DISCARDED

LIBRARY
WESTERN WYOMING COMMUNITY COLLEGE

The Primary Production of the Earth (from Chapter 5).

The energy of sunlight reaching the earth averages 700 cal/cm^2/day for all wave lengths outside the atmosphere, about 5.5×10^5 kCal/m^2/yr in the visible range at the earth's surface. A fraction of this energy is absorbed by chlorophyll and used in photosynthesis and primary productivity, the amounts of which vary widely in different kinds of communities.

Ecosystem Type	*Area*	*Net Primary Productivity per Unit Area Normal Range*	*Mean*	*World Net Primary Production*
	10^6km^2	*dry g/m^2/yr*		*10^9 dry t/yr*
Tropical forest	24.5	1000–3500	2000	49.4
Temperate forest	12.0	600–2500	1250	14.9
Boreal forest	12.0	400–2000	800	9.6
Woodland and shrubland	8.5	250–1200	700	6.0
Savanna	15.0	200–2000	900	13.5
Temperate grassland	9.0	200–1500	600	5.4
Tundra and alpine	8.0	10–400	140	1.1
Desert and semidesert	42.0	0–250	40	1.7
Cultivated land	14.0	100–3500	650	9.1
Swamp and marsh	2.0	800–3500	2000	4.0
Lake and stream	2.0	100–1500	250	0.5
Total continental	149.0		773	115
Open ocean	332.0	2–400	125	41.5
Continental shelf, upwelling	27.0	200–1000	360	9.8
Algal beds, reefs, estuaries	2.0	500–4000	1800	3.7
Total marine	361.0		152	55
World total	510.0		333	170

The energy content of organic matter averages 4.26 kCal/g of dry matter in land plants, 4.9 kCal/g in open-ocean plankton, and 4.5 kCal/g in other algae. Some mean values (excluding freshwater from land and estuaries from oceans):

	On Land Only	*In the Oceans*	*For the World*	
Mean net primary productivity	780	147	336	g/m^2/yr
Mean energy of NPP	3300	720	1500	kCal/m^2/yr
Chlorophyll/surface area	1.5	0.05	0.48	g/m^2
Efficiency of NPP				
Dry matter/visible sunlight	1.42	0.27	0.61	mg/kCal
Dry matter/chlorophyll	510	3000	700	g/g/yr
Energy/chlorophyll	2200	15,000	3100	kCal/g/yr
Energy/visible sunlight	0.60%	0.13%	0.27%	kCal/kCal

Gross primary productivity should average (approximately) 2.7 × NPP on land, 1.5 × NPP in the oceans, and 2.3 × NPP for the world.

COMMUNITIES AND ECOSYSTEMS

ROBERT H. WHITTAKER
Cornell University

Communities and Ecosystems

SECOND EDITION

MACMILLAN PUBLISHING CO., INC.
New York
COLLIER MACMILLAN PUBLISHERS
London

To Clara

Cover photo is of the southern Oregon coast at Gold Beach in the redwood belt, from *The Last Redwoods* by François Leydet, published by the Sierra Club. [Reproduced by permission of Philip Hyde, Box 220, Taylorsville, California.]

Copyright © 1975, Robert H. Whittaker

Printed in the United States of America

All rights reserved. No part of this book may be reproduced or transmitted in any form or by any means, electronic or mechanical, including photocopying, recording, or any information storage and retrieval system, without permission in writing from the Publisher.

Earlier edition copyright © 1970 by Robert H. Whittaker.

MACMILLAN PUBLISHING CO., INC.
866 Third Avenue, New York, New York 10022

COLLIER-MACMILLAN CANADA, LTD.

Library of Congress Cataloging in Publication Data

Whittaker, Robert Harding, (date)
Communities and ecosystems.

Includes bibliographies and index.
1. Ecology. I. Title.
QH541.W44 1975 574.5'24 74-6636
ISBN 0-02-427390-2

Printing: 1 2 3 4 5 6 7 8 Year: 5 6 7 8 9 0

Preface to the Second Edition

The suggestion that a second edition is intended to bring a book up to date surprises no one. Ecology is an active science, however, and especially in such areas of the field as species diversity and ecosystem function, knowledge has increased and new viewpoints have emerged since the time of the first edition. The sections of the book relating to these topics have been rewritten, and a new section has been added summarizing some recent knowledge of the biosphere.

Another purpose was improvement of the book as a working textbook. The first edition was written for an introductory course given in one academic quarter at the

University of California, Irvine. I have used a number of suggestions of teachers about what they would like in the book for use in courses elsewhere, especially for one-semester courses. A full chapter on populations has been added, so that the book can be used with or without a companion text on population ecology. The chapter will not suffice for courses with a strong emphasis on population biology, but there are now several good books for such courses. New sections on community classification and soils have been added in response to comments that these were slighted in the first edition and in other texts. The reference lists have been expanded and broken up by sections of the text, to make them more useful to students and others seeking further readings. I have used asterisks and daggers to indicate some of the readings that seem to me possible choices for class use, and some references recommended for further study of a topic. I thank many people for comments on parts of the book, but I should mention especially the reviews of the whole book by Robert McIntosh and Paul Zedler.

The first edition was criticized for asking more of the student than some others, and especially for its use of some mathematics in an introductory text. I have not really wanted the book to introduce ecology without at least indicating some ways measurements and mathematics are used by ecologists. In deference to the less mathematical reader, however, the mathematics has been segregated into tables that can be used like other tables—as supporting material for the reader to whom it is of interest—or in one case into bracketed sections of the text that can be passed over. The student should be assured that the tables and equations are simpler than they may appear to be at first. There are parts of the book where I have used concepts I think important to understanding the subject, even though the results will not be easy for all readers. I hope the reader will accept these as part of the over-all purpose: an introduction to some of the ideas of ecology in a book that is clear and short.

R. H. W.

Preface to the First Edition

On the surface of the earth living organisms and their environments form a thin film, the biosphere. A pervasive interrelatedness of living things and environments to one another characterizes the biosphere. Organisms form interacting systems or communities, these communities are coupled to their environments by transfer of matter and energy, and communities and environments of the biosphere as a whole are related by movements of air, water, and organisms. The biosphere is man's environment, and man is now altering the biosphere in ways disadvantageous to himself. The importance of understanding the natural systems formed by organisms and environments for its own sake and for the sake of man's future is not always granted in a civilization guided by technology, despite the increasing emphasis of functional systems in technology.

The study of living systems in relation to environment is the science of ecology. Because of the wide range of concerns of ecology it is difficult to treat in a single book. In developing the Macmillan series "Current Concepts in Biology," it was thought better to prepare two books representing major divisions of the science. Much ecological understanding can be integrated around populations as living systems and the manner in which populations function in relation to environment; this aspect of ecology is the subject of *The Ecology of Populations* by Arthur S. Boughey. Much ecological understanding can be integrated also around the concepts of communities as assemblages of different species which interact with one another, and ecosystems as functional systems formed by communities and their environments. These aspects of ecology are dealt with in *Communities and Ecosystems*. The two books are designed to complement one another and to serve as introductions to ecology either separately or together. An additional important part of ecology is represented in the series by David E. Davis's *Integral Animal Behavior*. The reader will find the present book concerns the structure of natural communities, the function of ecosystems, and the problems of man's relations to the biosphere.

R. H. W.

Contents

3

Community Structure and Composition 60

4

Communities and Environments 111

5

Production 192

List of Figures

List of Tables

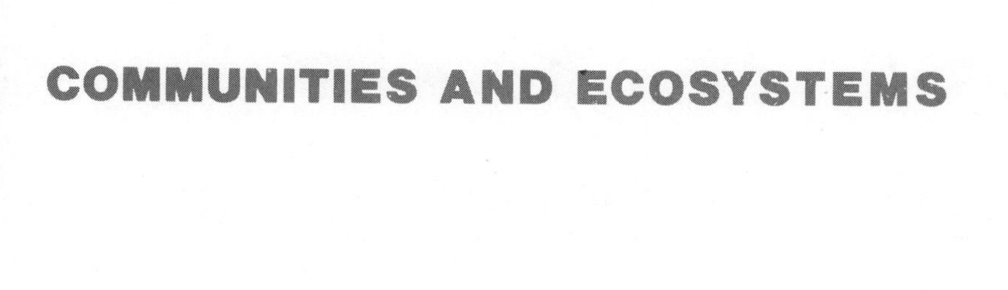

COMMUNITIES AND ECOSYSTEMS

1

Introduction

In northern California and southern Oregon there are forests like no others—the great coastal redwood forests. Some twenty-five years ago, this writer was drawn to study part of the ecology of these forests—their species composition and relation to other kinds of forests, their variation with topographic position and the climatic gradient from the fog belt inland, their evolutionary history and relation to the mixed, ancient forests that occurred across the continent in Tertiary time, and their dimensions compared with those of other forests. A redwood forest is a *natural community,* an assemblage of populations of plants, animals, bacteria, and fungi that live in an environment and interact with one another, forming together a distinctive living system with its own composition, structure, environmental relations, development, and function. A redwood forest or an oak forest, a piece of prairie or a patch of desert—each of these can be approached as a community, a system of organisms living together and linked together by their effects on one another and their responses to the environment they share.

In each case the community has a close-linked, interacting relation to environment, as climate and soil affect the community and the community affects the soil and its own internal climate or microclimate, as energy and matter are taken from environment to run the community's living function and form its substance, transferred from one organism to another in the community, and released back to environment. A community and its environment treated together as a functional system of complementary relationships, and transfer and circulation of energy and matter, is an *ecosystem.* Thus a redwood forest is an ecosystem that can be characterized as different in some degrees from other forests in its structure, its particular adaptation to a humid coastal environment and effects on its own microclimate, its manner of using sunlight energy in high productivity, and its way of circulating nutrients between the soil and the organisms of the community.

Offshore from the redwoods are very different communities. In the surface waters of the sea there is a community at the opposite extreme of size from the redwood forests—the plankton, an almost invisible community of microorganisms. The plankton is a full community of green plants, animals feeding on these, and predatory animals feeding on other animals, bacteria, and fungi (Figure 1.1), organisms that are suspended in the water and carried passively by currents. These organisms are in intimate chemical relation to the water as materials circulate from water through the organisms and back to water; the plankton with its aquatic environment is an ecosystem. It is possible, observing sea water without a microscope, to be unaware of the plankton in it, but the marine plankton is the most widespread kind of natural community on the earth's surface.

The area of this writer's study in the redwood belt has changed in the last twenty-five years. The area was then relatively remote and lightly used by man, reached by winding roads above the sea that passed extensive uncut forests and largely unoccupied shores. As we see it today the forests are small remnants of what they were, the towns are larger, and the human use of the area is ever heavier. There are other ecological changes: along much of the coast the richness of the shore life and inshore fishing is reduced; populations of some of the large predatory birds of the coast are in decline; smog appears at times in valleys with cities far smaller than Los Angeles; and the sardines that were fished by fleets of boats from Alaska to California and were the basis of one of the country's major fishery industries have decreased, overfished despite warnings about danger of depleting the population, to near disappearance.

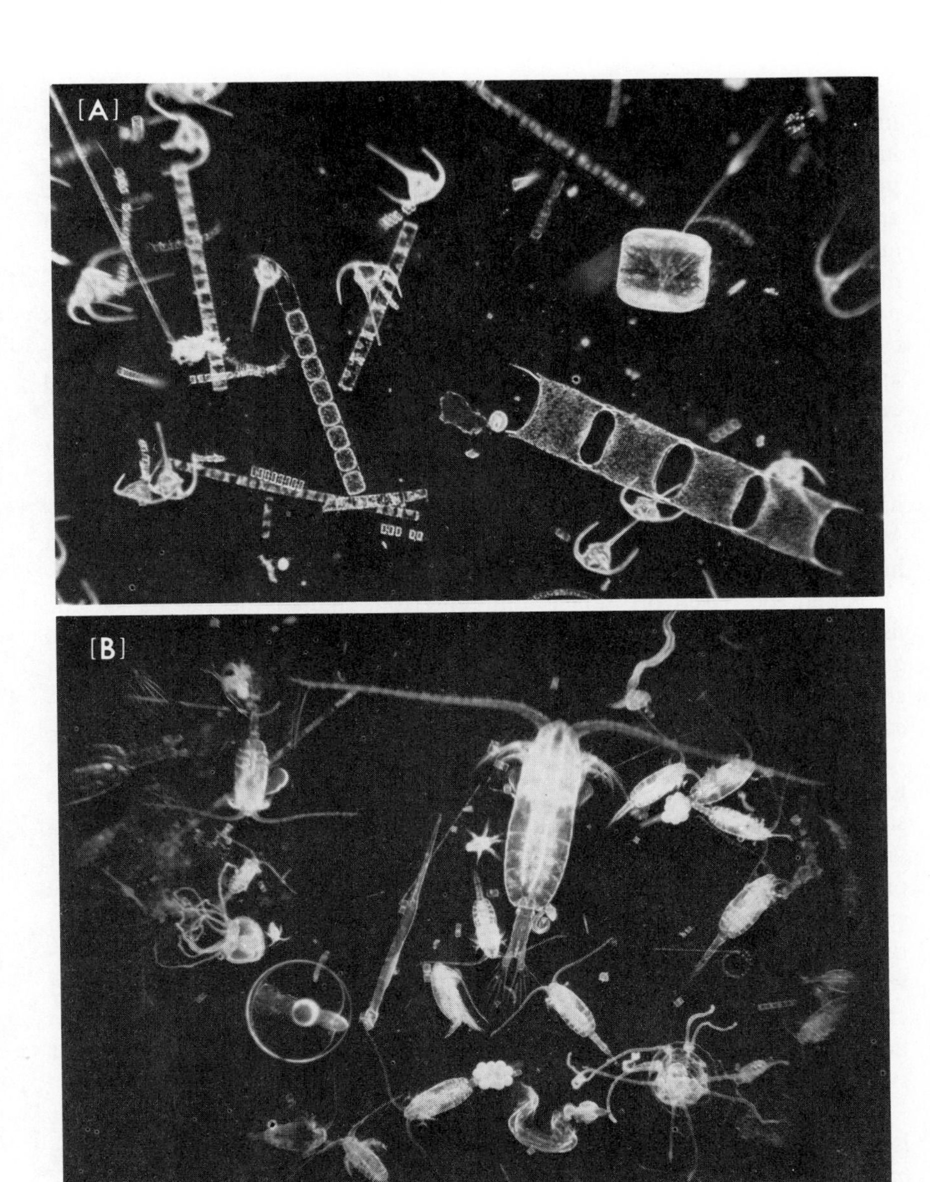

Figure 1.1. Marine plankton organisms. A: Plant plankton magnified 65 times. Most of the cells are diatoms (the large chain of four is *Biddulphia sinensis,* the smaller chains *Stephanopyxis turris* and *Rhizosolenia faeröense,* the single large cell *Coscinodiscus concinnus*); the cells with curved spines are dinoflagellates (*Ceratium tripos*). B: Animal plankton magnified 16 times. The shrimplike animals with long antennae are copepods (one large individual of *Calanus finmarchicus,* a number of smaller *Pseudocalanus elongatus,* and a larval copepod just to the left of the large *Calanus*). An arrow worm (*Sagitta*) parallels the large copepod on its left, and the figure includes also two small jellyfish, two tunicates (*Oikopleura,* curly organisms near the top and bottom), and a fish egg (circular object). Chains of diatom cells can be seen on the right.

[Both photographs are of organisms in the living condition concentrated in sea water, taken by electronic flash. Copyright by Douglas P. Wilson, see also A. Hardy, *The Open Sea.*]

If one visits a beach of the area now, in the evening, the glare of the lights of increasing traffic on a fast coastal highway contrasts with the softly luminous moving lines of the surf, which at times glows with the phosphorescence of a plankton organism, a dinoflagellate doubtless even more ancient as a type than the redwood. The seacoast, where many of the richest and most productive of natural communities are concentrated and where man is concentrating much of his population, thus juxtaposes not only the natural realms of the land and sea, but also two disparate orders: man's current ways of aggressively expanding technology and use of environment, and the evolutionarily old, vulnerable, natural order of communities and ecosystems.

Man necessarily uses environments and harvests from communities, but many of the changes brought by man in doing so are unforeseen and against his choice and interests. There is reason for concern now not only with unwanted changes in natural conditions, but with the implications to man himself of these changes. There is great need today that man's use of environment be based on models different from that of the Pacific sardine fisheries—exploitation expanding, uncontrolled by foresight, to the eventual destruction of the resource and of the institution exploiting it. The foresight should be directed by an understanding of ecosystems and of the world ecosystem that is man's environment.

Ecology is the area of the biological sciences that is concerned with living systems in their environmental contexts. In practice the living systems studied by ecologists are those of the highest levels of organization: individual organisms, populations, societies (as organizations of individuals of one species), communities (as systems of populations usually of many species), and ecosystems. Difficulties and challenges in ecology result from the effort to deal with the complexities of these higher organic systems, in which biological processes of lower levels are integrated with physical and chemical processes of environment into phenomena that are distinctive to the higher system and must be interpreted in terms of the function of that system.

Because of the breadth of the phenomena with which ecology deals, it is convenient to recognize two major divisions or levels. The first of these, *autecology,* concerns the ecology of individual organisms and populations and includes as fields of study physiological ecology, genecology, animal behavior, the study of symbioses, and population dynamics. (The study of populations is sometimes termed *demecology.*) These fields concern primarily one species at a time (but in symbiosis and often in population dynamics two or a few species). The other area of ecology concerns systems of many species—whole communities or major fractions of communities, and ecosystems. This area of study is termed *synecology* in the English-speaking countries, and *biocenology* or *biosociology* by many Europeans; it includes studies of terrestrial ecosystems, biological aspects of oceanography, limnology as the study of lakes and streams, biogeochemistry as the study of the circulation of materials in the world ecosystem, and applied problems of man's management and alteration of ecosystems. This book is designed as an introduction to synecology, and its discussion will proceed from population processes, to relations of species and communities to environment, to the structure of communities, to energy function in community production, to the circulation of materials in ecosystem function, to pollution processes and implications for human ecology.

References

BATES, MARSTON. 1960. *The Forest and the Sea.* New York: Random. 277 pp.

CLAPHAM, W. B., JR. 1973. *Natural Ecosystems.* New York: Macmillan. viii + 248 pp.

KORMONDY, EDWARD J. 1969. *Concepts of Ecology.* Englewood Cliffs, N. J.: Prentice-Hall. xiii + 209 pp.

KUCERA, CLAIR L. 1973. *The Challenge of Ecology.* St. Louis: Mosby. xiv + 226 pp.

LEOPOLD, ALDO. 1949. *A Sand County Almanac.* Oxford Univ. Reprint 1966, New York: Ballantine and Sierra Club. xix + 295 pp.

2

Populations

We begin with a simple observation. When the populations of species in natural communities are observed from year to year, these populations seem rather stable. The redwood forest gives a sense not only of spaciousness but of timelessness. There is no visible change in the redwood trees from year to year, and no noticeable change in the smaller plants of the forest floor. A record of bird songs in the forest during the breeding season of one summer is very similar to that of the previous summer. Samples of the small animals that live in the leaf litter on the forest floor are reasonably consistent from one summer to the next, although they do show changes with wetter and drier weather. One should not exaggerate the stability of natural populations; even in the redwoods quite different numbers of seedlings establish themselves from one year to another. We should thus speak not simply of stability but of *relative* stability. Nevertheless there is a question: How are we to interpret the relative stability of many species populations in many natural communities?

A Few Populations

It may be helpful to make the question more concrete with examples—four populations of plants and four of animals. To describe these populations we shall be using throughout this chapter the term *density* for the number of individual organisms of a species, as counted in a unit area of the earth's surface (or, in some cases, in a unit volume of water or of an experimental culture).

1. The white oak (*Quercus alba*) trees of a mature oak-hickory forest are counted and their ages determined by drilling to the centers of the trees. Density of the species is 20.7 trees over 6 inches (15 cm) in diameter at breast height per hectare (1000 m^2, 2.5 acres). Trees are grouped by 50-year age classes, and numbers of trees are plotted by these age classes in Figure 2.1. Note especially the curve on the right, where numbers of trees are plotted on a logarithmic scale. The numbers of trees in age classes then appear as a straight line that represents a simple geometric series. This straight line implies that from each age class to the next a nearly constant *fraction* (34.4 per cent) of the trees will die. The fraction of individuals dying per unit time, for a population or age class is a *death rate*. The death rate for white oak is 34 per cent per 50 years, hence 0.344/50 or about 0.08/decade. One can imagine a continuous flow of tree individuals through time, as a first age-group (1–50 years) is added by reproduction, this group loses 34 per cent of its individuals during the half century in which it ages to become the second age-group (51–100 years) then another 34 per cent as it becomes the third age-group and so on until the last of its individuals dies at an age of around 400 years. Meanwhile other age-groups follow it through the steps. Note that in each 50 year period the number of new individuals added to the first age-group is equaled by the number lost from all groups from the first to the last, and that the number being added to the second age-group is equaled by the loss from all groups from the second to the last, and so on. There is thus a rather constant *flow* of individuals through the age classes, and through the population as a whole. The fact that the steps do not perfectly fit the curve indicates that even in this relatively stable population there is some fluctuation in rates of birth and death, and some consequent fluctuation of density around a mean

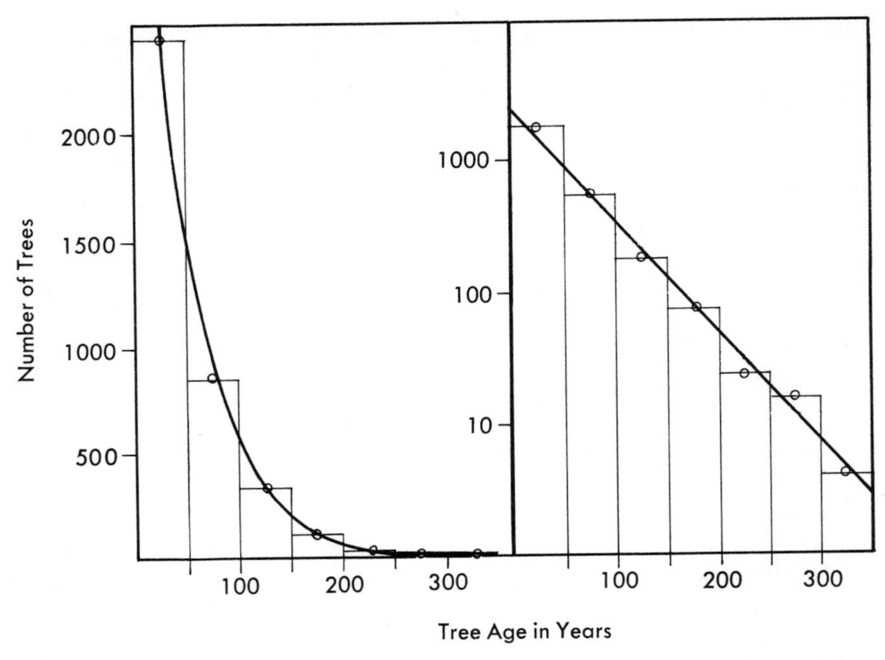

Figure 2.1. Age composition of white oaks (*Quercus alba*) in a mature oak-hickory forest. The graphs show numbers of trees in 50-year age classes, as these numbers decrease from younger to older age classes. Numbers of trees are on a linear scale on the left, producing a descending *J*-curve for age composition. Numbers of trees are on a logarithmic scale on the right; and on this scale the numbers of trees in classes approach a straight line, implying that a constant fraction of the trees die from each age class to the next. Numbers are based on 1525 tree diameters measured in 74 hectares, and interpretation of age classes from more limited age-diameter data (56 trees) [Data of Miller, *Illinois Nat. Hist. Surv. Bull.,* **14**:8 (1923)].

value. It is likely that we should find that some mature redwood populations have age compositions resembling that of Figure 2.1, if we could treat their giant trees by century age-classes. However, many forests (including many oak and redwood forests) do not have age compositions that fit geometric series.

2. Across the northern part of North America and Eurasia are extensive forests of spruce and fir. In some areas, such as the northern Appalachian Mountains, stands of these forests are periodically blown down by storms. In a given place young spruce trees (*Picea rubens*) establish themselves following a blow-down, and grow. While young they are less vulnerable to the wind; and as the trees grow older and taller some of them die because of the crowding of trees in the stand. A dense, mature forest develops in which few young trees can grow; and this forest is toppled, sooner or later, by wind. When the trees have fallen, other plants

replace them temporarily, and among these other plants young seedling spruces begin to grow. These trees in turn may become a mature forest that is blown down two centuries later. This population maintains itself not by continuous reproduction but in pulses, with bursts of reproduction followed by periods of growth and maturity during which the numbers of trees are more nearly constant. We can say that the population appears to be periodic or cyclic, if we observe that it is not periodic in the true sense of constant time intervals from one pulse of reproduction to the next.

3. The dominant plants of prairies are grasses that are more vulnerable to year-to-year fluctuation in climate than most trees. In a mixed-grass prairie in Nebraska, little bluestem grass (*Andropogon scoparius*) was observed to decrease to a fraction of its former population—from 50 to 1 per cent of ground cover—during a series of drought years, 1932–39. Other species, better adapted to drought, partly replaced it, but the prairie's total grass cover dropped from 61 to 17 per cent during the drought. Some of the bluestem plants survived, however; and when more favorable years of greater rainfall returned, the bluestem became dominant again. The bluestem is a persistent dominant of the prairie, but one whose population fluctuates in an irregular way, rising and falling as climatic conditions become more or less favorable for it. It is less stable than the oak trees, but we can say that it is relatively stable in the sense that it is always present, although fluctuating widely about its mean density.

4. Fireweed (*Epilobium angustifolium*) is an herb common in areas of spruce-fir forest in the Rocky Mountains. These forests are less affected by extensive blow-downs than those of the northern Appalachians, but they are vulnerable to fire. When a forest is destroyed by fire, the wind-carried seeds of fireweed that are brought to the burn area germinate, and a fireweed population establishes itself and reproduces. A few years after the fire the area may be covered with fireweed, and in summer the mountain slopes may be purple with its flowers. As the spruces and other plants grow back, the fireweeds fail to reproduce, and eventually disappear. By this time, however, the wind is likely to have carried some of their abundant seeds to a new burn area. Fireweed is a "fugitive" species that leads a nomadic existence of appearance here and disappearance there, depending on fire. Yet in the absence of fires the species survives—in smaller numbers along streams, on landslides, and in some of the openings formed where individual trees have fallen.

5. The true locusts are species of grasshoppers subject to population outbreak and swarming. The locust *Chortoicetes terminifera* can be found in small numbers, at least, in most parts of Australia. In most areas the death rate is high, and the local populations are probably maintained only by immigration. In southern Australia, however, there are outbreak areas, dry grasslands where the locusts find combinations of soils and vegetation favorable for breeding and population growth. In these areas the locusts

are persistent, and in periods of favorable climate they can multiply to form vast populations of millions. These can emigrate in swarms that descend on other areas, and in some cases multiply further there, to become locust plagues. The plagues are in time ended by population catastrophe through drought and starvation or disease. In South Australia plagues have occurred every 30 or 40 years since the first one recorded in 1845; in eastern Australia they are more frequent. In the latter area disturbance by man, particularly through grazing by sheep, has much increased the area where soil and vegetation characteristics permit population outbreak. The locust is relatively unstable in a different sense from fireweed; it is an eruptive pest species.

6. The bay checkerspot (*Euphydryas editha* ssp. *bayenis*) is a butterfly that occurs in local areas of serpentine, a distinctive kind of rock, in northern California. The species forms local populations that do not occur on all serpentine areas and that do not occupy the full extents of the areas in which they occur. The sizes of populations fluctuate widely, and apparently independently, in different serpentine areas for unknown reasons; and at times the species becomes extinct in some of its areas. The serpentine areas that lose the species appear to regain it eventually by recolonization from other areas where the species has survived. The population as a whole is persistent in a broad geographic area, but in an individual serpentine area it is unstable.

7. The snowshoe hare (*Lepus americanus*) in the Canadian Arctic increases for several years to maximum population density, and then decreases to a minimum density that is a small fraction of the maximum. The period from one maximum to the next is about ten years on the average, and the hare consequently seems to fluctuate in a ten-year cycle, as does its predator the lynx (Figure 2.2). A number of other arctic mammal and and bird species fluctuate in similar ways. These fluctuations have fascinated ecologists and produced extensive research and discussion of whether they are truly cyclic and what factors determine their periods.

8. The red-eyed vireo (*Vireo olivaceus*) is the principal small bird species of the deciduous forests of the eastern United States. Successful breeding pairs occupy territories, and the number of territories is much the same from year to year in a given forest; 43 pairs per 40 hectares (or 100 acres) were counted in one oak-hickory forest. There is a surplus of nonbreeding birds that move about as they feed, but the population of breeding pairs is relatively stable. When one of the breeding birds is killed, one of the nonbreeding birds quickly takes over its territory and mate. Losses from age classes (after the youngest ages) are relatively constant at about 50 per cent per year. Thus the vireo, like the white oak, has a relatively stable population with continuous replacement, though it has a much shorter life span.

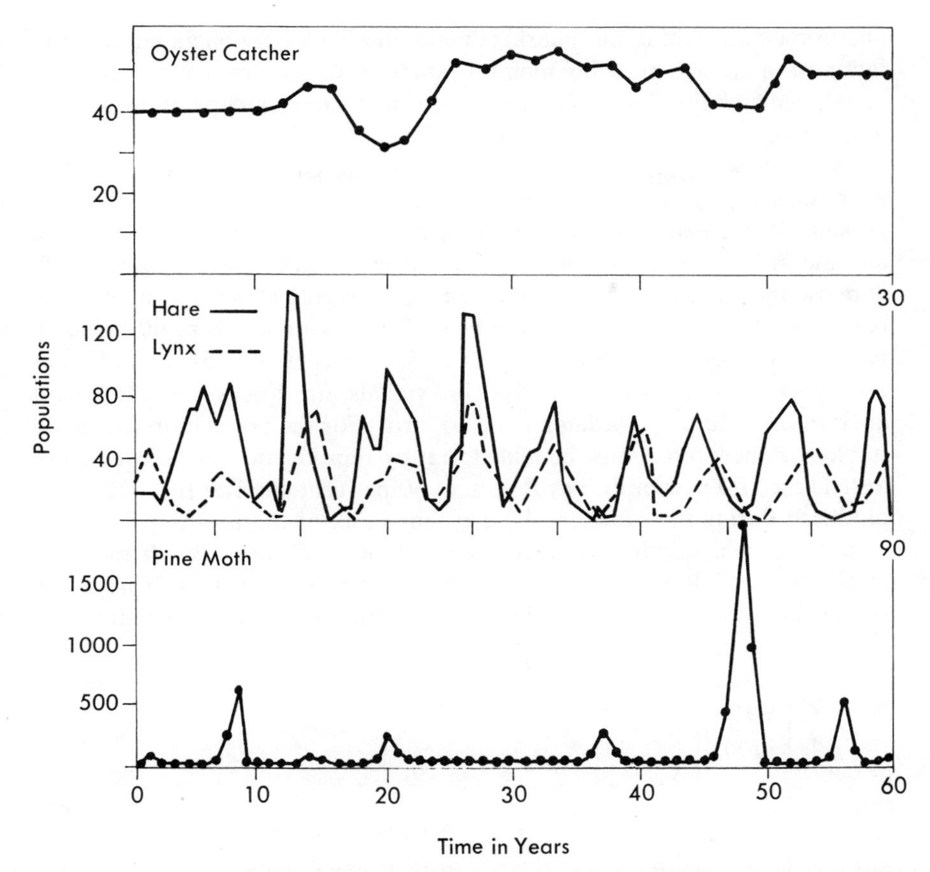

Figure 2.2. Types of fluctuation in animal populations. Top: Highly stable population of a bird, the oyster catcher (*Haematopus ostralegus*), numbers of breeding pairs on the island of Skokholm (96 hectares) off southwest Great Britain [Lack, *J. Anim. Ecol.,* **38**:211 (1969)]. Middle: Less stable populations of a predator, the lynx (*Lynx canadensis*), and its prey, the snowshoe hare (*Lepus americanus*), based on numbers of pelts (in thousands) received by the Hudson's Bay Company from trappers, 1845–1935 [MacLulich, *Univ. Toronto Stud. Biol.,* **43** (1937)]. Bottom: The widely fluctuating population of a pine moth (*Bupalus piniarius*) in Germany subject to eruption at irregular intervals, wintering individuals per 100 m^2 [Schwerdtfeger, *Z. Angew. Ent.,* **28**:254 (1941); *Demökologie,* p. 305 (1968)].

These cases permit us to make some general comments on populations. The first is the variety of ways populations behave (Figure 2.2). Some are relatively stable, some are conspicuously unstable in a local area, and many are in between—far from constant but persistent, fluctuating around a mean density that stays the same. Among these intermediate cases some are clearly irregular or aperiodic in their fluctuation, whereas others may be more regular, periodic or cyclic. It would be a mistake to pursue our question about stability of natural communities with only one kind of population in mind. We must broaden our question to ask not just the meaning of relative stability, but of both narrower and broader population fluctuations.

Second, we ask how to define and measure "stability." A number of definitions of stability are used, but our concern here is with extent of population fluctuation. On this basis we can define the concept as an ideal, and measure it in actual cases. For the ideal concept, note population 8. Suppose the vireo has a population density averaging about 200 individuals per square kilometer of forest, of which 120 are individuals in 60 breeding pairs. If the death rate for the birds is 50 per cent or 0.5/yr; each year 200 individuals times 0.50, or 100 individuals, are lost to the hazards of migration, capture by predators, and so forth. For the population to remain stable, 100 new birds must be added to it by reproduction, to replace those lost. These 100 new birds may represent 50 per cent survival from 200 eggs produced during the summer. We can thus think of the bird population as a "pool" to which individuals are added and lost (Figure 2.3). If the rates of addition and loss are the same, the pool remains constant. This is the special kind of stability with which we are concerned; it is a condition of

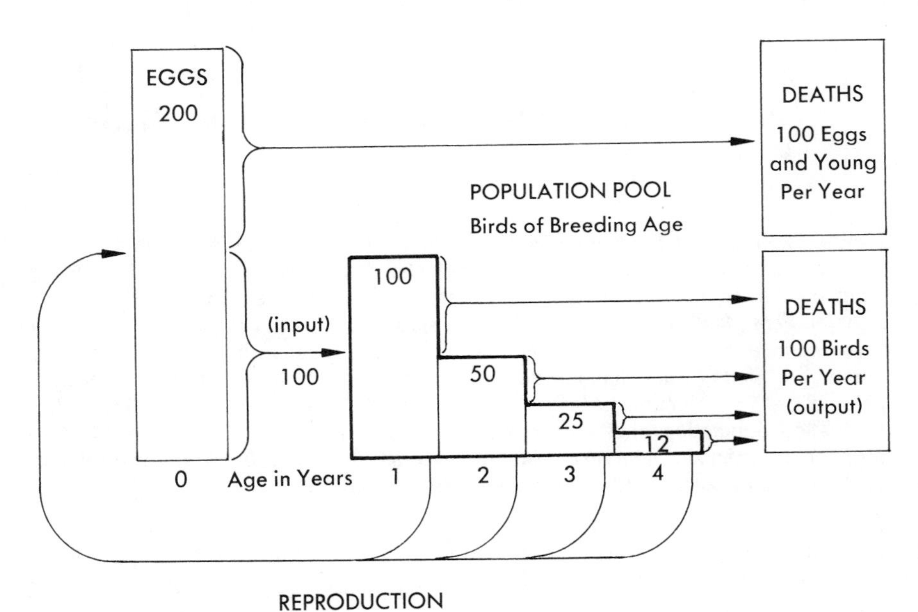

Figure 2.3. Population steady state for a bird. The population averages about 200 birds of breeding age per 100 hectares of forest; of these birds 120 are nesting in 60 pairs, producing 200 eggs each year. A death rate of 50 per cent per year is assumed for each age class, and for the eggs and young. The "population pool" of birds of breeding age thus gains 100 young birds and loses 100 older birds each year, and the population remains relatively constant.

relative constancy (of the population as a pool) superimposed on a *flow* of individuals through the population. Such constancy of a system based on flow through the system (with input and output equal) is termed a *steady state*. Ideally, population "stability" implies a population steady state, with births and deaths equal on the average. "On the average" allows for some fluctuation of birth and death rates, and consequently of population size, while we still interpret the population as relatively stable.

We should like a measure to express degree of stability, or extent of fluctuation (Table 2.1). (The non-mathematical student should not be dis-

Table 2.1 Measuring Relative Population Stability

$N_1, N_2, N_3 \ldots N_t$ are measurements of population density at t different times.

$$\overline{N} = (N_1 + N_2 + N_3 \ldots + N_t)/t = \sum_{x=1}^{t} N_x/t = \text{arithmetic mean density} \quad (1)$$

$$\sigma = \sqrt{\sum_{x=1}^{t} (N_x - \overline{N})^2/(t-1)} = \text{standard deviation} \quad (2)$$

$$CV = \sigma/\overline{N} = \text{coefficient of variation} \quad (3)$$

$$\tilde{N} = \sqrt[t]{(N_1 \times N_2 \times N_3 \ldots \times N_t)} = \text{antilog of} \left[\left(\sum_{x=1}^{t} \log N_x \right) / t \right] = \text{geometric mean density} \quad (4)$$

$$D_l = \sqrt{\sum_{x=1}^{t} (\log N_x - \log \tilde{N})^2/(t-1)} = \text{logarithmic deviation} \quad (5)$$

$$CF = \text{antilog } D_l = \text{coefficient of fluctuation} \quad (6)$$

couraged by this and other tables of equations, but should read the text and use the equations as aids to the text to the extent that the equations are informative to him.) If we measure a population's density at a number of different times, the degree of spread or scatter of these measurements away from the mean expresses the intensity of fluctuation, the degree of contrast between peaks and troughs in the population graph (Figure 2.2). This spread on each side of the mean value is a dispersion, for which a standard deviation (or, better, coefficient of variation) might be a natural expression (Table 2.1, equations 2 and 3). However, populations are best treated on a logarithmic scale because of the manner of population growth (by fractions or ratios, not by particular numbers of individuals added). If we use logarithms of the densities of the population at different times, then we can characterize the population by its geometric mean (Table 2.1, equation 4) rather than its arithmetic mean, and by a dispersion measure we can call the coefficient of fluctuation (equation 6). The coefficient of fluctuation (*CF*) permits us to characterize the wide range of relative stabilities in our examples. Records of breeding oyster catchers (Figure 2.2) give a year-to-year *CF* of breeding pairs of 1.153, implying that the densities in different years are mostly not more than 15 per cent above or below the mean. (*CF* expresses the range not as plus or minus values, but as $N \times 1.153$ to $N/1.153$.) *CF* values for the hare and lynx are about 2.8 and 2.2. The relatively unstable insect *Bupalus,* in contrast, has a *CF* of 10.2, implying that tenfold fluctuation of its population above and below the mean is frequent. The coefficient of fluctuation can express relative stability, but the densities used should not include zero values. Zero densities imply that the species becomes extinct, and that it is not relatively but absolutely unstable for the area and time-period in which densities have been measured.

We may comment finally on the nature of populations. From here on we shall often talk about populations as if they were sets of identical individuals, but they are not. Populations differ in their degrees and kinds of genetic variability, in kinds of reproduction that affect variability, and in manners in which variability relates to environment and permits survival. Populations differ in age composition, in birth and death rates at different ages, and in ways age composition changes with change in environment. We have illustrated only one of the simplest possible forms of age composition in populations 1 and 8. Figure 2.4 illustrates a wide range of death rate and survival curves. Actual populations show curves of diverse forms mostly different from and intermediate to those illustrated. These features of populations are given extensive treatment in books on population biology and demography. We should observe, though, that a population is a fairly complex living system, with its own distinctive structure and function, and its particular way of adapting to environment and other populations.

Population Growth

The study of populations may best start with simplified, experimental situations. *Paramecium caudatum* is a protozoan, a "slipper animalcule" experimented with by the Russian ecologist, G. F. Gause. *Paramecium* reproduces by binary fission: every few hours a paramecium divides itself in two. Gause set up an aquarium with a broth of bacterial cells suspended in water to provide food for paramecium; he then introduced one paramecium and followed the growth of the population it produced. Let us first observe this population while no limit affects its growth. If a paramecium divides every three hours, then after 3, 6, 9, 12, and 15 hours there will be 2, 4, 8, 16, and 32 individuals. After 30 hours there will be 1024 paramecia. Such population growth is described by an expanding geometric series (Table 2.2, equation 2) with a rate of increase (*R*) per generation of 2 per 3-hour time unit. The steps of a geometric series are points along

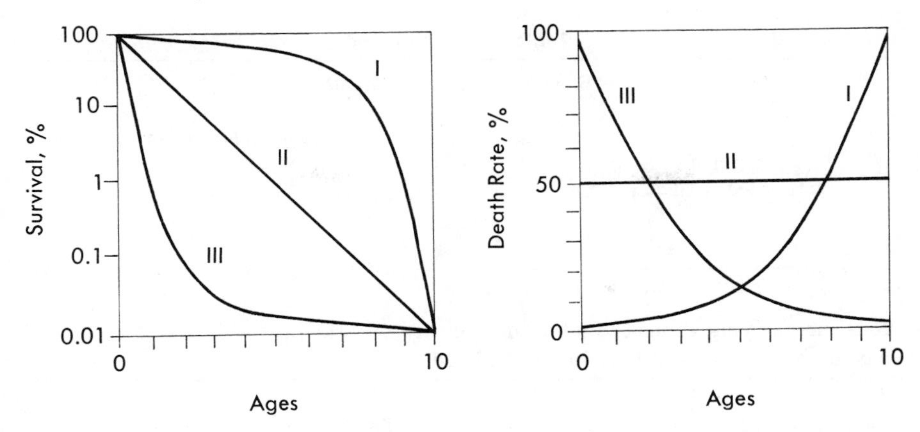

Figure 2.4. Survival- and death-rate curves. The survival curves, on the left, indicate the per cents of individuals born in a given time period that are surviving as they age through the 10 age and time units on the horizontal axis, while subject to death rates at different ages as indicated on the right. If the population is fully stabilized, then the curves on the left also indicate age composition at a given time—relative numbers of individuals in age classes corresponding to the same ten units. A logarithmic scale is used for survival per cent and linear scales for death rate and age.

Type I is for a species with a fairly definite life span, such as man in societies with well-developed health and medical services. The death rate is low and survival high in youth and middle age; but in old age the death rate increases, and survival per cent rapidly decreases.

Type II is for a species with a constant death rate at different ages, as in some forest tree and bird populations. With constant fractions of individuals dying in each age class, the survival per cent becomes a straight line. (See also Figure 2.1.)

Type III is for a species with a high death rate early in the life cycle, such as oysters and other animals producing large numbers of eggs and young of which few survive. The survival curve drops rapidly to small numbers of individuals that survive into reproductive age, but the death rate is lower for these survivors.

Table 2.2 Population Growth Equations

A. Growth Without a Limit

N is a population density. N_0 is an initial population density, N_1 is density after a unit time ($t = 1$) of population growth, and N_t is density after any time period t, with growth at a constant rate of increase.

Geometric Formulas

$\frac{N_1}{N_0} = R$	rate of increase per time unit	(1)
$N_t = N_0 R^t$	density after time units t	(2)

Exponential Formulas

$\frac{dN}{dt} = N_0 r$	rate of increase	(3)
$N_t = N_0 e^{rt}$	density after time t	(4)

Table 2.2—Continued

B. Growth Within a Limit		
	Logistic Formulas	
$\frac{dN}{dt} = rN(1 - \frac{N}{K})$	rate of increase	(5)
$N_t = \frac{K}{1 + [(K - N_0)/N_0]e^{-rt}}$	density after time t	(6)

K is the carrying capacity or limiting density.
e is the base of natural logarithms, 2.71828.

an ascending J-curve on a linear plot (Figure 2.5, left), and along an ascending straight line on a logarithmic plot (Figure 2.5, right). Note that the increase in numbers of individuals in a single time unit is N_0R, in which N_0 is the number of individuals at the beginning of that time unit. Capital R is then a ratio by which any density, N, can increase per unit time. It is because populations generally increase and decrease by ratios applied to N, that treatment of population data on a logarithmic scale is appropriate.

The paramecia will not really divide all at the same time every three

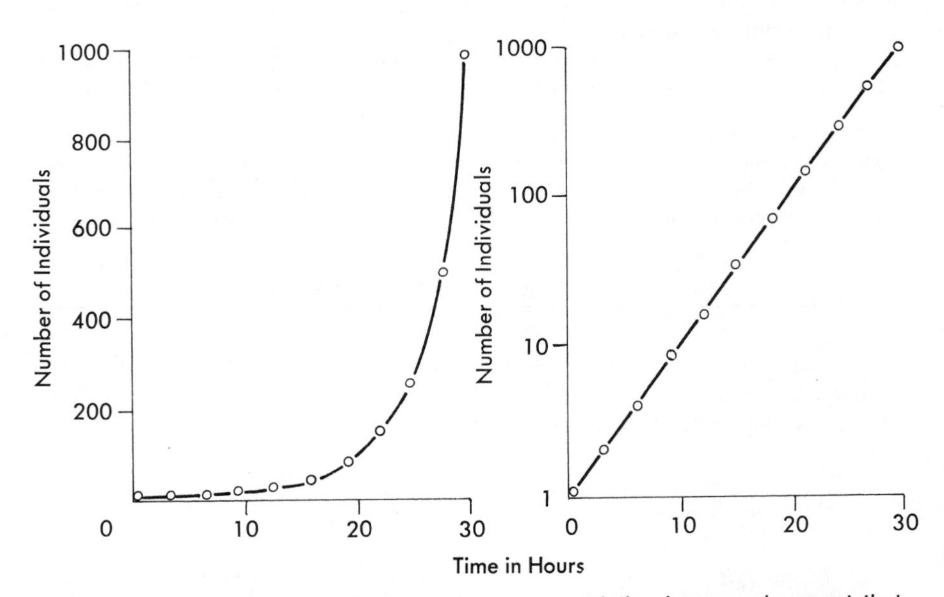

Figure 2.5. Exponential population growth, for a population in an environment that has not limited its growth. On the left, numbers of individuals are plotted on a linear scale, and exponential growth becomes a *J*-curve (in form the mirror image of that in Figure 2.1, left). On the right, numbers of individuals are plotted on a logarithmic scale, and the exponential increase becomes a straight line.

hours; some will divide more rapidly and some more slowly. After a few generations the rhythm of division will be lost, and the population growth will be continuous, not step-wise by powers of two. We can generalize from the step-wise growth to a continuous growth, as in equation (4). This equation describes the *exponential* curve of growth that is normal to populations when that growth is not somehow restricted. The most interesting symbol in the equation is r, the intrinsic rate of increase for the population. Little r is the natural logarithm of capital R, $R = e^r$. Little r does not necessarily describe the growth of a population in a natural community; it is a potential rate of increase when neither food, nor space, nor any other limitation of environment restricts that increase. Also, r varies within a species with environmental factors that do not limit growth but affect rate of growth—factors such as temperature.

If food and space were unlimited, and after 30 hours there were 1024 paramecia, then after 60 hours there would be more than a million, after 90 hours more than a billion, and after 120 hours more than enough paramecia. Unchecked exponential growth is eruptive, explosive, a force toward population instability. Since organisms must die, all populations must have r values that permit population increase, if they are not to decline to extinction. Yet that capacity for increase implies the hazard of explosive growth that consumes and destroys the resources that support the population. A kind of conflict—between the intrinsic tendency to increase, and the limits that increase must encounter—is characteristic of population function. The varied population behaviors we have observed express different ways the limits and hazards affecting populations contain or cut back their growth.

Population growth is always, sooner or later, checked. In some experimental situations a population of paramecia, or of yeasts, will smoothly approach and stabilize itself at a population ceiling, K, as in Figure 2.6. K is termed the *carrying capacity* of the environment for this population; it is the population density that the environment can support on a continuing, steady-state basis. In Gause's experiments with paramecia (Figure 2.6) K was determined primarily by the rate at which food was added to the culture to support growth and reproduction of new paramecia. As the paramecia become more crowded in the culture, the food becomes too limited to support them all, and the death rate rises. The equilibrium condition of the right-hand part of Figure 2.6 is a steady state in which, by definition, the birth rate and death rate are equal.

The curve in the upper part of Figure 2.6 is of the stretched-S form that is termed sigmoid. In sigmoid population growth we can recognize three phases (even though these are continuous with one another). In the early phase the population is little affected by limitations of food and space and grows at a rate close to rN, as is shown more clearly in the logarithmic plot in the lower part of Figure 2.6. In the middle phase environmental limita-

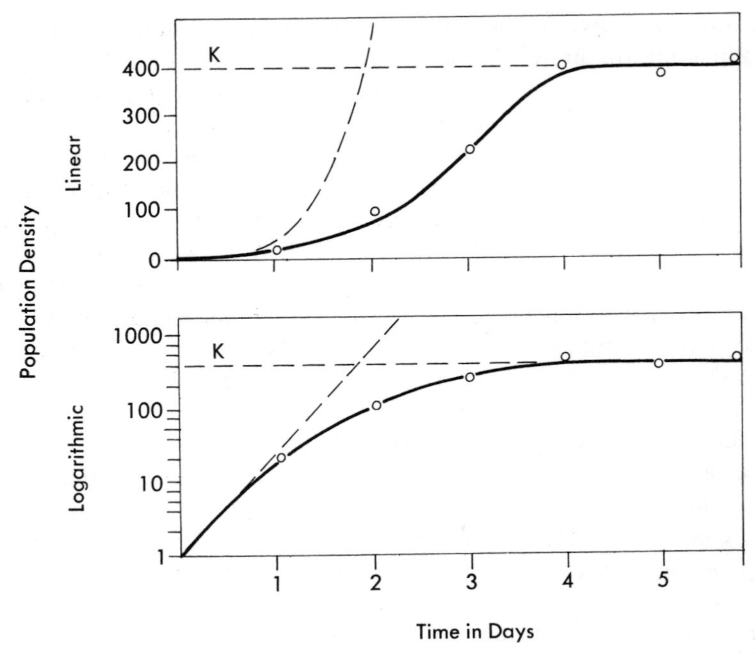

Figure 2.6. Sigmoid population growth, for a population in an environment that limits its growth. Above, with population density on a linear scale, the densities form a sigmoid curve to which the logistic equation (Table 2.2) has been fitted. The dashed line is the exponential growth of a population with the same rate of increase, *r*, in an environment that does not limit its growth. Below, density is on a logarithmic scale, showing for the same population the decrease in relative rate of population growth. Relative growth $(dN/dt)/N$ at first equals the *r* for exponential growth, as indicated by the dashed line, but later decreases to zero as *K* is reached and the population is stabilized.

tions begin to slow the growth by increasing the death rate (or decreasing the birth rate, or both). In the later phase the population approaches as an asymptote, and stabilizes at, the carrying capacity *K*. We want a simple expression of the manner in which population growth slows to zero as *K* is approached. We can choose, for example, the factor $(1 - N/K)$ and multiply equation (3) by this to obtain equation (5), which is called the *logistic* equation.

Some populations grow and stabilize in a sigmoid pattern that the logistic equation fits reasonably well, but many do not. The factor, $(1 - N/K)$, is an algebraic device that says that population growth shall be immediately, directly, and proportionately limited according to the difference between N/K and 1.0. For many populations, it will not be so. To interpret these we can apply to populations the concept of *feedback*. When information about a process returns to a control mechanism or center for that process and alters the rate of the process, this return of information is called feedback. In some cases, as the rate of the process increases, feedback information returns to the control center, which transmits control information to the process to reduce its rate; this is the negative feedback by which many control mechanisms work. Populations do not have distinct control centers, or circuits, or coded regulatory messages. Stabilization in the logistic pattern can be interpreted, however, as involving negative feedback. As the density of the population approaches K and reduces the resources available, "information" on (that is, effects of) that reduction acts on the population as feedback, slowing its growth by increasing deaths, or decreasing births, or both.

The effect of this feedback does not have to be proportional to $(1 - N/K)$. A population that is close to its carrying capacity may not slow its growth, because there may still be enough resources to permit reproduction. The effect of the exhaustion of resources on population growth is then delayed until the population has already overshot its carrying capacity. Population growth equations can be variously modified to describe such overshoot, and to produce behaviors ranging from an oscillating approach to K (Figure 2.7A), through a periodic fluctuation around K (Figure 2.7B), to an irregular pattern of eruption and crash in which the population seems to pay no attention to its carrying capacity (Figure 2.7C). It is probably uncommon for a natural population to be so directly controlled by feedback effects of N/K that it will grow to and be stabilized at the carrying capacity in the manner of Figure 2.6.

The equations in Table 2.2 do not describe natural populations, but simplified ideal situations that we can approximate in experiments. We can, however, abstract interesting characteristics of populations from these equations. Some relatively stable populations (1 and 8) spend much of their time at or near their carrying capacities. (For this statement we may have to include not just resources, but competition and predation by other populations as determining the actual carrying capacity for a given species.) In these populations, individuals that survive and reproduce under conditions of competition with other organisms pass on their genetic characteristics. Populations of such species tend to be selected toward tolerance of competition, or toward specialization in the community, or both. Thus the seedlings of the white oak are tolerant of soil effects and shade of older trees, and the red-eyed vireo is specialized to feed on forest insects in a way that is different from those of other bird species in the forest.

Relatively unstable populations, in contrast, may be selected for reproductive success—particularly their capacity for wide dispersal and rapid population growth in new environments where they are less affected by competition. The fireweed, with its wind-carried seeds and rapid multiplication in a burn area, is an example. We can use the symbols of Table 2.2 to say that adaptations of the first group of species emphasize *K-selection,* and the second group *r-selection*. Species do not fall into separate r and

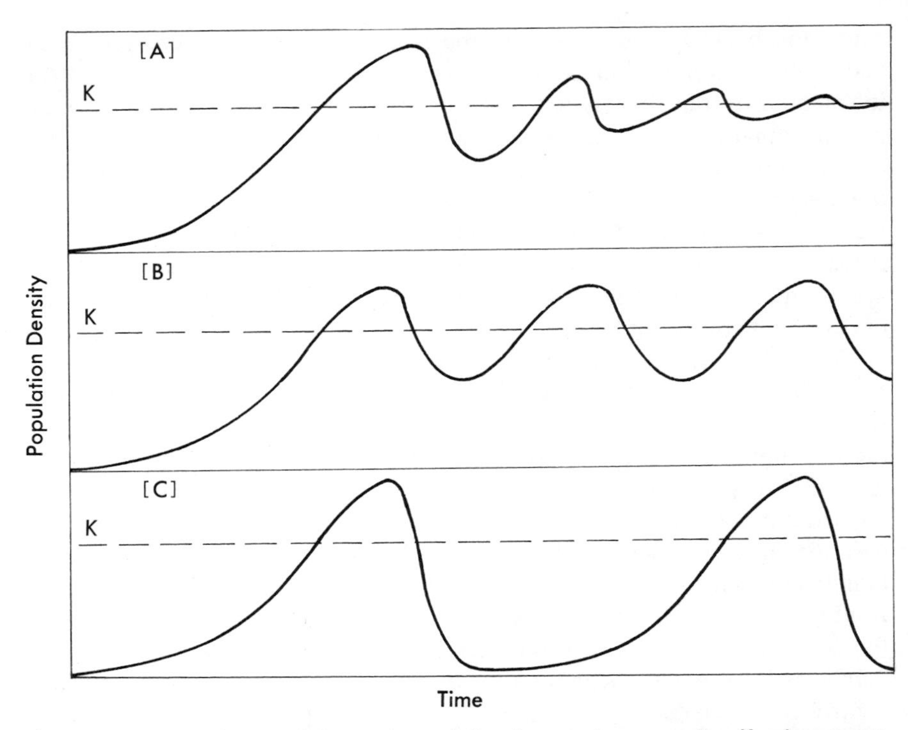

Figure 2.7. Three forms of fluctuation relative to carrying capacity, *K,* when population growth overshoots carrying capacity. **A:** Initial overshoot following which the population drops back and again increases, with decreasing fluctuation until stabilized at *K.* **B:** Initial overshoot followed by fluctuations of constant amplitude. **C:** Overshoot and crash pattern of a relatively unstable population.

K groups. These are ideal types, characteristics of most species are somewhere between them, and we shall observe a third type below. Nevertheless, the different selective conditions for survival in relatively stable and occupied, versus unstable and open environments are part of the reason for wide differences in behavior among our example populations. The different ways populations function are products of evolution; they result from selection for different ways of surviving as populations.

Competition and Crowding

By competition we mean a situation in which there is not enough of an environmental resource for both of two individuals, or two species populations. Use of the resource by one individual or species reduces the resource available to the other, and the growth or survival of the other is affected by shortage of the resource. (Or both may be affected by the competition.) Thus a young tree growing in the shade of a larger tree suffers from competitive limitation of the light available to it; the young tree will in consequence grow slowly, and it may die. A shrub population growing in the shade of a canopy tree population is limited in its growth because most of the sunlight is used or intercepted by the trees, and only a fraction of it reaches the shrubs. It may be true both that the shrubs are physiologically adapted to grow and reproduce successfully at this light intensity and that the shrub population could be much denser in the absence of trees competing with the shrubs for light.

Competition can thus occur both among the individuals of a species population, and between individuals of two or more species populations. Within-species competition increases when a population grows until it presses upon, or overshoots, its carrying capacity. Gause found, for example, that competition for food was limiting *Paramecium aurelia* at a density of 240 individuals in 0.5 cc of standard culture medium, whereas a culture with twice the standard concentration of food supported twice that density of *P. aurelia.* A larger species, *Paramecium caudatum,* requires more food; its carrying capacity with the standard culture medium was 58 individuals per 0.5 cc. In these cultures competition for food, shortage of which exerts a direct and immediate feedback on the growth and reproduction of individuals, acts to stabilize the population of either species. Another basis of this population stability should, however, be observed. Gause was adding constant amounts of food to the culture every day, and at the same time was removing excretory wastes. The paramecia were thus living on a steady-state flow of food through their system, while the rates of that flow were unaffected by the paramecia. This situation is in contrast with that of paramecia introduced into a culture provided with a fixed amount of food at the beginning of the experiment. In the latter case the paramecia will consume this limited supply of food, and then starve and crash to extinction in the culture.

We infer that there are different possible relations to food supply, or other resources, in communities.

1. The species feeds directly on the population of another species and reduces that population by feeding on it. Competition may not become effective until late, when the carrying capacity has been overshot, and a crash results. The fact that the population is able to use and reduce the "capital"—the size of the food population supporting it—contributes to instability. We shall consider such relationships further in the next section, on predation.
2. The species does not feed directly on the food population, but feeds instead on something it produces. The feeding species is living on "interest" from the food population and is not able to reduce the latter. When the feeding species is provided with a relatively constant flow of food, such as the leaf fall in a tropical rain forest, then the population

may be stabilized by competition between individuals for a steady food supply.

3. The species feeds on the population of another species, either directly or indirectly, but the population of the feeding species is limited by some factor other than competition. The feeding population may then be relatively stable, but it is not stabilized by competition.

Relation 2 can produce relative population stability for some species, and it may apply to the fungi of the forest floor. It may apply also to some animals, such as scavengers, that do not attack their food species directly. For animals that do, it may apply in a less simple way, if the feeding population has some effect on the food population but does not destroy the latter's capacity to reproduce and maintain itself. Competition may then occur among individuals of the feeding population as they use only the "interest," or surplus above maintenance, of the food species.

For competition within the species to stabilize its population, competition must reduce resources available to individuals, thereby changing the death rate or birth rate. If in a growing population the birth rate and death rate are constant—whether the population is below, or at, or above its carrying capacity—then the population will overshoot its carrying capacity. If, however, the death rate from competition increases as the population density increases, then effects of competition become negative feedback that may stabilize the population. Effects on a population that change in relative intensity as population density changes are termed *density-dependent.* Competition can contribute to the stabilization of a population if competition acts in a density-dependent manner to increase the death rate (or decrease the birth rate) as density increases.

Two types of competition were distinguished by A. J. Nicholson. Unrestrained competition, in which each individual pre-empts as much of a resource as it can, is *scramble competition.* Some species, however, have their own devices for population limitation, and these spare the reproducing individuals from adverse effects of competition. Territorial birds regulate their own densities, for the males establish territories and post these territories against intruders by singing. The territories are large enough to provide food for the male, female, and young in a year of normal food supply. Competition involving an encounter or contest between individuals, from which the victor is assured sufficient resources, is *contest competition.* The birds that are not able to establish territories apparently drift about, entering less favorable environments in which they may be more exposed to predators, and suffer a higher proportion of deaths than the territorial individuals. In many animal species competition may imply dispersal of "surplus" individuals into less favorable environments. In such species the combination of competition with emigration of and predation on the overflow individuals may function as density-dependent population control.

For species with scramble competition, the consequence of population growth may be an increased death rate, as already observed, or decreased growth by individuals, or both. Some species have a fixed or determinate size for adult individuals. In these species carrying capacity may affect mainly the numbers of individuals, as in paramecium cultures, and not the size of individuals. Other animals have indeterminate growth that does not cease at some age of reproductive maturity; for these there is no fixed size for adults. Sizes of fish in ponds can be determined in large part by the food available. Of two small ponds stocked with 100 and 50 young fish, the fish in the former may be stunted while those in the latter grow twice as large, so that total weight of fish supported can be about the same in both ponds. Sizes of plants are even more adaptable to competition and resources available. Of two trees of a given species, one in the shade of other trees may spend a century of slow growth to a stem diameter of 5 cm; the other in an opening of the forest, without competition for light, may grow in the same time to a stem diameter of 50 cm. Annual plants in deserts are adapted to their unstable environment by flexible growth to different sizes. According to the amount of rain that has fallen and the competition of adjacent plants, a desert annual may grow to a height of 2 cm and produce seeds from a single flower, or to a height of 20 cm and produce seeds from 30 flowers.

The flexibility of plant size-growth makes possible a paradoxical result of competition—decreased density with increased resources. Suppose plants are grown in two plots of the same area in a greenhouse. The plots are identical and have equal numbers of planted seeds, but one plot receives twice as much fertilizer as the other. The plants in the high-fertility plot grow faster to larger sizes, and the most successful of these plants exert strong competition (that may involve light and water as well as nutrients) on the slower-growing individuals. The death rate from this scramble competition is high, and at maturity only a few, large plants remain. In the low-fertility plot the plants grow more slowly, the death rate from competition is lower; and at maturity there are a large number of small plants. Density and mean plant weight are linked together by an exponential or logarithmic relationship. The mean area per plant (which is the reciprocal of density) varies as the square of a linear dimension such as diameter or height, whereas the mean weight per plant varies as the cube of a linear dimension. Mean weight per plant then varies as the 3/2 power of the mean area per plant, or the —3/2 power of density (Table 2.3, A). The manner in which density and weight change through time is illustrated in Figure 2.8, with five rather than two levels of fertilizer. At maturity the densities and mean weights for the five plots form, on a double-logarithmic graph, a straight line with a slope approaching —3/2.

Some other effects of crowding act in a density-dependent manner. By crowding we mean a density at which adverse effects of individuals on one

Table 2.3 Competition Equations

A. Within-Species Competition in Plants (Yoda Relationship)

Mean weight and area per individual

$$w = Ca^{3/2} = C/d^{3/2} \qquad d = 1/a \tag{1}$$

$$y = wd = C/\sqrt{d} \tag{2}$$

Mean area per plant (a) is the reciprocal of density (d); mean weight per unit area (y) is mean weight per plant (w) times density. C is a constant.

B. Between-Species Competition (Lotka-Volterra Equations) for Survival of One or Both of Two Competing Species

Given the logistic equation for change of density (N) with time (t),

$$\frac{dN}{dt} = rN\left(\frac{K - N}{K}\right) \tag{3}$$

Let species 1 and 2 each have its rate of increase (r_1, r_2) and carrying capacity (K_1, K_2), and a competition coefficient (α_1, α_2) expressing the effect of the other species on it; then

$$\frac{dN_1}{dt} = r_1N_1\left(\frac{K_1 - N_1 - \alpha_1N_2}{K_1}\right) \tag{4}$$

$$\frac{dN_2}{dt} = r_2N_2\left(\frac{K_2 - N_2 - \alpha_2N_1}{K_2}\right) \tag{5}$$

The outcome of competition will be determined by the competition coefficients and carrying capacity:

Condition		Outcome	
$\alpha_1 < K_1/K_2$,	$\alpha_2 > K_2/K_1$	Only species 1 survives	(6)
$\alpha_1 > K_1/K_2$,	$\alpha_2 < K_2/K_1$	Only species 2 survives	(7)
$\alpha_1 > K_1/K_2$,	$\alpha_2 > K_2/K_1$	One or the other survives	(8)
$\alpha_1 < K_1/K_2$,	$\alpha_2 < K_2/K_1$	Both survive	(9)

another become detrimental or limiting to the population; not all such effects are from competition. In some experimental cultures of animals waste products accumulate until they limit population growth or cause population decline. Gause observed such effects with his paramecium cultures, if the medium was not frequently replaced with fresh medium. Wastes can also limit growth in cultures of fungi and bacteria, and some plant species release substances that accumulate in the soil until growth of the plants is inhibited. In experimental cultures of flour beetles (*Tribolium*) studied by Thomas Park, the larvae and adults cannibalize the eggs and pupae. As the density of individuals in the culture increases, the frequency with which eggs and pupae are encountered and eaten increases, until a density is reached at which the death rate equals the birth rate. Cannibalism is in this case density-dependent and acts to limit and stabilize the populations. Can-

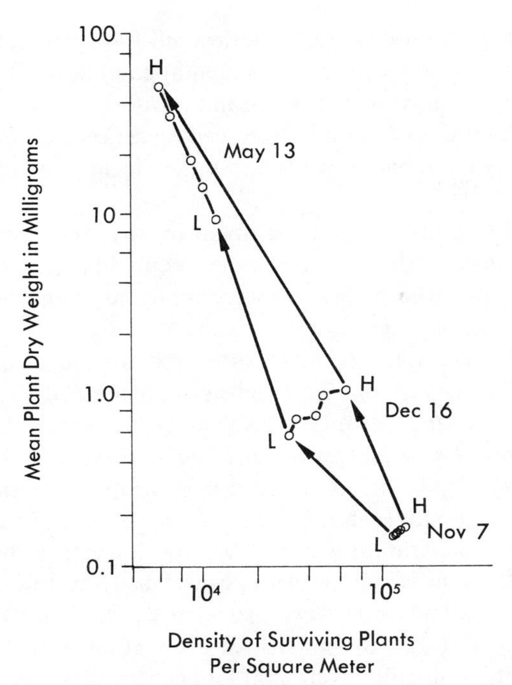

Figure 2.8. Plant density and mean weight. Effects of competition are shown for populations of an herb (*Erigeron canadensis*) growing at five levels of soil fertility from high, *H,* to low, *L,* using N–P–K–Mg fertilizer in amounts with ratios of 5/4/3/2/1. Both axes are logarithmic. [Modified from Yoda *et al., J. Biol. Osaka Univ.,* **14:**107 (1963).]

nibalism of the young is not uncommon in natural populations of carnivorous animals and can reduce the height of population peaks in some of these. In some vertebrate animals, crowding in laboratory experiments produces symptoms of stress, including endocrine changes, increased aggressiveness, and neglect or killing of the young. It is not clear that these laboratory observations can be applied to natural populations, which generally have much lower densities than the laboratory cultures. However, a variety of penalties—not only self-impoverishment in resources with stunted growth or increased death rate, but toxication of environment, self-aggression or cannibalism, and social or reproductive failure—can affect a population that grows until it is crowded.

Some of Gause's experiments also concerned between-species competition. Two species of *Paramecium* were introduced in equal numbers into a culture with bacteria for food and an essentially constant environment. Both populations grew until there was not enough food to support the growth of both. No two species have identical population functions. One species will have a higher rate of obtaining food, or of population growth or of survival, than the other. One species will have competitive advantage over the other, in a given set of environmental conditions. The experimental culture can support only so many paramecia. Given two species in direct competition, within this essentially fixed ceiling set by resources, one species grows to its carrying capacity while the other declines to extinction in the culture (Figure 2.9).

Gause sought to modify the experiment in ways that would permit both populations to survive. In one set of experiments fixed fractions of the culture were withdrawn, the paramecia were removed from this fluid, and the fluid minus paramecia was returned to the culture. This removal of part of the population resembles the effect of a predator that is density-independent in its prey consumption. This density-independent predation did not stabilize the culture and permit survival of both species. In other experiments *Paramecium bursaria,* instead of *P. caudatum,* was placed in competition with *P. aurelia. P. bursaria* could survive in the bottom part of the culture, even though only *P. aurelia* persisted in the rest of the culture. Thus a degree of spatial separation of these two species within the culture could permit a coexistence in which neither species was forced to extinction.

Equations for competition were provided by A. J. Lotka and V. Volterra (Table 2.3, B). Let the growth of each population be according to the logistic equation already given, and let each species have its own carrying capacity (K_1 and K_2), its equilibrium density in the culture in the absence of the other species. The growth of population 1 is now limited not just by $(1 - N_1/K_1)$, but by this and the effects of the other species, hence by $(1 - N_1/K_1 - \alpha_1 N_2/K_1)$, or $1 - (N_1 + \alpha_1 N_2)/K_1$. The last expression states that as the density-equivalent of (and resource utilization by) species 1 and 2 together approaches K_1, population growth of species 1 must cease. The α_1 is a competition coefficient that expresses the effect of species 2 in limiting the growth of species 1; α_2 is the corresponding effect

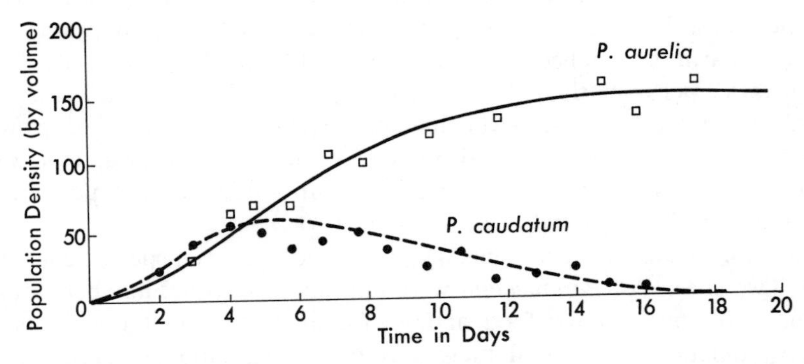

Figure 2.9. Competition between two *Paramecium* species. The two species are direct competitors in an experimental culture, and one of them declines to extinction in the culture. [After Gause, 1934.]

of species 1 in limiting species 2. These equations, like the logistic on which they are based, are much too simple to be realistic for many natural populations; but their implications are interesting.

The equations do not lead to a single outcome, but different results of competition depending on the relations of α_1 and α_2 to K_1 and K_2 as indicated in the lower part of Table 2.3B. Of the four possible relationships, three imply that one of the species will go extinct as in Gause's first experiments. The last case is more interesting; given this relationship the two species can survive at equilibrium in one another's presence. In this case each species is limiting its own population growth more than that of the other species. It can do this if, for example, the species are spatially separated in the culture, so that one species is limited by its own within-species competition in one part of the culture, and the other in the other part of the culture. Persistent competition is then possible because the species have divided the space of the culture between them, so that each species is the more successful competitor in its own part of the culture. Such was the case with *Paramecium aurelia* and *P. bursaria. P. aurelia* and *P. caudatum,* in contrast, did not divide the space or resources of the culture between them, and the less effective competitor (*P. caudatum*) went extinct.

Between-species competition is thus not, in its simplest form, a means of population control and stability; it is often a force toward instability, that is, extinction. Yet it is also an evolutionary force toward survival through divergence in the way species relate to environment and each other. If a new species is added to a community, by extension of its range or by transport to an island where it had not occurred, the new species may be a close competitor to one already in the community. If the new species is the stronger competitor, it may make the weaker native species extinct there. The weaker competitor need not become extinct, however, if it also occurs in a different community to which the stronger competitor is not adapted, or if the weaker competitor so evolves that its population becomes adapted to a different community from that now occupied by the stronger species. In these cases competition produces not extinction but difference in species distribution.

On the rocky shores of Scotland two barnacle species occur, one at higher and the other at lower tide levels. One of the species, *Balanus balanoides* is able to grow throughout the intertidal zone; the other, *Chthalamus stellatus,* can occur through a wide range of tide levels, but its population is centered at higher levels than that of *Balanus.* Where the two species occur together, *Chthalamus* is restricted to the upper tide levels by competition with *Balanus,* which is the stronger competitor at middle and lower tide levels. The two barnacle species can survive on the same shore because each is better adapted to some part of the shore environment than the other.

Such adaptive differences appear among plants also. The soils formed from serpentine rock are peculiar in chemical composition, as will be discussed further in Chapter 6. Many plant species cannot grow on these soils; some that grow both on and off serpentine have genetically different populations (ecotypes) that are adapted to the two soils. Some species grow only on serpentine. Experiments with some of the latter have shown that they grow well in gardens on non-serpentine soils, but fail to grow on non-serpentine soils occupied by other plants because of the competition with these plants. *Chthalamus* and the serpentine plants are able to occur in special (and relatively extreme) environments to which they are adapted, and which exclude or limit the occurrence of other species. *Chthalamus* and the serpentine species are excluded from other, apparently more favorable environments by competition. Competition thus regulates species distribution in a way that could not be recognized except by the behavior of the species in the absence of competition.

Predation

Predation typically occurs when one animal (the predator) catches, kills, and eats another and less fortunate animal (the prey). For discussion of population interactions in general, we need a broader definition of predation: any direct killing for food, of organisms of one species by those of another. The capture of a rabbit by a fox is predation, but so is the capture of flies by a pitcher plant, the grazing of aquatic animals on algal cells, and the consumption of acorns by squirrels. Predators intergrade with parasites, and many insects are intermediate. A parasitic wasp may lay an egg on a caterpiller, the egg becomes a larva that lives in the flesh of the caterpiller and feeds on its tissues until the caterpiller is killed. The wasp larva becomes a pupa from which a wasp later emerges. The wasp is both a parasite and a predator with a single prey. Insects with this pattern of predation by way of a parasitic larva are referred to as *parasitoids*. We are interested in the way any kind of predation affects the death rate of a population.

It is again convenient to use Gause's work for illustration. Into a culture of *Paramecium caudatum* he introduced individuals of *Didinium nasutum,* a barrel-shaped protozoan predator that attaches itself to a paramecium and sucks the latter dead. Once introduced into the paramecium culture didinium feeds on and kills the paramecia, and multiplies, and the additional didinia feed on and kill paramecia, until paramecium is extinct in the culture. Didinium is then rewarded for its predatory efficiency by starving to extinction (Figure 2.10A). Gause sought to produce a more stable interaction by using cultures with sediment at the bottom, where some paramecia were protected from didinium. A stable interaction did not result: the didinia multiplied and destroyed all the paramecia except those in hiding; the didinia then starved to extinction, and the paramecia began a sigmoid growth in the absence of the predator (Figure 2.10B).

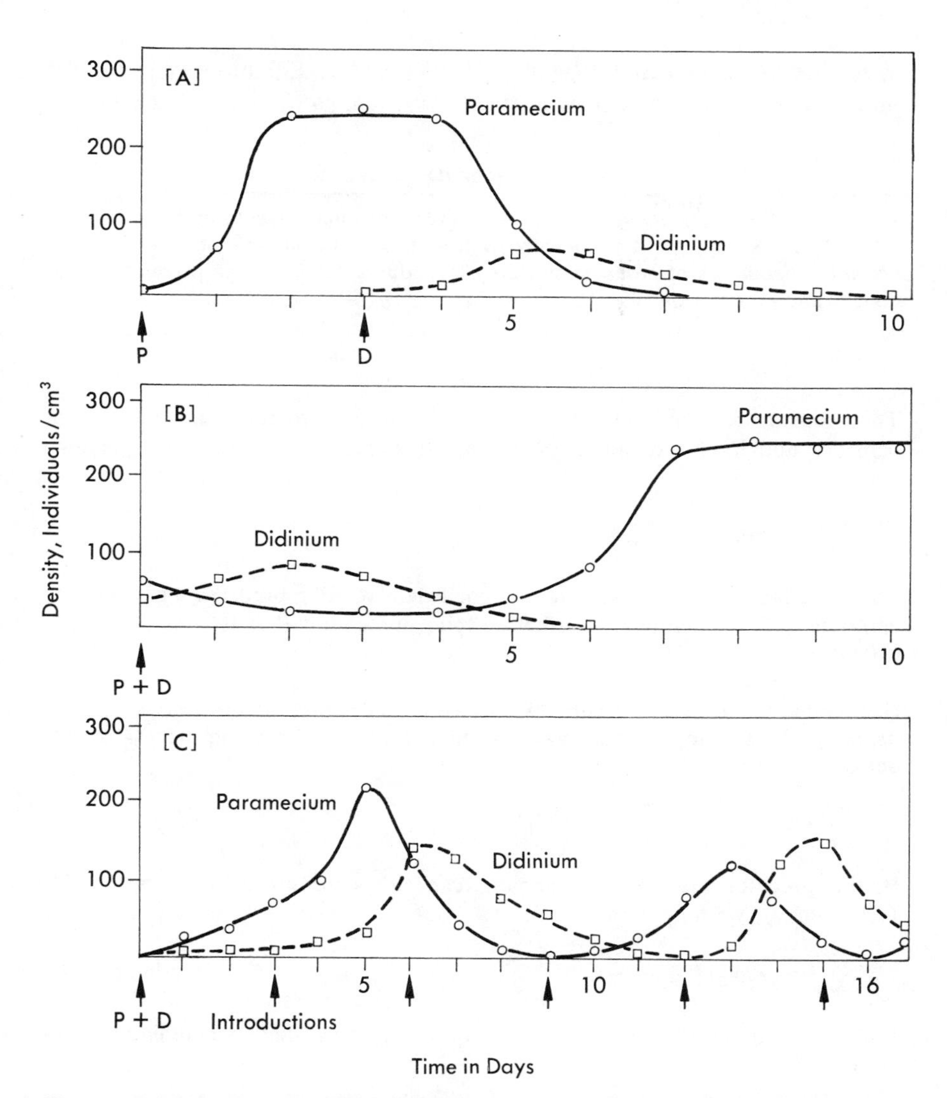

Figure 2.10. Predator-prey interactions between *Paramecium* and *Didinium.* **A:** Introduction of *Paramecium* (at P) and *Didinium* (at D) into a culture without a refuge for *Paramecium.* **B:** *Paramecium* and *Didinium* are introduced together (P + D) into a culture with sediment as a refuge for *Paramecium.* **C:** *Paramecium* and *Didinium* are introduced together every three days, preventing permanent extinction of either and producing predator-prey cycles. [Modified from Gause, 1934.]

Gause was able to produce a sustained interaction (with marked fluctuations in numbers) only by periodic additions of both predator and prey to the culture (Figure 2.10C). Only as part of a larger system from which there was continual "immigration" was the interaction of paramecium and didinium relatively stable.

The traditional equations for interaction of predator and prey are those developed by Lotka and Volterra (Table 2.4A). Equation (1), for the prey, states that this can grow at the exponential rate (r_1N_1), but will also

Table 2.4 Predation Equations

A. The Lotka-Volterra Equations. The prey adds individuals according to its intrinsic rate of increase (r_1) times its density (N_1), and loses individuals at a rate proportional to encounters of predator and prey individuals, hence to the product of prey density and predator density (N_2),

$$\frac{dN_1}{dt} = r_1N_1 - PN_1N_2 \qquad \text{prey} \qquad (1)$$

The predator adds individuals at a rate proportional to this same product of densities and loses individuals according to a death rate (d_2) times its own density,

$$\frac{dN_2}{dt} = aPN_1N_2 - d_2N_2 \qquad \text{predator} \qquad (2)$$

P is a coefficient of predation and a a coefficient relating predator births to prey consumed. The equations describe a cyclic fluctuation that is unstable if disturbed.

B. Carrying Capacities. It is more realistic to consider that the prey has a carrying capacity, K_1, and to give the prey equation the form of the logistic (Table 2.2, equation 5). Then,

$$\frac{dN_1}{dt} = r_1N_1(1 - \frac{N_1}{K_1}) - PN_1N_2 \qquad \text{prey} \qquad (3)$$

If the predator also has a carrying capacity determined by prey density ($K_2 = bN_1$), we may write,

$$\frac{dN_2}{dt} = r_2N_2(1 - \frac{N_2}{bN_1}) \qquad \text{predator} \qquad (4)$$

Interactions described by (3) and (2), or (3) and (4) have stable equilibrium points.

C. Predator Saturation. It may also be more realistic to consider that there is a limit on the number of prey that the predator individuals can consume, and to express decreasing consumption of prey (per prey individual) as this limit is approached by a saturation factor, C. (For C we can use such forms as $(1 - e^{-cN_1})/N_1$ of Ivlev or $D/(D + N_1)$ of Holling. Then,

$$\frac{dN_1}{dt} = r_1N_1(1 - \frac{N_1}{K_1}) - PN_1N_2C \qquad \text{prey} \qquad (5)$$

$$\frac{dN_2}{dt} = aPN_1N_2C - d_2N_2 \qquad \text{predator} \qquad (6)$$

Equations (5) and (6) describe interactions that will generally either have a stable equilibrium point, or fluctuate in a stable limit cycle, depending on the coefficients.

Table 2.4—Continued

D. Prey Refuge. The effect of a refuge for the prey may be most simply described by giving the prey a number (B) of individuals protected from predators, and replacing N_1N_2 with $(N_1 - B)N_2$ in both predator and prey equations. Only the prey in excess of the protected number $(N_1 - B)$ are consumed, and $(N_1 - B)N_2$ is replaced by zero when N_1 is less than B. Using this refuge expression in equations (1) and (2) gives an interaction with a stable equilibrium point.

[See also May, 1973 (equations 3.1a and b, 4.1a and 4.3b and 4.4 or 4.5), and Maynard Smith, 1974.]

lose individuals to predation $(-PN_1N_2)$. The loss to predation is assumed to bear a simple, linear relation to the number of encounters between predator and prey as both move at random. The frequency of encounter is a function of the product of prey density (N_1) times predator density (N_2); P is a coefficient of predation relating encounters to actual number of prey killed. The predator population (2) grows at a rate directly proportional to the same prey consumption (PN_1N_2) times a, a coefficient relating the number of new predator individuals born to prey killed. Predators are assumed also to die $(-d_2N_2)$ at a rate proportional to the predator density (N_2). These equations are much too simple for real predator–prey interactions, but it is of interest to see what they imply.

They do not imply stable populations. Instead they describe a cyclic fluctuation of predator and prey (Figure 2.11). In each cycle, as the prey population increases, the predator population follows, overtakes, and overconsumes it, producing a decline in the prey. The predator population now follows the declining prey supply downward until the predators reach so low a level that the prey population begins again to increase; whereupon the predator begins again to follow it upward. Assuming the environment is constant, then the intensity or amplitude of the cycles—the contrast between peak and bottom densities—is determined by the initial densities and

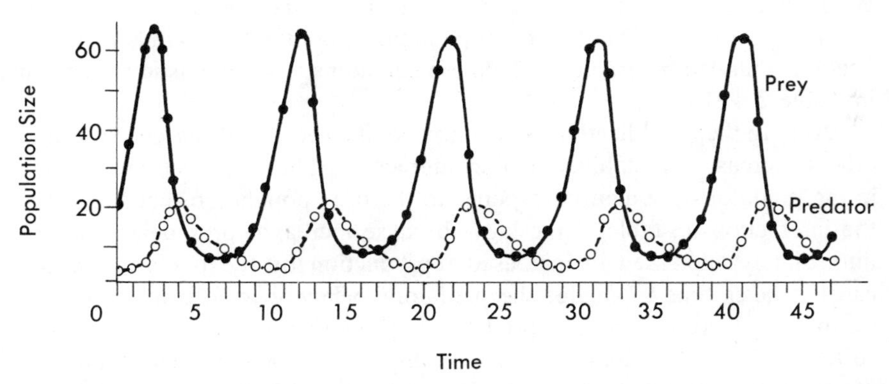

Figure 2.11. Population cycles for a hypothetical predator and prey, as predicted by the Lotka-Volterra equations (Table 2.4A). [Krebs, 1972, p. 249.]

remains the same. The predator and prey populations can thus cycle forever, in a computer.

It has been tempting to use these equations to interpret natural cycles such as that of the snowshoe hare, referred to above, and its predator the lynx (Figure 2.2). There are problems in such interpretation. The lynx's age at first reproduction is greater than the hare's and its number of young per brood is smaller. In this situation the lynx has a smaller intrinsic rate of population increase, *r,* than the hare; and it is probably impossible for a lynx population to grow so fast as to overtake a growing hare population. The lynx consequently cannot overtake and control the hare population. The same difficulty applies to other interactions of small mammal prey with predators that are larger and slower in reproducing than the prey. Some of these interactions appear to involve a prey population that increases out of control by the predator (while the predator increases, but more slowly) until the prey encounters a limit to its growth (for reasons unknown). A rising predator population may now feed on a declining prey population and hasten its decline. The predator population may cycle in response to the prey, but the prey population is not really being controlled by the predator.

For description of these and other interactions the Lotka-Volterra predation equations are quite inadequate. All populations are subject to some kind of environmental limit or carrying capacity, but none is included in these equations. The equations make no allowance for the effect of refuges in protecting part of the prey population, for the fact a predator can only eat so many of the prey it encounters, for change in the predator death rate with food supply, and so on. Moreover, the cycles of equations (1) and (2) are "stable" only in a highly artificial way. In a constant environment the amplitudes remain constant; but environments are not, in fact, constant. Any environmental fluctuation increases the amplitude of the cycles. In an environment of continual fluctuation, the amplitudes of the cycles increase until the populations described by equations (1) and (2) go extinct. The Lotka-Volterra predator-prey equations are thus so unnatural as to be irrelevant to real populations. Some of the kinds of modifications that can be used to make these equations less unrealistic are shown in Table 2.4 B to D.

Among the complications that must be allowed for, some decrease and others increase the stability of the interaction. The introduction of a lag in the predator's population response to the prey population tends to make the interaction unstable. Time lags (because increased prey density is not immediately expressed in increased reproduction by the predator, or because it takes time to change the predator's behavior to feed more heavily on the prey) are common in predator-prey interactions. Other effects suggested in Table 2.4 contribute to stability of interactions. The limitation of the prey population by a carrying capacity tends to increase the stability of an interaction. If we give the prey a carrying capacity as in Table 2.4 (3) and use (2) for the predator, these equations describe a stable interaction. The densities of the predator and prey in this interaction will remain constant if the environment remains constant. If that stability is disturbed by environmental fluctuation, the prey and predator populations will return from their altered densities to the same constant densities. This interaction has a *stable equilibrium point* to which the densities converge following disturbances. If we give the predator a carrying capacity determined by prey density, Table 2.4 (4), the interaction is stable whether the prey has a carrying capacity (3) or does not (1). Instead of the logistic we may modify equation (2) to imply that the predators are increasingly starved to death as the ratio between their density and the prey density increases. If, to express this effect, we substitute a death term using the ratio of the predator and prey densities ($-d_2N_2/N_1$) for the death term ($-d_2N_2$) in equation (2), the combination of (1) with this modified (2) has a stable equilibrium point. If part of the prey population is protected from the predators by refuges (Table 2.4D), the interaction may be stable (even if it was not in Gause's experiment with sediment as a refuge for paramecium). If immigration re-establishes local populations that have fluctuated to extinction, then over a larger area that includes many local populations, instability may be decreased by the effects of migration (note Gause's experiment with continued introductions).

Other effects may result from characteristics of predator behavior. The number of prey individuals that a predator can consume in a given time may be limited. The effect of the saturation of the predator's appetite as this limit is approached can be expressed as in Table 2.4C. Interactions in the form of equations (5) and (6) may have stable equilibrium points, or they may show cyclic fluctuations. Such cycles are different from those of the Lotka-Volterra equations (1) and (2), however. Cycles such as described by (5) and (6) may have constant amplitudes and mean densities while environment is constant; and after disturbance they may return to the same amplitudes and mean densities as before. Such cycles that recover from perturbation are *stable limit cycles*. The hare and lynx interaction may be a stable limit cycle; it is not a Lotka-Volterra cycle.

Thus a number of ways real populations behave may contribute to relative stability of predator-prey interactions. These damping effects are opposed by the effects of environmental fluctuation in perturbing the interactions, and the disparity of intrinsic rates of increase of many predator and prey populations. The equations of Table 2.4 include neither lag effects, nor, except in (4), any indication of predators' intrinsic rates of increase (r_2). Except in (4) it is assumed that the rate of predator population growth is determined directly by the rate of encounters of predator and prey, aPN_1N_2, or by this encounter term as modified by predator carrying capacity or saturation, or prey refuges. Many of the kinds of interactions in Table 2.4B to D become less stable if the predator is limited by an intrinsic rate of increase smaller than that of the prey. We note again the essence of population limitation: density-dependent effects on the population. The prey population is not stabilized if the predator takes 50 per cent of the prey at both high and low densities (density-independent predation). Much less is a prey stabilized if a slow-growing predator population takes 50 per cent of the prey at its low densities and only 30 per cent at its high densities (which is an inverse density dependence). The prey population may be stabilized if the predator takes 20 per cent of the prey at low density and 80 per cent at high density. Stability is possible if the *relative* (percentage) loss to predation increases as the prey density increases. In density-dependent predation the prey is receiving a negative feedback "message" in the form of an increasing death rate as its population density rises.

Density-dependent predation may result from refuge effects as suggested in Table 2.4D. It may also be based on two effects that are not modeled in the table and that may sound more appropriate to railroading than ecology: tracking and switching. Control of the prey by the predator may be possible if the predator can "track" the prey and not merely follow the prey population as it increases but gain on and overtake the prey population by more rapid growth. Control may be possible if a sufficiently large predator population, that does not track the prey, switches from feeding on other species to feeding on the prey as the prey population increases. Control may be possible if the predator both increases more slowly than the prey *and* switches to feeding on it.

Tracking is possible for some of the large group of predators we termed parasitoids. These parasitic insects, many of them small wasps and flies, have life cycles that may be of the same length as, or shorter than, those of their hosts. The female parasitoids seek out the hosts to lay eggs on them, and each female may lay tens or a few hundreds of eggs. It thus may be possible for the parasitoid to overtake the prey, if the latter is not itself increasing too rapidly. Figure 2.12 illustrates population fluctuations of the same host species with a slowly reproducing parasitoid, above, and a rapidly reproducing parasitoid, below. Coefficients of fluctuation for the host (and parasite) are 3.38 (and 1.93) in the former case, 1.42 (and 1.30) in the latter.

Parasitoids are not unimportant in the natural world. Most free-living insect species have one or several parasitoid species attacking them, and some of these parasitoids have their own hyperparasitoids. Most parasitoids are narrow specialists, attacking one or a few prey species. It is possible, therefore, that there are more parasitoid than prey species and, in fact, more insect parasitoids than all other species of the living world together. There are probably more than a million parasitic wasps, ranging in size from that of free-living wasps down to minute egg parasitoids, smaller than some

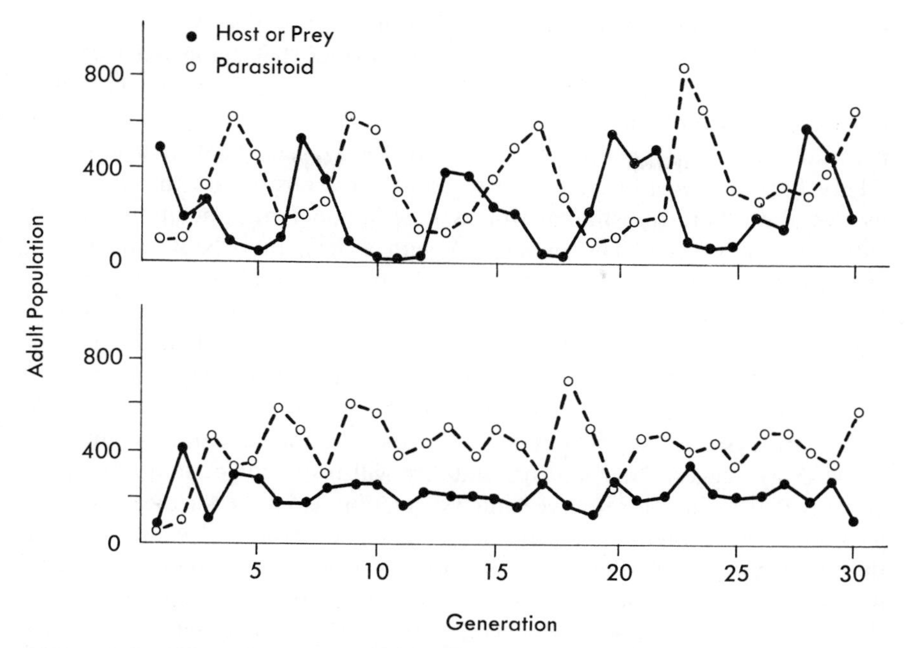

Figure 2.12. Effect of predator reproductive rate on the prey. In the interaction above, the parasitoid reproduces slowly, and the delay in its response to increase in the prey produces well-defined predator-prey cycles. (See Figure 2.11.) In the interaction below, the parasitoid reproduces rapidly and acts as a control that stabilizes the prey population. The host or prey is the azuki bean weevil (*Callosobruchus chinensis*) in both cases, as affected by two wasp parasitoids: *Heterospilus prosopidus* above, and *Neocatolaccus mamezophagus,* below. [Utida, 1957.]

of the largest protozoans. Parasitoids probably are responsible for, or contribute to, the control of many insect populations. Parasitoids do not necessarily control a species they attack, however, and many insects are apparently not controlled by them.

Vertebrate predators are more flexible in food choice and more likely to switch. A predator that had been feeding on other species can respond to an increasing prey population by learning a search-image for that prey and concentrating on it, sometimes with immigration to areas of denser prey population. The switch will not necessarily control the prey. If prey numbers exceed the appetite or capacity of the predator, the prey eruption is not controlled.

There is evidence, however, that some vertebrate predator-prey interactions are fairly stable. On the Kaibab Plateau of Arizona the predators on the deer (cougars and wolves) were largely killed off by man. The response of the deer population—eruption, to the destruction of its food supply by overgrazing, followed by a crash—is one of the classic observations of ecology. Similar eruptions and declines have been observed in some mammal populations released on islands where they were free from predation. The removal of many cattle and sheep, that competed with the deer for food, from the Kaibab Plateau may have contributed to the eruption there. It is probable, however, that killing the predators on the plateau permitted a deer population that had previously been kept at a low and relatively stable level by predation to erupt. In contrast with this eruption, in several northern forest areas wolves exist in apparent population balance with large prey (deer, elk, or moose) with a mass ratio between the two of about 1 kg of predator to 150–300 kg of prey. These balances may seem to contradict what we have said about the necessity of tracking and switching, for control of these herbivores may be based on neither of these.

We expect a two-part predator-prey interaction with a lag in the predator's response to increase in the prey to be unstable, taken by itself. A three-part system can, however, be stable. Mathematical modeling by R. M. May suggests, in fact, that stability will result for certain three-part systems that include a resource limit (vegetation as food) for the herbivore and a long-lived predator on the herbivore. The herbivore population can then be regulated at a level that is normally below the resource limit and is not destructive to the food resource. The regulation is based not on any single density-dependent mechanism, but on the combined effect of available food and predation on the herbivore population. The relative stability results from the function of the three-part interaction.

Certain other examples bear mention. In an experiment by R. T. Paine a major predator species, a starfish, was removed from a rocky ocean shore. The removal had an unstabilizing effect on the community as it had existed. Certain species, particularly mussels, increased in densities when relieved of predation, whereas others, apparently affected by competition with the mussels, declined to disappearance in the study plots. When grazing animals are removed from a grassland, certain grass species may become more strongly dominant, and the number of plant species in the grassland may decrease. Such observations suggest a regulatory function of predation that, by restraining the populations of some species that are strong competitors, makes possible a greater number of species in the community.

The Klamath weed (*Hypericum perforatum*) is a plant toxic to cattle that was introduced from Eurasia into the western United States. It became a major pest, occupying millions of acres of range land. A flea beetle (*Chrysolina quadrigemina*) that feeds on the weed was introduced from Europe, multiplied in the infested range lands, and reduced the Klamath weed to virtual extinction in them. The weed did not become extinct, however, but persisted at low densities in shaded areas in forests, where the beetle does not effectively prey on it. The flea beetle is now uncommon. These observations make a number of points:

1. The plant prey survives because it has a refuge from the predator.
2. The relative stability of the plant is maintained by the predator, which prevents the plant population from erupting into the range lands.
3. An animal predator which is well able to overtake the plant population as it grows is the effective control for this plant population in the range land.
4. This predator, not simply environmental tolerance, determines the present distribution of the plant.
5. The control mechanism is almost invisible. If eruption and decline of the plant had not been observed, we should not know what restrains its population or why it is largely limited to forests.

Symbiosis

Symbiosis refers to various lasting, close associations between organisms of different species. Three kinds of relations are usually distinguished: mutualism, in which both organisms benefit; commensalism, in which one benefits and the relation is largely neutral for the other; and parasitism, in which one benefits to the detriment of the other. These relations intergrade; and parasites both intergrade with predators and include pathogens (disease-producing parasites).

Paramecium bursaria has green algal cells scattered through its protoplasm, and there are a number of other protozoans and multicellular animals that have symbiotic algal cells in their cells. The algae in *P. bursaria* are capable of producing enough food by photosynthesis for themselves and the paramecium; but the paramecium is also capable of feeding itself as a predator on yeasts or bacteria. Probably the relationship between the two fluctuates between mutualism in the light and commensalism or parasitism in the dark. It is likely that chloroplasts and mitochondria, and possibly some other organelles, in the cells of higher organisms evolved long ago from cellular symbionts. The most conspicuous symbionts in many communities are lichens. A fungus forms the structure of the lichen and contributes to the partnership water-holding capacity, nutrient uptake, and chemical defense; the cells of a green or blue-green alga are held among the filaments of the fungus and produce by photosynthesis the food for both partners. The lichen can be regarded as a fungus with a symbiotic alga, or as a new organism formed of both. The success of the partnership permits lichens to be major organisms in the arctic tundra, and on rock and bark surfaces in many climates.

Some other mutualistic relations are important in communities. Wood is one of the major biological resources of the world, but very few higher animals are able to digest the cellulose and lignins that are major components of wood. In cool-temperate climates wood decay must be accomplished primarily by higher fungi. In warm-temperate and tropical climates much dead wood is consumed by termites, which have in their intestines strange flagellated protozoans able to use wood as food. From this partnership the protozoans obtain a home and a supply of wood particles as food, while the termites are supported by the surplus of sugars from wood digested beyond the protozoans' own needs. Large grazing mammals are dependent on symbiotic bacteria that live in the rumen, a special part of the stomach, for adequate digestion of plant tissues. Some higher plants (especially legumes, species of the pea family) form a partnership with nitrogen-fixing bacteria in their roots, with the plant providing food to the bacteria and the bacteria providing nitrates to the plant.

Symbiotic relations termed *mycorrhizal* are formed also by higher plants and the fungi on their roots. The root fungus forms a sheath around smaller roots, and fungal filaments extend into the root tissues between the cells in some cases, or into the root cells in other cases. Other fungal filaments or hyphae extend out into the soil or the litter of dead leaves and wood on the soil surface. In plants with mycorrhizal roots the fungal filaments provide the surface through which inorganic nutrients are taken up by the fungus and passed on to the plant, while the fungus may in return receive food from the plant. Some mycorrhizal fungi digest dead organic matter on or in the soil and obtain from it inorganic nutrients that are passed on to the plant. In some cases the plant as a seedling, or even as an adult, receives from the fungus food derived from decomposition of dead organic matter by the fungus. Most higher land plants are in partnership with root fungi; and members of all three major groups or phyla of higher fungi (conjugation fungi, sac fungi, and club fungi) have evolved to participate in such partnerships.

A variety of relations of mutual benefit involve plants, and animals that either pollinate them or disperse their seeds. These relationships are not symbioses in the usual sense, for there is no close and continued living together of individuals of the two species. The two species may be to some extent dependent on each other, however, and may be adapted to each other. A squirrel both preys on acorns and accidentally plants oaks by burying and forgetting some of the acorns. The squirrel depends in part on the oaks for food; while the oaks gain more effective seed dispersal. Many striking adaptations relate the color, form, and food offering (nectar or pollen) of flowers to the behavior of pollinating animals. In the Tropics particularly, many of these relationships are highly specialized. Pollination is a kind of mutualism of organisms that do not live in close contact—the animal gains food and the plant gains fertilization. Pollination is subject also, however, to a kind of parasitism. Some bees and other animals feed on nectar or pollen without transferring pollen from one flower to another.

The plant species is then supporting, with the food supply it has evolved to produce as a part of the interaction, both pollinators and free loaders.

In commensal relations one organism generally serves the other as a surface for attachment, or means of shelter, without being fed upon. Many plants are attached to the surfaces of other plants that support them; the plant being supported is termed an *epiphyte*. Many lichens are epiphytic on the bark of trees. Some aquatic animals are borne on the surface of other animals, and in aquatic communities the surfaces of both plants and animals are coated with films of bacteria, microscopic algae, and other microorganisms. Commensalism includes some remarkably specialized relationships. Certain small leech-like worms live on the pincers of crayfish; some mosquito larvae live only as commensals in the water held by the leaves of plants that are commensal epiphytes on tropical rain forest trees; a small fish (*Carapus*) is intimately commensal with a sea cucumber and swims in and out its anus. Certain cleaner shrimps (*Periclimenes*) of coral reefs are commensal with sea anemones, living in the shelter offered by the tentacles and stinging cells of the anemones. These shrimps also, however, have relations of mutual benefit with fish that come to the shrimp to be cleaned: the fish gains by removal of parasites from its skin, and the shrimp gains food without having to leave its shelter to search for it.

Parasitism is pervasive; most or all individuals of most species bear or contain parasites. Since a parasite is typically dependent on its host for food, it is to the parasite's disadvantage to kill the host. When a parasite species kills some, but not all of the host individuals attacked, two genetic implications are possible. First, some of the more vulnerable host individuals will die or fail to reproduce, and will then fail to pass on their genetic characteristics. The parasite exerts a differential selection in favor of the less vulnerable host individuals and their genetic characteristics. The host species consequently evolves toward increased resistance to infection or toward tolerance of the parasite (or a combination of greater resistance and tolerance). Second, the virulent parasite that kills its host (and thus itself) before the parasite reproduces fails to pass on its genetic characteristics. A milder parasite of the same species does reproduce and pass on its characteristics. There is consequently differential selection toward reduced virulence in the parasite; this selection can be considered a kind of feedback affecting the genetic characteristics of the parasite. These selective processes imply that host and parasite evolve toward a reasonable accommodation, one in which the host tolerates the parasite and the parasite limits its effect on the host, which then can support it without serious detriment. (Parasitoids are set aside from this statement as specialized predators.)

We can express symbiotic relations, in a simplified way, with equations comparable to the Lotka-Volterra equations for competition and predation. Suppose first that two populations are present in a community without interaction and that their population behaviors are reasonably described by logistic equations (Table 2.5, equations 1 and 2). If, now, species 1 evolves toward partial symbiosis, its new carrying capacity becomes the sum of the old carrying capacity (K_1) independent of species 2, plus a new resource use based on species 2 (bN_2). A modified equation (3) results for species 1, while the effect on species 2 is expressed by the term aN_1 in equation (4). The sign and magnitude of the coefficient a now define the character of the relationship. When a is positive, it expresses the number

Table 2.5 Symbiosis Equations

Let two species behave by logistic equations with separate carrying capacities (K_1 and K_2) and no interactions,

$$\frac{dN_1}{dt} = r_1N_1\left(\frac{K_1 - N_1}{K_1}\right) \qquad (1)$$

$$\frac{dN_2}{dt} = r_2N_2\left(\frac{K_2 - N_2}{K_2}\right) \qquad (2)$$

Species 1 now evolves toward partial symbiosis, using species 2 as an additional resource, so that the new carrying capacity for species 1 is $K'_1 = (K_1 + bN_2)$,

$$\frac{dN_1}{dt} = r_1N_1\left(\frac{K_1 - N_1 + bN_2}{K_1 + bN_2}\right) \qquad \text{symbiont} \quad (3)$$

and the equation for species 2 becomes

$$\frac{dN_2}{dt} = r_2N_2\left(\frac{K_2 - N_2 + aN_1}{K_2}\right) \qquad \text{host} \quad (4)$$

In these, b is a coefficient for support of individuals of species 1 by individuals of species 2, and a is a coefficient for the effect of species 1 on the population of species 2.

If species 1 evolves to become wholly dependent on species 2, K_1 approaches zero, and the carrying capacity for species 1 is now defined by the population of species 2 and the coefficient of support, b. The relation is then described by equations (4) and (5),

$$\frac{dN_1}{dt} = r_1N_1\left(\frac{bN_2 - N_1}{bN_2}\right) \qquad \text{symbiont} \quad (5)$$

A saturation factor may be added, as a limit on the benefit per host individual from the interaction with the symbiont, in the form $C = D/(D + N_2)$. A stable mutualism is then expressed by equations (5) and (6)

$$\frac{dN_2}{dt} = r_2N_2\left(\frac{K_2 - N_2 + aN_1C}{K_2}\right) \qquad \text{host} \quad (6)$$

[I am indebted to Robert May for discussion of these and related equations.]

of individuals (or the fraction of an individual) of N_2 supported per individual of N_1 and the relation is mutualistic. When a is zero or a very small value, species 1 is commensal on species 2. When a is negative, it expresses a loss of individuals from N_2 per individual of N_1; and equations (3) and (4) then describe parasitism of species 1 on species 2.

[The resemblance of some of the symbiosis equations to these in Tables 2.3 and 2.4 may be noted. Equations (3) and (4) in Table 2.5 and the Lotka-Volterra competition equations (Table 2.3B) differ, apart from the signs of a and b, in the inclusion of the interaction term bN_2 in the denominator of (3). The form of equation (3) (with $-\alpha_1$ and $-\alpha_2$ replacing $+b$) is, in fact, an alternative form for competition equations. If an interaction term is placed in the denominator, this implies the interaction will affect the stable density but not the initial growth rate of the population described by that equation; if the interaction term is not placed in the denominator, this implies the interaction will affect both the stable density and the initial growth rate. Equations with or without the interaction term in the denominator may be the more appropriate ones for different species interactions. Equation (5) in Table 2.5 has the same form as one of the predation equations, Table 2.4 (4).]

Species 1 may evolve further, from partial symbiosis to complete dependence on species 2. When it does so, it no longer has an independent carrying capacity, K_1; and its use of species 2 as a resource, bN_2, becomes its carrying capacity. A full or obligate symbiosis is then described by equations (4) and (5).

If the relationship in equations (4) and (5) is mutualistic, it is stable only if a times b is less than one; the two populations would otherwise grow without effective limit. We can specify a stable mutualistic relation by complicating equation (4) with a saturation factor (see also Table 2.4 C) to obtain equation (6). Equations (5) and (6) describe a stable mutualism. The stability, however, does not result from the symbiotic relationship but from our use of limiting factors (K_2 and D) in the equations. Despite the evolutionary interest of mutualism and commensalism, these do not appear to be important means, by themselves, of stabilizing populations in communities. Rather than this, they are ways species are added to communities. For both commensals and mutualists, the addition is possible because the use of one species by another as a resource (bN_2) increases the number of kinds of resources, and corresponding carrying capacities, available to the species in the community.

When a is negative, equations (4) and (5) can describe a stable relationship of a parasite and host. The stability does not result from the parasite-host interaction, but from the limit (K_2) set on the host population. Parasitism, like commensalism and mutualism, permits the addition to the community of species using other species as resources. Parasites do not seem to stabilize their host populations in most cases, whether or not parasite and host are accommodated to each other. A fungus, the chestnut blight (*Endothia parasitica*), is parasitic on chestnut trees in Eurasia without evident effect on the chestnut population, to which it is accommodated. A parasite not thus accommodated can be destructive and unstabilizing. The same chestnut blight fungus was introduced from Eurasia to North America and a chestnut species not accommodated to it. The blight was then a virulent pathogen that has made the American chestnut (*Castanea dentata*), probably the most important tree species of the eastern United States, almost extinct. There may be some parasites that kill (or reduce reproduction by) some host individuals, and affect an increasing proportion of the host population as host density increases and transfer of the parasite between host individuals becomes more frequent. Such an effect is density-dependent, but it has not been observed as an important means of population regulation. Disease epidemics that strike fluctuating populations appear to be one of the sources of fluctuation, rather than a mechanism limiting that fluctuation.

There are other possibilities, however, of parasitic effects between virulence and harmlessness. Two species of flour beetles, *Tribolium confusum* and *T. castaneum,* have been used for experiments on population growth and interaction. In competition experiments *T. confusum* commonly wins and *T. castaneum* goes extinct, although in some environmental conditions *T. castaneum* has the advantage or the outcome is a matter of chance. In these experiments, however, *T. castaneum* is affected by a protozoan parasite (*Adelina*) that has little effect on *T. confusum.* In competition cultures from which the parasite is excluded, *T. castaneum* more frequently wins. For some environmental conditions the parasite thus reverses the competitive advantage between the two beetle species. *T. castaneum* was also cultured with and without the parasite, in cultures without *T. confusum.* The mean density of *T. castaneum* in cultures with the parasite was about half that in cultures without the parasite. Presence or absence of the parasite thus determines different relations of the flour beetle's population to *K,* the carrying capacity. The effects of parasites have been called the community's "hidden taxes"; in this case the hidden tax significantly lowers the beetle population's standard of living. Many parasites are internal and their effects invisible; such effects generally cannot be recognized in field populations. Yet the parasites may be significantly affecting population densities and competitive advantages; they may be important in the system of influences affecting a population.

Community Stability

We return to our guiding question: How are we to interpret the relative stability of populations in natural communities? The question has aroused its share of controversy. Argument has been permitted by the difficulty in determining how natural populations function and are controlled, and encouraged by different authors' choices of different kinds of populations to generalize from. It seems clear that no one kind of mechanism governs all populations. Simple questions need not have single answers.

One position is that natural populations predominantly fluctuate in response to changing environment, and are controlled primarily by density-independent mortality. The publications of H. G. Andrewartha and L. C. Birch in particular emphasize irregular fluctuation in response to change in environment, and population limitation by shortage of time during which favorable environment permits population increase. An alternative view is that a population that fluctuates in a random and density-independent increase and decrease in response to change in environment must sooner or later fluctuate down to its lower limit. That limit is a density so low that individuals fail to meet and mate, or pollination fails, or the last few individuals fall to some environmental hazard, and the population becomes extinct. In principle, a population that randomly walks in time, without some density-dependent limitation, must walk randomly to extinction. In this view density-independent population control is a contradiction in concepts. Influences limiting fluctuation are necessary to the long-term survival of populations.

We have observed that environmental limits and competition, parasitism and predation, and self-regulating mechanisms *can* exert density-dependent influences; but all these except the last are two-sided in effect, either stabilizing or unstabilizing. Moreover, the effort to identify specific density-dependent mechanisms regulating populations seems often defeated. Not only are we often ignorant, but density-dependent population behavior may result from linkages of several mechanisms, or from the ways sets of more than two species function as interacting systems. Populations fluctuate, but mostly do not fluctuate to extinction. The meaning of population stability is not to be sought simply in the wide fluctuations that characterize many populations. These fluctuations are, in a sense, the noise of population behavior; the theme that must be listened for is survival. Discussion of populations has often concentrated on fluctuations and upper limits, to some neglect of the behavior of populations toward their lower limits.

It might seem that the exponential curve for density-independent population increase, if applied in reverse to population decrease, could provide a kind of insurance against extinction. Suppose mean daily temperature in July decreases from 30°C to 25°C and this decrease reduces a population by 50 per cent, a further decrease of 5° could leave 25 per cent of the population alive, and a still further decrease of 5° could leave 12.5 per cent, and so on. At this pace, at a July mean daily temperature of 0°C, 1.5 per cent of the population would still be alive. This manner of population decrease is suggestive of a variation of Zeno's paradox of Achilles in pursuit of the tortoise. Suppose that in a first time unit Achilles reduces the distance between the tortoise and himself by one half, in a second (half) unit of time he reduces the remaining distance by one half, in a third (quarter) unit of time the remaining distance is reduced by half. Never, in any number of time units (of decreasing lengths), will Achilles pass the tortoise. Similarly, if a population decreased by a constant fraction for each unit of increasing unfavorableness of an environmental factor, the population would never reach zero, its point of extinction.

The exponential curve of population growth in a favorable and constant environment does not imply any such reverse relationship as this. Given a population in a homogeneous but changing environment, the further environment departs from the optimum for the population the greater the reduction in population a unit change in environment will cause. If the temperature optimum of the species is 30°, and a decrease to 25° produces a reduction of population by 50 per cent, a decrease to 20° may cut the population by 90 per cent and a decrease to 15° by 100 per cent. This population suffers not a constant, but an accelerating death rate as environment becomes more unfavorable. Figure 2.13A illustrates, for an experi-

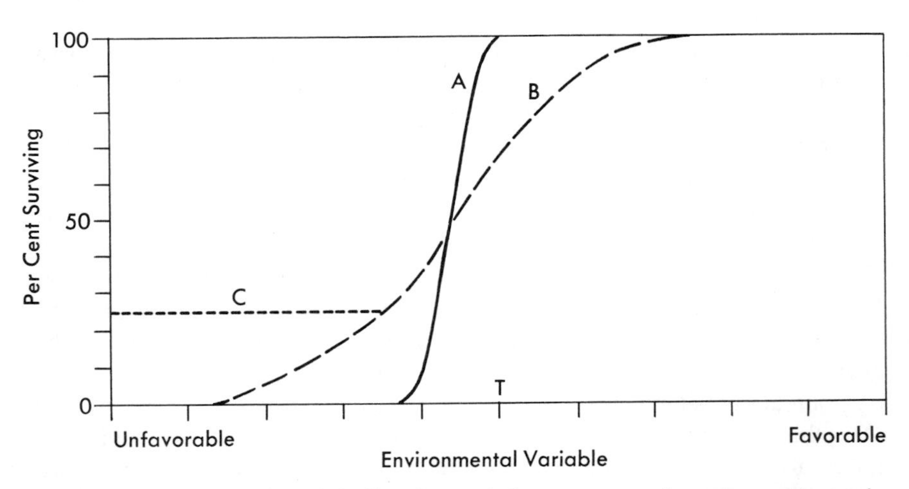

Figure 2.13. Unbuffered and buffered population response to unfavorable environment. **A:** An experimental, genetically homogeneous population for which an environmental variable (such as temperature, drought, or oxygen level) becomes increasingly unfavorable past a physiological tolerance limit *T*. Past this limit the population declines rapidly to extinction. **B:** A buffered, gradual population decline in which some individuals are more vulnerable and others less so, because of genetic differences among them, or difference in the microenvironments they occupy, or both. Genetic and microenvironmental heterogeneity can permit the population to survive environmental fluctuations that would make the population extinct in a homogeneous environment. **C:** A further buffering in which a part of population **B** is converted into resting stages that are scarcely affected as the environmental factors become increasingly unfavorable.

mental population, the abrupt increase in death rate that may occur as an environmental factor falls below a tolerance limit. Many environmental factors do not act with so sharply defined a tolerance limit, but produce increasing death rates with increasing departure from conditions that are optimal for the species.

To interpret population survival of unfavorable conditions we need different assumptions from those of the exponential and logistic curves. We need to consider how death rates may change not only with change in density, but also with unit changes of environment toward less favorable conditions. We need population mechanisms that, as environment becomes increasingly unfavorable, act in ways that prevent accelerating loss from the population. We shall refer to processes that reduce the losses from a population as environment becomes more unfavorable, as population *buffering*. Density-dependent factors can buffer the population, but there are other processes that protect a population against accelerating loss, processes for which the concept of density dependence may be inappropriate (Figure 2.13B and C).

A negative feedback may limit population extremes by slowing or reversing either population increase above the mean, or decrease below the mean. A density-dependent effect may either kill larger fractions (or reduce reproduction) of the population as density increases, or kill smaller fractions (or increase reproduction) as density decreases (Figure 2.14). Some relatively stable populations seem to be controlled primarily by density-dependent limits on population increase beyond their carrying capacity. Many populations, however, are not relatively stable, and these may survive because of factors that set lower limits on population fluctuation. A number of factors may act in this way. As the species becomes scarce, predation on it may decrease. Predators may forget a search image of the population and neglect it as food. Even if the predator continues the search, the search becomes more difficult as density decreases. Consider one hundred colored marbles, scattered in a weed patch. The first 60 may be found with little effort, the next 30 may be found only by search, and the last 10 may not be found. The lower the density of a population, the greater the percentage of its individuals likely to be in especially favorable situations—situations that provide concealment against predators or protection against environmental fluctuation. When environment becomes unfavorable, the population may be reduced to these individuals in especially favorable situations, and may survive because of them. Also, as environment becomes unfavorable, the population may be reduced to those individuals that are genetically best adapted to the adverse conditions. The presence of these individuals in small numbers, in the range of genetic variability of the population, can permit survival during an adverse period.

Many populations, moreover, have special buffering mechanisms for persisting through unfavorable periods. Fresh-water environments are on the whole unstable ones, and inactive resting stages—spores, cysts, pro-

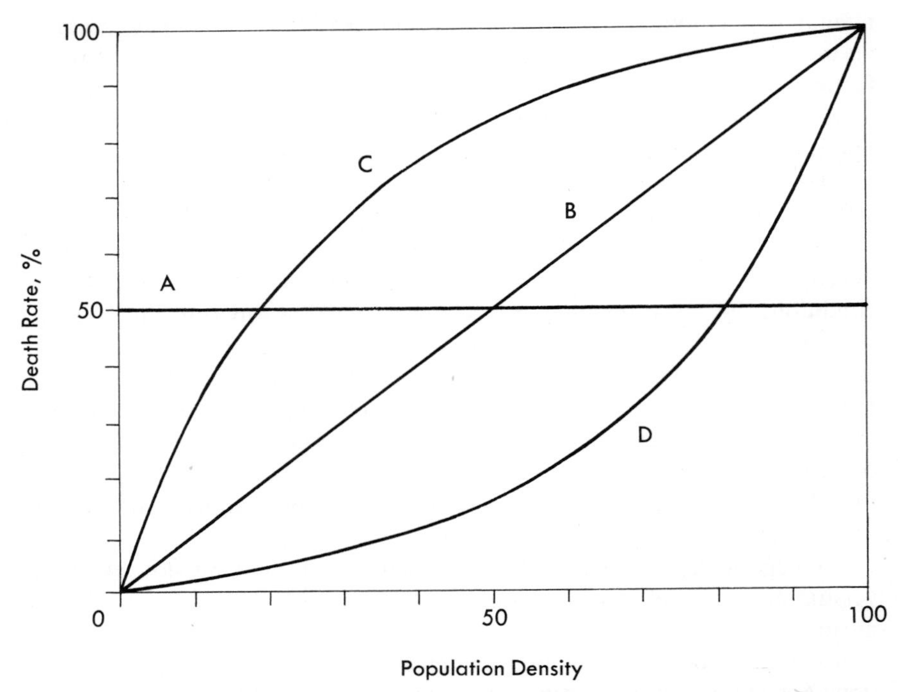

Figure 2.14. Density-dependent and density-independent death rates. Line **A** represents a death rate that is density-independent—constant at all population densities. Line **B** is density-dependent, with a linear increase in death-rate per cent as population density increases. Curve **C** is a convex density dependence that would tend to keep the population at a low density; curve **D** is a concave density dependence that would tend to keep the population at a high density.

tected eggs, gemmules, statoblasts, and so forth—are common among freshwater organisms. Plankton populations fluctuate widely, and most disappear during part of the seasonal cycle. Inactive resting stages can remain on the bottom of the lake or stream; from these an active population can emerge and reproduce when conditions are favorable again. The resting stages are relatively invulnerable to environmental change beyond that which stimulates their formation; they consequently buffer the population against loss as the environment becomes increasingly unfavorable. The small annual plants of deserts similarly survive unfavorable seasons—or a series of unfavorable years if necessary—as seeds. Many bacteria have relatively unstable populations that survive fluctuations as spores, or survive by reduction of the population to a few colonies in most favorable microenvironments. Fungi with relatively unstable populations probably survive in much the same ways. The mycorrhizal and other higher fungi that appear as mushrooms on a forest floor may have more stable populations, with long-lived mycelia that can expand and contract as conditions change, while reproducing sexually only when conditions are favorable. Certain desert shrubs survive drought by reducing the number of living stems they support; free-living flatworms can digest their own tissues and shrink in size, and thus survive a period without food.

We can represent a kind of population buffering, based on a population's relation to environment, with a model. The model may seem complex in detail, but its purpose may be simply stated. It should serve as an example of how heterogeneity of habitat, or difference in microenvironments within a community, may affect population behavior and survival. In the model the species population is affected by two environmental variables: (1) a local environmental gradient that varies in space (such as soil moisture), and (2) a variable in time (such as rainfall and humidity) that as it changes determines what part of the local gradient the species population can occupy. From the way the variable in time affects the population we ask: (1) what kinds of population fluctuation or stabilization may result, and (2) how differences in microenvironment may make possible the population's survival of unfavorable periods. The model will be described for a microenvironmental gradient within a given community, but the reasoning of the model applies also to population behavior along broader habitat gradients such as a topographic moisture gradient from a moist river valley to a dry hill top.

Let a species population occupy a range of microenvironments within a community with a bell-shaped frequency distribution of population density in relation to some environmental gradient (Figure 2.15A). The gradient might be, for example, soil moisture conditions that vary from wetter to drier in small depressions and rises in the microrelief of a forest floor. The population will have an optimum along this gradient, determined by its genetic characteristics (and those of the species with which it must compete). The population may also be genetically variable, with the individuals occurring in the driest and wettest environments genetically different from those in the optimum. We expect that the genetic composition of a population may change continuously along an environmental gradient to which its individuals are adapted, although a population within a given community may not be differentiated in this way.

Consider now the same population curve in a different form, one summing the number of individuals encountered as one proceeds from left to right. This curve (Figure 2.15B) is sigmoid and has an upper asymptote K, but the assumptions are different from the logistic. The horizontal axis here is not time, and the curve is not intended to describe population growth in a constant environment. We are interested instead in the population response per unit change in environmental favorableness affecting species distribution along the horizontal axis. In our example, the change would be periods of wetter or drier weather that determine in which of the microenvironments the population can survive. As the weather becomes wetter,

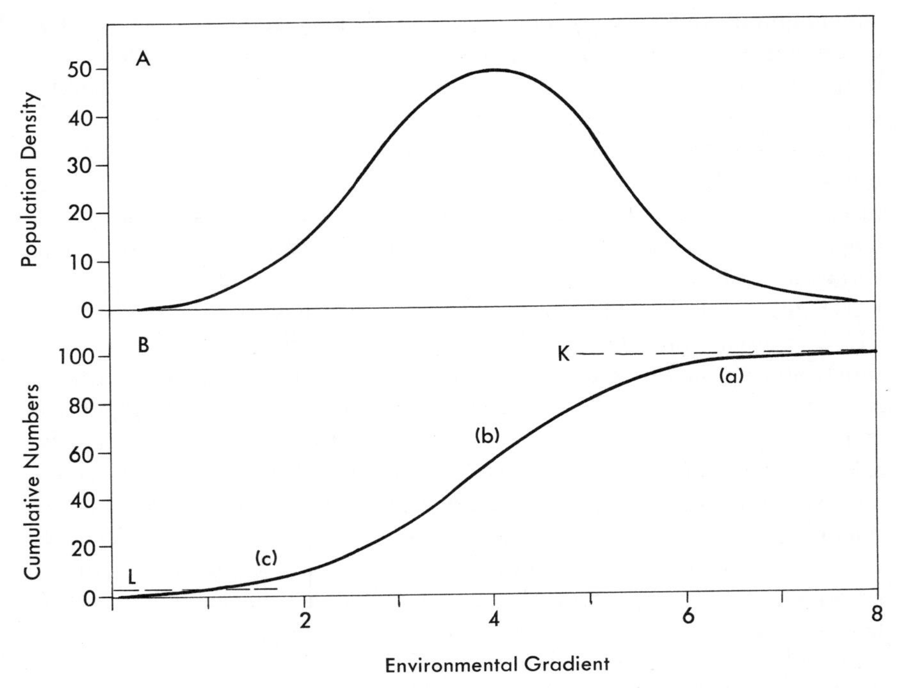

Figure 2.15. Population response to change in environment, assuming a normal distribution along an environmental gradient. Above: The population is shown to form a bell-shaped distribution along an environmental gradient (which could be a gradient of microenvironment within a community). Only during periods of very favorable weather does the population occupy the full range of environments that curve **A** indicates it is able to occupy. Peak density is then at point 4 along the gradient, and on each side of this the density is lower because the environment is less favorable (to the right of the peak), or environment is more favorable but the population is kept at low density by more intense competition (to the left of the peak). Below: Cumulative numbers (all the individuals already encountered as one proceeds along the gradient from left to right) are plotted against the same environmental gradient. Cumulative numbers form a sigmoid curve. As weather becomes increasingly unfavorable, individuals in the middle part of the distribution (2–6) die, with rapid decrease in total population along the middle part (b) of the sigmoid curve. As the weather becomes still more unfavorable, only the individuals in the most favorable part of the gradient (0–2) survive. The death rate among these per unit change of weather conditions is lower, and population decrease follows the lower tail (c) of the sigmoid curve. This lower death rate for the individuals in the most favorable situations tends to protect the population against extinction, which occurs when its density is reduced to a threshold level or lower limit, *L,* which may be zero or, as suggested in the figure, a point above zero. [D. Goodman and R. H. Whittaker, unpublished data.]

the population can expand into normally drier microenvironments, and the population increases, following curve *B* up and to the right. As the weather becomes drier, individuals in the drier microenvironments die, and the population decreases, following curve *B* down and to the left. (The model assumes that all individuals to the right of a given point on the horizontal axis die as weather becomes less favorable, but the function of the model remains the same if we assume that the deaths contract the distribution of curve *A* from the right while preserving its bell-shaped form.) Even in a dry period, some individuals on the left tail of the curve survive in the most favorable microenvironments. The presence of these favored individuals tends to buffer the population against increase in its death rate as environment becomes more unfavorable.

The population response per unit change in environment differs in different parts of curve *B*. The middle of the curve (b) is steep, and a unit change in weather conditions produces a relatively large change in population density. The upper part (a) of curve *B* is flatter, and the response to a unit change in environment is a much smaller change in density. We shall assume that this population is subject to a definite upper limit, because the microenvironments on the right of the gradient are less favorable for it (and may be occupied, even during periods of wetter weather, by persistent competing species populations). The lower part (c) of curve *B* is also flatter, and response to a unit change in environment is a small change in density. The population thus has a response that slows its approach to the lower limit, *L,* a zero population or a threshold population below which the species must decline to extinction.

We speak sometimes in human affairs of a principle of diminishing returns. As an effort is made, say, to increase food production by expanding farms onto less and less favorable lands or by adding more and more fertilizer, the amount of food gained per unit effort decreases. For the population in curve *B* we can speak of a principle of diminishing returns, or diminishing gains. As the population increases past its middle range, from (b) to (a), the gain in numbers per unit improvement in environment decreases. For the other end of the curve we can speak of a principle of diminishing losses. As the curve slopes down from its middle range, from (b) to (c), a unit change for the worse in environment reduces the population by smaller and smaller numbers. Populations can be stabilized by one, or the other, or both of these principles. The principles themselves express what we have called density-dependent limits and buffering. The principle of diminishing gains implies density-dependent reduction of population growth as *K* is approached or overshot; the principle of diminishing losses implies reduction of population decrease as environment becomes less favorable or as *L* is approached. The implication of the two principles is not symmetrical. A population that frequently and widely overshoots *K* upward is relatively unstable. A population that "overshoots" *L* downward is, in contrast, absolutely unstable; it has become extinct.

Curve *B* (like the equations discussed above) represents a simplified, idealized view of population function. We expect real populations to differ from the model in various ways (K overshoot, departure from symmetrical bell-shaped distribution, population movement along the gradient, response to several environmental variables, effects of other species, and so on) that affect their behaviors but need not invalidate the essential point of the model. We can, however, illustrate some typical manners of population fluctuation on the basis of curve B. In Figure 2.16 population (a) is

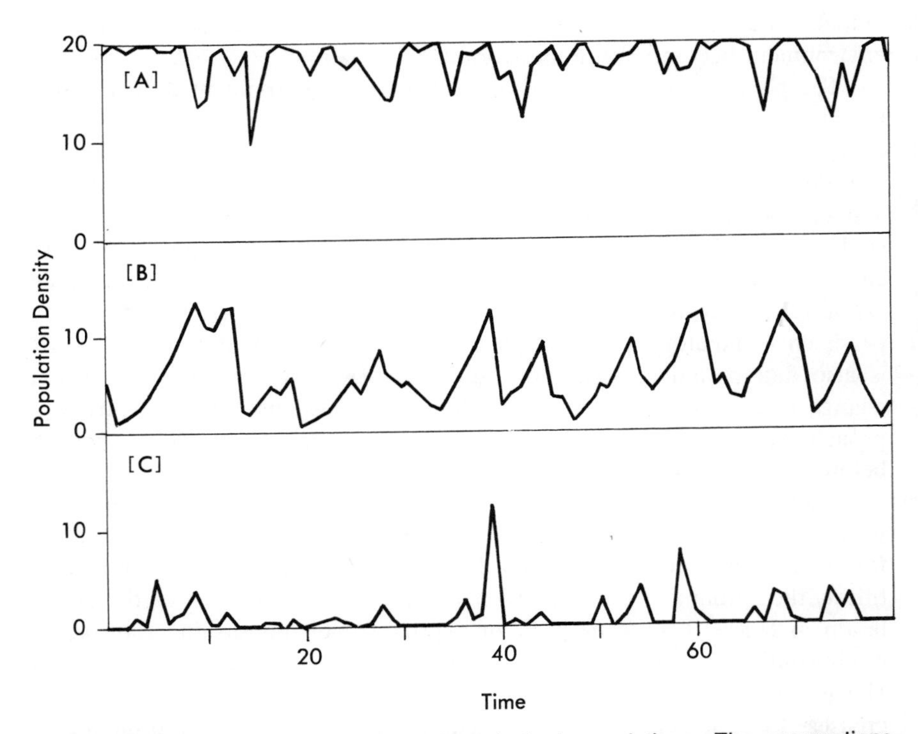

Figure 2.16. Modes of fluctuation in hypothetical populations. The assumptions about environmental relations illustrated in Figure 2.15 were used in a computer simulation to generate population fluctuations in response to random environmental change. **A:** Population that fluctuates near its carrying capacity ($K = 20$) (compare sections of this trace, such as that between time 45 and 65, with the oyster catcher in Figure 2.2). In this case the population is affected by environmental fluctuation within a predominantly favorable range of values that permits the population to occupy most of the potentially suitable microenvironments. The fluctuations occur within part (a) of the sigmoid curve of Figure 2.15. **B:** Population that fluctuates more widely at a range of densities between K and L (compare with the hare and lynx in Figure 2.2). In this case the population is affected by environmental fluctuation causing population increase and decrease in the middle part of the range of microenvironments available to it, part (b), in Figure 2.15. **C:** Population that fluctuates near lower limit ($L = 0$) and is rare much of the time, but is subject to periodic outbreaks (compare the pine moth in Figure 2.2). In this case the population normally exists at low densities and is affected by environmental fluctuation acting on individuals in the most favorable part of the microenvironmental gradient, (c) in Figure 2.15. [D. Goodman and R. H. Whittaker, unpublished.]

a species with a relatively stable population that fluctuates near its carrying capacity without exceeding it. In this species K selection acts on the mechanisms by which the species maintains its population in the face of competition and other interactions in an occupied environment; the reproductive rate may or may not be high. We have mentioned some forest trees and territorial birds as subject to K selection. Population (b) is a relatively unstable one that fluctuates widely while seldom approaching L or K (though natural populations may of course overshoot the latter). In such a population selection produces a high reproductive rate and effective dispersal; the species is subject to r selection as we characterized this above, with fireweed as an example. Such a population may not have highly developed adaptations for surviving extended unfavorable periods. Population (c) is a normally uncommon species that fluctuates near L (and may at times erupt). In this species selection tends to perfect the mechanisms for surviving unfavorable periods, and the reproductive rate may or may not be high. Desert annuals and animals that are adapted to temporary ponds are examples of species subject to L selection.

It is tempting to speak of K, r, and L species, but this might be misleading. In all species selection must act to produce a sufficient r, and most species will be affected by L selection at least at some times and in some environments. Furthermore, selection does not simply act to produce a maximum value of r (or of K or L). Among all species, including those we called r-selected, values of r will result from selective trade-offs between the potential advantage of a higher r (producing a larger number of progeny), and the expense or penalty to the individual (or its progeny) of an increased reproductive effort. The values of r vary widely within each of the three groups, and some K- and L-selected species have high r values. The three types cannot simply be characterized by r values or by the proportion of an individual organism's energy that is directed into reproduction. It seems best to focus attention on the different circumstances of selection affecting these populations, and to regard K, r, and L selection, if these terms are used at all, only as labels for these circumstances. The circumstances we recognize are: (1) saturation selection or interaction selection in an environment that is predominantly favorable, relatively stable, and saturated in the sense of full occupation by other organisms; (2) exploitation selection in an environment that is intermittently favorable and unfavorable, or in environments that are only temporarily favorable in changing locations, such as the burn areas occupied by fireweed, and that are consequently less fully occupied by other species; and (3) adversity selection in environments that are predominantly unfavorable and restrictive, and that only at times permit population growth. Species do not form three distinct groups corresponding to these, but differ in the relative effects of these circumstances on their adaptive characteristics. Species may form a continuum of relative emphases of saturation, exploitation, and adversity selection in their adaptations, as characterized by the proportions of their genes selected for species interactions in occupied and stable environments, for wide dispersal and rapid growth in unstable environments, and for survival of periods of environmental adversity, respectively.

On this basis we can answer our question on the meaning of population and community stability. The most general reason species in natural communities persist is that species lacking some kind of buffering that ensures their persistence are mostly extinct. Some less stable communities are dominated by relatively unstable (but persistent) species populations. Other communities, such as the redwood forest, are dominated by more stable populations. In these many individual species—birds and herbs as well as redwoods—may have relatively stable populations regulated by density-dependent factors. The relative stability of the redwood forest is a consequence, first, of the long-lived and continuously reproducing tree populations that dominate the forest, second of the adaptation of other species to this relatively stable community and occupied environment.

One other idea on community stability should be mentioned. Relatively unstable, and apparently cyclic, fluctuations of mammal and bird populations occur in some extreme environments in which the number of species is small, particularly in the Arctic. As we have indicated, these fluctuations are not Lotka-Volterra cycles, though they may be stable limit cycles. The rather even peaks of Figure 2.16B, produced by random fluctuation of environment, are quite suggestive of the "cycles" of arctic animals. In some more favorable climates, however, where there are larger numbers of species, these seem to have more stable populations. The observation suggests an attractive idea: that community stability results from a larger number of species that act on one another in density-dependent ways. Relative stability then would be a product of the numbers of species and complexity of interactions in communities.

The idea that complexity produces stability may be more attractive than true. Some experiments indicate that adding another predator species to a set of interacting prey and predator (or parasitoid) species can either increase or decrease stability. R. M. May has used mathematical modeling of communities to study the effect of increasing numbers of interacting species. The results did not show that increasing numbers of species interactions increased community stability. In fact, increased number of species and complexity of interactions produced increasing instability and vulnerableness of the community models to perturbation. Relative stability in communities apparently results from the population function of individual species and small sets of interacting species, not from richness in species of the whole community. Environments that are stable may permit many species to survive in interaction with one another in a complex community. Individual species may then have relatively stable populations because of density-dependent relationships, but the complex community may be fragile if its environment is perturbed. May's work suggests not that community complexity produces community stability, but that environmental stability permits the evolution of community complexity.

This supports our theme: The relative stability of natural populations results from the accumulation in communities, through evolutionary time, of species affected by buffering mechanisms and density-dependent relationships. To this there is a complementary theme—the diversity of manners of population behavior and means of population control and survival. Communities are mixtures of species that differently respond to environmental fluctuation and differently interact with one another. We need to explore further the implications of differences among species in a given community.

SUMMARY

Some of the species populations in natural communities fluctuate widely in densities, or numbers per unit area, whereas others seem relatively stable. If unaffected by limits on them, populations grow at accelerating rates following exponential curves. Normal population growth is thus a force toward instability. Populations are ultimately subject to limits of the environmental resources to support them, and some populations follow a sigmoid (logistic) curve of increase to, and stabilization at, an environmental limit or carrying capacity. Many populations, however, are relatively unstable, growing until they exceed their carrying capacities and then crashing to much lower densities. Environmental limits do not by themselves stabilize populations.

Relative population stability may result from density-dependent effects. These density-dependent effects either (1) kill an increasing fraction of the individuals (or decrease the rate of births per individual) as the population grows, thus tending to set an upper limit on it, or (2) kill a smaller fraction of individuals (or increase the rate of births per individual) as the population decreases toward a low density, thus tending to set a lower limit on its fluctuation. Many populations possess special buffering mechanisms (such as resting stages) that cut the losses from the population as environment becomes increasingly unfavorable.

Competition between individuals of the population as resource limits are approached can result in density-dependent effects that stabilize the population in some cases. Competition between two species, in contrast, can imply either extinction of one species, or change in distribution or behavior so that the two species survive in different environments, or by using different resources. Predators can act in a density-dependent manner by taking an increasing fraction of a prey population as it increases, but many predator-prey interactions are unstable. Symbiotic relations (mutualism, commensalism, and parasitism) are important aspects of carrying capacity for many species, but these relations do not seem widely important in limiting population fluctuation. Some species, such as territorial birds, have self-regulating mechanisms that stabilize their populations, but most species do not. Some density-dependent controls involve combinations of effects, such as: competition within resource limits, that results in dispersal of individuals that lose in the competition, which dispersal exposes these displaced individuals to a high death rate from predation.

There is thus no single mechanism responsible for the relative stability of populations. Some species are relatively stable near an upper limit by the effects of one or more of these mechanisms. Many species fluctuate widely but persist because of density-dependent effects or buffering mechanisms that act near the lower limit of the population and prevent extinction. Such relative stability of populations in natural communities as we observe results from the accumulation through evolutionary time of species subject to mechanisms that limit population fluctuation.

References

General, Populations and Population Growth

ANDREWARTHA, H. G. 1961. *Introduction to the Study of Animal Populations.* London: Methuen. xvii + 281 pp.

ANDREWARTHA, H. G. and L. C. BIRCH. 1954. *The Distribution and Abundance of Animals.* Univ. Chicago. xv + 782 pp.

BOER, P. J. DEN and G. R. GRADWELL, editors. 1971. *Dynamics of Populations:* Proceedings of the Advanced Study Institute at Oosterbeek, 1970. Wageningen: Centre for Agric. Publ. and Docmt. 611 pp.

BOUGHEY, ARTHUR S. 1973. *Ecology of Populations.* 2nd ed. New York: Macmillan. x + 182 pp.

COLLIER, BOYD D., G. W. COX, A. W. JOHNSON, and P. C. MILLER. 1973. *Dynamic Ecology*. Englewood Cliffs, N. J.: Prentice-Hall. 563 pp.

COLINVAUX, PAUL A. 1973. *Introduction to Ecology*. New York: Wiley. ix + 621 pp.

CONNELL, JOSEPH H., D. B. MERTZ, and W. W. MURDOCH, editors. 1970. *Readings in Ecology and Ecological Genetics.* New York: Harper & Row. viii + 397 pp.

EMLEN, J. MERRITT. 1973. *Ecology: an Evolutionary Approach.* Reading, Mass.: Addison-Wesley. xiv + 493 pp.

GAUSE, G. F. 1934. *The Struggle for Existence.* (Reprint, 1964) New York: Hafner. ix + 163 pp.

HAZEN, WILLIAM E., editor. 1970. *Readings in Population and Community Ecology*. 2nd ed. Philadelphia: Saunders. ix + 421 pp.

*KREBS, CHARLES J. 1972. *Ecology: The Experimental Analysis of Distribution and Abundance*. New York: Harper & Row. x + 694 pp.

LACK, DAVID. 1954. *The Natural Regulation of Animal Numbers*. Oxford: Clarendon. viii + 343 pp.

MCNAUGHTON, S. J. and LARRY L. WOLF. 1973. *General Ecology*. New York: Holt, Rinehart & Winston. x + 710 pp.

PIANKA, ERIC R. 1974. *Evolutionary Ecology*. New York: Harper & Row. xi + 356 pp.

POOLE, ROBERT W. 1974. *An Introduction to Quantitative Ecology*. New York: McGraw-Hill. xii + 532 pp.

*RICKLEFS, ROBERT E. 1973. *Ecology*. Newton, Mass.: Chiron Press. x + 861 pp.

SLOBODKIN, LAWRENCE B. 1962. *Growth and Regulation of Animal Populations*. New York: Holt, Rinehart & Winston. viii + 184 pp.

VOLTERRA, V. 1931. Variations and fluctuations of the number of individuals in animal species living together, pp. 409–448 in *Animal Ecology, with Especial Reference to Insects*, by Royal N. Chapman. New York: McGraw-Hill.

WILSON, EDWARD O. and W. H. BOSSERT. 1971. *A Primer of Population Biology*. Stamford, Conn.: Sinauer. 192 pp.

Population Behavior

BRUSSARD, P. F. and P. R. EHRLICH. 1970. The population structure of *Erebia epipsodea* (Lepidoptera: Satyrinae). *Ecology* **51:**119–129.

*BRUSSARD, P. F. and P. R. EHRLICH. 1970. Contrasting population biology of two species of butterfly. *Nature* **227:**91–92.

CISNE, J. L. 1973. Life history of an Ordovician trilobite *Triarthrus eatoni*. *Ecology* **54:**135–142.

COLE, L. C. 1951. Population cycles and random oscillations. *Journal of Wildlife Management* **15:**233–252.

COLE, L. C. 1954. The population consequences of life history phenomena. *Quarterly Review of Biology* **29:**103–137.

DEEVEY, E. S. JR. 1947. Life tables for natural populations of animals. *Quarterly Review of Biology* **22:**283–314.

DEMPTSTER, J. P. 1963. The population dynamics of grasshoppers and locusts. *Biological Reviews* **38:**490–529.

EHRLICH, P. R. 1965. The population biology of the butterfly *Euphydryas editha*. II. The structure of the Jaspar Ridge colony. *Evolution* **19:**327–336.

*EHRLICH, P. R. and L. E. GILBERT. 1973. Population structure and dynamics of the tropical butterfly *Heliconius ethilla*. *Biotropica* **5:**69–82.

HARPER, J. L. and J. WHITE. 1974. The demography of plants. *Annual Review of Ecology and Systematics* **5:**419–463.

KREBS, J. R. 1971. Territory and breeding density in the great tit, *Parus major* L. *Ecology* **52:**2–22.

LACK, DAVID. 1966. *Population Studies of Birds.* New York: Oxford Univ. v + 341 pp.

NIERING, W. A., R. H. WHITTAKER and C. H. LOWE, JR. 1963. The saguaro: a population in relation to environment. *Science* **142:**15–23.

PIANKA, E. R. 1970. On *r*- and *K*-selection. *American Naturalist* **104:**592–597.

REYNOLDSON, T. B. 1966. The distribution and abundance of lake-dwelling triclads—towards a hypothesis. *Advances in Ecological Research* **3:**1–71.

WEAVER, JOHN E. and F. W. ALBERTSON. 1956. *Grasslands of the Great Plains, Their Nature and Use.* Lincoln, Nebr.: Johnsen. 395 pp.

WELLINGTON, W. G. 1964. Qualitative changes in populations in unstable environments. *Canadian Entomologist* **96:**436–451.

Competition

*CONNELL, J. H. 1961. The influence of interspecific competition and other factors on the distribution of the barnacle *Cthalamus stellatus. Ecology* **42:**710–723.

*HARPER, J. L. 1967. A Darwinian approach to plant ecology. *Journal of Ecology* **55:**247–270.

Miller, R. S. 1967. Pattern and process in competition. *Advances in Ecological Research* **4:**1–74.

MILTHORPE, FREDERICK L., editor. 1961. Mechanisms in Biological Competition. *Symposia of the Society for Experimental Biology* **15,** vi + 365 pp. Cambridge Univ.

NICHOLSON, A. J. 1954. An outline of the dynamics of animal populations. *Australian Journal of Zoology* **2:**9–65.

*PARK, T. 1962. Beetles, competition, and populations. *Science* **138:**1369–1375.

PARK, T., D. B. MERTZ, W. GRODZINSKI and T. PRUS. 1965. Cannibalistic predation in populations of flour beetles. *Physiological Zoology* **38:**289–321.

SHARITZ, R. R. and J. F. MCCORMICK. 1973. Population dynamics of two competing annual plant species. *Ecology* **54:**723–740.

WHITE, J. and J. L. HARPER. 1970. Correlated changes in plant size and number in plant populations. *Journal of Ecology* **58:**467–485.

WILLIAMSON, M. H. 1957. An elementary theory of interspecific competition. *Nature* **180:**422–425.

WYNNE-EDWARDS, VERO C. 1962. *Animal Dispersion in Relation to Social Behavior.* New York: Hafner. xi + 653 pp.

WYNNE-EDWARDS, V. C. 1965. Self-regulating systems in populations of animals. *Science* **147:**1543–1548.

Predation

BROOKS, J. L. and S. I. DODSON. 1965. Predation, body size, and composition of plankton. *Science* **150:**28–35.

CAUGHLEY, G. 1970. Eruption of ungulate populations, with emphasis on Himalayan thar in New Zealand. *Ecology* **51:**53–72.

CONNELL, J. H. 1970. A predator-prey system in the marine intertidal region. I. *Balanus glandula* and several predatory species of *Thais*. *Ecological Monographs* **40:**49–78.

ERRINGTON, PAUL L. 1963. *Muskrat Populations.* Ames: Iowa State Univ. x + 665 pp.

HOLLING, C. S. 1959. The components of predation as revealed by a study of small-mammal predation of the European pine sawfly. *Canadian Entomologist* **91:**293–320.

HUFFAKER, C. B. 1958. Experimental studies on predation: dispersion factors and predator-prey oscillations. *Hilgardia* **27:**343–383.

HUFFAKER, C. B. 1959. Biological control of weeds with insects. *Annual Review of Entomology* **4:**251–276.

JORDAN, P. A., D. B. BOTKIN and M. L. WOLFE. 1970. Biomass dynamics in a moose population. *Ecology* **52:**147–152.

*LUCKINBILL, L. S. 1973. Coexistence in laboratory populations of *Paramecium aurelia* and its predator *Didinium nasutum*. *Ecology* **54:**1320–1327.

MAY, R. M. 1972. Limit cycles in predator-prey communities. *Science* **177:**900–904.

MAY, R. M. 1973. Time-delay versus stability in population models with two and three trophic levels. *Ecology* **54:**315–325.

MURDOCH, W. W. 1969. Switching in general predators: experiments on predator specificity and stability of prey populations. *Ecological Monographs* **39:**335–354.

PAINE, R. T. 1966. Food web complexity and species diversity. *American Naturalist* **100:**65–75.

PIMLOTT, D. H. 1967. Wolf predation and ungulate populations. *American Zoologist* **7:**267–278.

UTIDA, S. 1957. Cyclic fluctuations of population density intrinsic to the host-parasite system. *Ecology* **38:**442–449.

*UTIDA, S. 1957. Population fluctuation, an experimental and theoretical approach. *Cold Spring Harbor Symposia in Quantitative Biology* **22:**139–151.

*ZARET, T. M. and R. T. PAINE. 1973. Species introduction in a tropical lake. *Science* **182:**449–455.

Symbiosis

*ARDITTI, J. 1966. Orchids. *Scientific American* **214**(1):70–78.

BATRA, S. W. T. and L. R. BATRA. 1967. The fungus gardens of insects. *Scientific American* **217**(5):112–120.

*COLWELL, R. K. 1973. Competition and coexistence in a simple tropical community. *American Naturalist* **107:**737–760.

CROLL, NEIL A. 1966. *Ecology of Parasites.* Cambridge: Harvard Univ. 136 pp.

CULBERSON, W. L. 1970. Chemosystematics and ecology of lichen-forming fungi. *Annual Review of Ecology and Systematics* **1:**153–170.

FAEGRI, KNUT and L. VAN DER PIJL. 1971. *The Principles of Pollination Ecology.* London: Pergamon. xii + 291 pp.

HACSKAYLO, E. 1972. Mycorrhiza: the ultimate in reciprocal parasitism? *BioScience* **22:**577–583.

†HARLEY, J. L. 1968. Mycorrhiza, pp. 139–178 in *The Fungi: An Advanced Treatise,* ed. G. C. Ainsworth and A. S. Sussman, Volume **3:**139–178. New York: Academic.

†HENRY, SYDNEY M., editor. 1966. *Symbiosis.* New York: Academic. 2 vols.

JANZEN, D. H. 1966. Coevolution of mutualism between ants and acacias in Central America. *Evolution* **20:**249–275.

LIMBAUGH, C. 1961. Cleaning symbiosis. *Scientific American.* **205**(2):42–49.

MARGULIS, L. 1971. Symbiosis and evolution. *Scientific American* **225**(2):48–57.

PIMENTEL, D. 1961. Animal population regulation by the genetic feed-back mechanism. *American Naturalist* **95:**65–79.

PODGER, F. D. 1972. *Phytophthora cinnamomi,* a cause of lethal disease in indigenous plant communities in Western Australia. *Phytopathology* **62:**972–981.

STEWART, W. D. P. 1967. Nitrogen-fixing plants. *Science* **158:**1426–1432.

Stability

EHRLICH, P. R. and L. C. BIRCH. 1967. The "balance of nature" and "population control". *American Naturalist* **101:**97–107.

GOODMAN, D. 1975. The theory of diversity and stability in ecology. *Quarterly Review of Biology* (in press).

HOLLING, C. S. 1973. Resilience and stability of ecological systems. *Annual Review of Ecology and Systematics* **4:**1–23.

MACARTHUR, R. 1955. Fluctuations of animal populations, and a measure of community stability. *Ecology* **36:**533–536.

MAY, R. M. 1971. Stability in multispecies community models. *Mathematical Biosciences* **12:**59–79.

†MAY, ROBERT M. 1973. *Stability and Complexity in Model Ecosystems.* Princeton Univ. ix + 235 pp.

†MAYNARD SMITH, J. 1974. *Models in Ecology.* Cambridge Univ. xii + 146 pp.

Nicholson, A. J. 1957. The self-adjustment of populations to change. *Cold Spring Harbor Symposia in Quantitative Biology* **22:**153–173.

SLOBODKIN, L. B. 1968. Toward a predictive theory of evolution, pp. 187–205 in *Population Biology and Evolution,* ed. Richard C. Lewontin. Syracuse Univ.

*SLOBODKIN, L. B., F. E. SMITH and N. G. HAIRSTON. 1967. Regulation in terrestrial ecosystems and the implied balance of nature. *American Naturalist* **101:**109–124.

*SMITH, F. E. 1972. Spatial heterogeneity, stability, and diversity in ecosystems. In *Growth by Intussesception,* ed. Edward S. Deevey. *Transactions of the Connecticut Academy of Arts and Sciences* **44:**309–335.

SOLOMON, M. E. 1949. The natural control of animal populations. *Journal of Animal Ecology* **18:**1–35.

WATT, K. E. F. 1965. Community stability and the strategy of biological control. *Canadian Entomologist* **97:**887–895.

* Suggested student reading.

† Recommended reference.

3

Community Structure and Composition

A community consists of species—many species with different kinds of population fluctuation and interaction with one another. We can say something about the community by giving a list of its species composition, but a community is poorly described by such a list alone. We want to know more than composition: how different species contribute to the community's structure, how species fit together to make up the community as a whole, what the relative importances of the different species mean, what determines the numbers of species that make up different communities, and so on. We study community structure and composition in the effort to understand how a community, as a living system of interplaying species populations, is organized.

Growth-Forms and Life-Forms

The study of the forms and structures of organisms is the science of morphology. It is an important area of biology, more important than is often recognized in current writing that takes for granted the knowledge of morphology built by generations of biologists. It is primarily by form and structure that living things are classified, by which their adaptations to environment are recognized, and by which evolutionary relationships are known or surmised. It is by the movement of research through the study of structure into the study of function related to that structure that much of the development of physiological and chemical biology has occurred. In our concern with natural communities it is appropriate that we first consider aspects of their structure, and then proceed to their environmental relations and functions. The study of form and structure in natural communities is termed not morphology but *physiognomy*.

The structure of the plankton community is usually invisible, which is not to say nonexistent. In adaptation to the free-floating life, most plankton organisms are microscopic and short-lived or rapidly multiplying. There is no accumulation of massive structure, and the physiognomy of a plankton community is limited to a rather sparse and changeable dispersion of microorganisms in water. More impressive physiognomy appears in communities of organisms on or attached to the bottom of the sea—in giant kelp beds, the elaboration of forms and colors of coral reefs, and the patterns of distinctive starlike, plumelike, fanlike, and flowerlike animals on the deep ocean bottom. It is in the study of communities on land, however, that physiognomy has been most discussed and most rewarding.

To describe the forms of communities on land one needs to characterize major kinds of form in plants, for physiognomy results from the forms of the plants that make up the community. The classes or kinds of form in plants are referred to as *growth-forms*. A number of characteristics of plants—height, woody versus herbaceous or nonwoody growth, stem form, leaf form, and leaf deciduousness or evergreenness, and so on—are used to define growth-forms. The growth-forms do not (with a few exceptions in some systems) correspond to the units into which taxonomists classify plants. There are a number of systems of growth-forms, one of which is outlined in Table 3.1. The list is not complete, but is limited to the growth-forms that are most important in determining community structure.

Another system of plant forms was designed by the Danish botanist C. Raunkiaer. Instead of the mixture of characteristics by which growth-forms are characterized, Raunkiaer used a single principal characteristic—the relation of the perennating tissue to the ground surface. "Perennating tissue" refers to the embryonic (meristematic) tissue that remains inactive during a winter or dry season, and then resumes growth with return of a favorable season. Perennating tissues thus include buds, that may contain

Table 3.1 Major Plant Growth-Forms on Land

Trees, Larger Woody Plants, Mostly Well Above 3 M Tall

- Needle-leaved (mainly conifers—pine, spruce, larch, redwood, and so on)
- Broad-leaved evergreen (many tropical and subtropical trees, mostly with medium-sized leaves)
- Evergreen-sclerophyll (with tough, evergreen, mostly smaller leaves)
- Broad-leaved deciduous (leaves shed in the Temperate Zone winter, or in the tropical dry season)
- Thorn-trees (armed with spines, in many cases with compound, deciduous leaves)
- Rosette trees (unbranched, with a crown of large leaves—palms and tree-ferns)
- Bamboos (arborescent grasses)

Lianas (Woody Climbers or Vines)

Shrubs, Smaller Woody Plants, Mostly Below 3 M in Height

- Needle-leaved
- Broad-leaved evergreen
- Evergreen-sclerophyll
- Broad-leaved deciduous
- Thorn-shrubs
- Rosette shrubs (yucca, agave, aloe, palmetto, and so on)
- Stem succulents (cacti, certain euphorbias, and so on)
- Semishrubs (suffrutescent, that is, with the upper parts of stems and branches dying back in unfavorable seasons)
- Subshrubs or dwarf-shrubs (low shrubs spreading near the ground surface, less than 25 cm high)

Epiphytes (Plants Growing Wholly Above the Ground Surface, on Other Plants)

Herbs, Plants Without Perennial Above-Ground Woody Stems

- Ferns
- Graminoids (grasses, sedges, and other grasslike plants)
- Forbs (herbs other than ferns and graminoids)

Thallophytes

- Lichens
- Mosses
- Liverworts

miniature twigs with leaves that expand in the spring or rainy season, and seeds. Since the perennating tissue makes possible the plant's survival during an unfavorable season, the location of this tissue is an essential feature of the plant's adaptation to climate. The harsher the climate, the fewer plant species are likely to have buds far above the ground surface, fully exposed to the cold or the drying power of the atmosphere. Among land plants positions of the perennating tissues define five major types of plants, termed *life-forms* (Figure 3.1).

Phanerophytes are woody plants that have their buds well above the ground surface, fully exposed to the atmosphere. Phanerophytes include trees, shrubs down to an arbitrary minimum height of 25 cm, and lianas and epiphytes supported by trees and shrubs.

Chamaephytes are various plants with their buds above the ground surface, but below 25 cm. The buds may be somewhat less exposed to cold or dry winds than are the buds of phanerophytes, and in cold climates they may be protected in winter because they are covered by snow. Chamaephytes include dwarf-shrubs and semishrubs (see Figure 3.1), small succulents and rosette-shrubs, and in some uses of life-forms mosses and lichens.

Hemicryptophytes are perennial herbs with their perennating tissues at the soil surface. Not only snow in a cold climate, but leaf litter or dead plant remains may give the buds of these plants some protection.

Geophytes are perennial herbs with underground perennating tissues

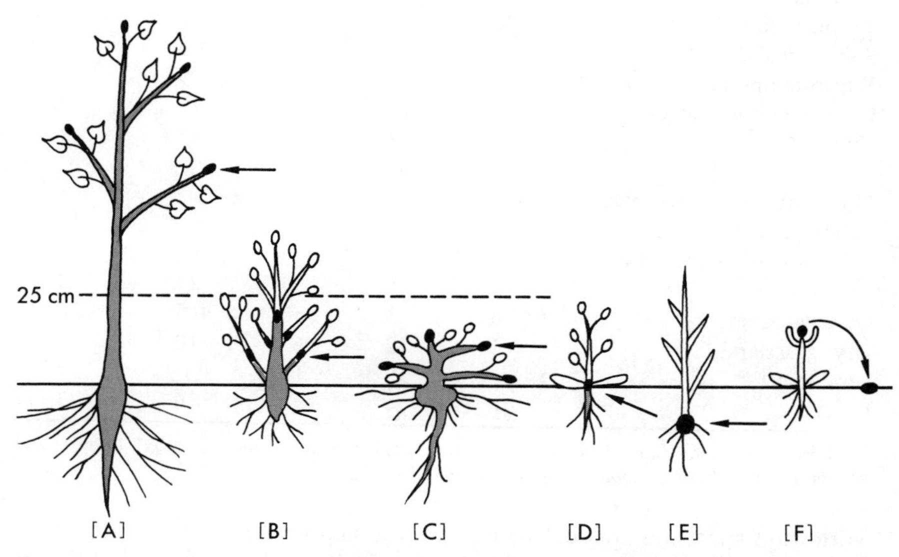

Figure 3.1. Plant life-forms of Raunkiaer. Perennating tissues are shown in black, woody tissues in gray, and deciduous tissues unshaded. **A:** Phanerophyte (tree or tall shrub) with buds more than 25 cm above the ground. **B:** Chamaephyte, semishrub (suffrutescent low shrub) with buds less than 25 cm above the ground. **C:** Chamaephyte, subshrub, with buds less than 25 cm above the ground. **D:** Hemicryptophyte, perennial herb with its bud at ground surface. **E:** Geophyte, perennial herb with a bulb or other perennating organ below the ground surface. **F:** Therophyte, annual plant surviving unfavorable periods only as a seed.

(such as bulbs, corms, tubers, or rhizomes) that are more fully protected from the above-ground climate.

Therophytes are short-lived annual or ephemeral herbs that survive unfavorable seasons (or in some cases a number of unfavorable years) only as seeds.

Growth-forms are used to characterize community structure by the fact that certain growth-forms are dominant, or most conspicuous, in the community. The use of life-forms, in contrast, is not structural but floristic in the sense of species composition of the community—in this case numbers of species in different life-forms. When the numbers of species in life-forms in a community, or a geographic area, are converted to per cents, these per cents form a *life-form spectrum* (Table 3.2). The "normal" or

Table 3.2 Life-Form Spectra

	Phanerophytes	*Chamaephytes*	*Hemicryptophytes*	*Geophytes*	*Therophytes*
World or normal spectrum	46	9	26	6	13
Latitudinal series (humid climates)					
Tropical rain forest	96	2		2	
Subtropical forest	65	17	2	5	10
Warm-temperate forest	54	9	24	9	4
Cold-temperate forest	10	17	54	12	7
Tundra	1	22	60	15	2
Humidity series (temperate latitudes)					
Mid-temperate mesophytic forest	34	8	33	23	2
Oak woodland	30	23	36	5	6
Dry grassland	1	12	63	10	14
Semidesert		59	14		27
Desert		4	17	6	73

[Spectra, taken from Cain and Castro (1959), Whittaker, *Ecol. Monogr.*, **30**:317 (1960); and Whittaker and Niering, *Ecology,* **44**:446 (1965).]

worldwide spectrum was calculated from a sample representing the whole vascular plant flora of the world. Departures from its pattern in different directions reflect the effects of environment, and especially climate, on plant adaptation in communities. Predominance of phanerophyte species, especially large numbers of tree species, is characteristic of tropical rain forest. The high proportion of phanerophytes in the world-wide spectrum results from the great numbers of tree species in different tropical rain forest areas. Toward cooler climates the numbers of tree species decrease. In cool temperate forests and many grasslands hemicryptophytes include the largest numbers of species. In still colder climates chamaephytes (both dwarf-shrubs and thallophytes) may predominate over hemicryptophytes or share predominance with them. In desert climates therophytes predominate, or share predominance with semishrubs and phanerophytes (shrubs).

Life-form spectra are only one among many ways of describing community composition. As one of these ways, however, they suggest that:

1. A community is a mixture of differently adapted species.
2. The characteristics of this mixture express community environment and the relative prevalence of different adaptations in the community.
3. These characteristics change along environmental gradients in ways that are usually interpretable, and in some cases relatively predictable.

Vertical Structure

The reader may note that growth-forms and life-forms have the same principal axis of differences among plants: plant height. Most communities show vertical differentiation or stratification—different species occur at different heights above the ground, or depths below the water surface. The species have different positions along a vertical gradient of depth in the community and decreasing light intensity. Intensity of light necessarily decreases from the surface of the community, which is in full sunlight, downward; light absorption by the organisms themselves is a principal reason for this extinction of light with depth. In a forest one may observe a several-storied physiognomy with a number of growth-forms (in each of which there may be a number of species) occurring one above the other to form the community's vertical structure. Shrublands and grasslands too are mixtures of species of different heights with some species shaded at least part of the time, as the position of the sun changes, by other species.

In a forest the trees, with their upper foliage in full sunlight, form the canopy or uppermost layer (Figure 3.2). The leaves and branch surfaces of the canopy trees may absorb and scatter more than half of the sunlight energy, but beneath the canopy there is a lower layer of smaller trees using some of the remaining light. This lower tree stratum usually contains both younger individuals of the canopy tree species, and mature trees of other, smaller species that do not normally reach canopy height. Less than 10 per cent of the sunlight reaching the upper canopy may penetrate through the tree foliage of both levels, and the spectral composition of the remaining internal light of the forest is changed from that of sunlight. Species of a third level of vegetation, shrubs, are adapted to photosynthesize using this weaker light within the forest, and they further reduce the light that reaches herbs beneath the shrub layer. The remaining light (1 to 5 per cent of

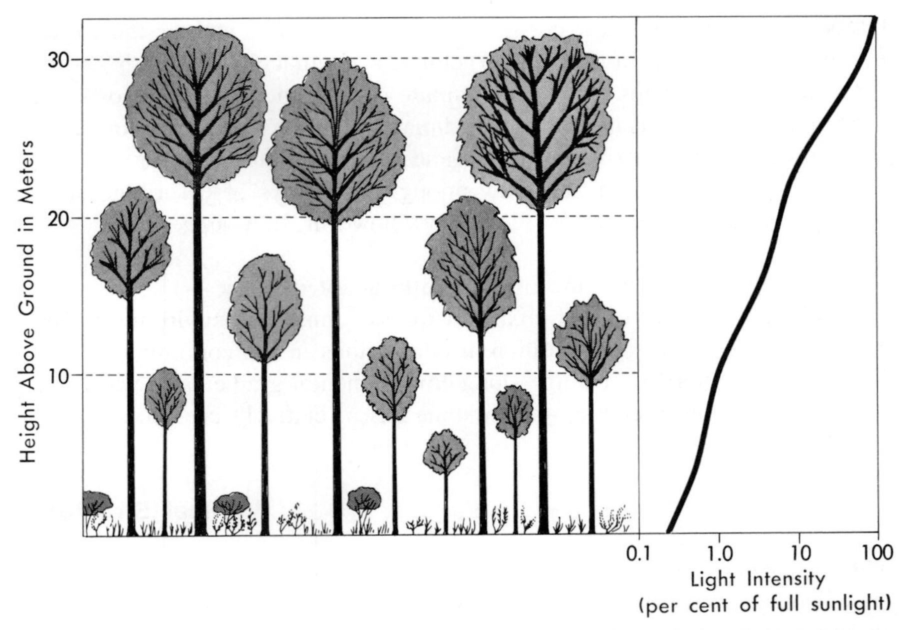

Figure 3.2. Stratification and light extinction in a forest. Different species of trees, shrubs, and herbs bear foliage at different heights above ground (left) and are adapted to life at the different light intensities (right) that result from the absorption of sunlight by that foliage.

incident sunlight in many forests) supports the growth of the herb layer. Beneath the herbs, mosses on the ground may form still another vegetation layer. In some dense forests only a fraction of 1 per cent of incident sunlight reaches the forest floor.

A forest tree may gain advantage by reaching the canopy, where abundant sunlight supports photosynthesis, but the tree must spend much of the energy of photosynthesis in growing the woody tissue of stem and branches to support the foliage in the canopy. There may be apparent disadvantage in the low light intensities in which the forest herbs must live, but the herb need not spend its more modest photosynthetic profit on woody supporting tissue. Forest structure involves a gradient of growth-forms—upper and lower trees, upper and lower shrubs, upper and lower herbs, and soil-surface mosses—in adaptation to the gradient of light intensity. Along the gradient growth-form designs change from one extreme (the upper tree with foliage in full sunlight, massive supporting stem and branch structure, and a root system smaller in mass than the above-ground structure) to herbs with adaptations at the other extreme (photosynthesis at low levels of light intensity, small investment in above-ground supporting structure, and accumulation of reserve food in a root system more massive than the above-ground structure).

Even as different plant species are adapted to different positions in this vertical gradient, so different animal species also occupy different levels in the forest. Different groups of bird species, for example, can be found feeding and nesting near the ground, in the shrub and small-tree foliage beneath the canopy, and in the canopy itself. Different arthropod species occur at different levels from the canopy downward to the herb stratum and to and below the ground surface. A group of animals—mites and springtails, millepedes and centipedes, ground beetles, and so on—occur primarily in the leaf-litter on the soil surface; these animals, which are seldom seen on the surface by daylight, are the *cryptozoans*. Other animals occur at different depths in the soil, in which different plant species also have their roots extending to different depths.

In the plankton too there is a degree of vertical differentiation in the adaptation of different species to different levels and light intensities. Vertical movements affect the distribution of these species, however, and vertical differences in the community are less evident than in the forest. Communities of lighted zones below tide levels on the ocean floor show a differentiation determined in part by light intensities. The species of green algae are concentrated in shallow water and the species of brown algae in somewhat deeper water, while some of the red algae occur at depths below those of the brown algae. The brown and red algae possess supplementary pigments, in addition to chlorophylls and carotenoids, that adapt these algae to use of light lower in intensity and different in spectral composition than that in shallow water. Vertical differentiation is thus a common feature of natural communities. So also is a degree of horizontal differentiation.

Horizontal Pattern

We say that plants form a carpet on the forest floor, but this is a carpet with its own kind of pattern. Suppose we lay out one hundred randomly located quadrats or plots, each 1 meter square, on the forest floor and record the undergrowth plants present in each of these. We can then ask two kinds of questions about the pattern.

First, are the individual plants scattered on the forest floor at random, or are they to some degree grouped or clustered? As a basis for answering the question, consider occurrence in plots of a hypothetical species whose individuals are randomly distributed and whose foliage covers a small fraction of the forest floor. A Poisson distribution is appropriate to describe the numbers of individual plants in plots—61 plots with no plants of the species, 30 with 1, 8 with 2, and 1 with 3, say, for 50 individual plants in 100 plots. (The Poisson distribution for random distribution of individuals may be described by the first equation in Table 3.3. Given a mean density of 0.5 individuals per square meter, a Poisson distribution for plants in plots can be calculated as shown in Table 3.3A.)

The numbers of individuals in plots will fit the Poisson distribution only if the distribution of individuals is random—if the location of each individual is determined by factors independent of those determining the

Table 3.3 Poisson Distribution and Contagion Tests

A. Poisson Distribution

$$f = e^{-m}, \quad me^{-m}, \quad \frac{m^2e^{-m}}{2!}, \quad \frac{m^3e^{-m}}{3!}, \ldots$$

for plots containing 0 individuals, 1 individual, 2 individuals, etc. In this f is the relative (decimal) frequencies of plots containing 0, 1, 2, 3 . . . individuals, m is the mean number of individuals per plot, e is the base of natural logarithms, and ! indicates a factorial. For a sample of 100 plots with $m = .5$,

$$F = 100f = 60.6,\ 0.5\times60.6,\ 0.25\times60.6/(1\times2),\ 0.125\times60.6/(1\times2\times3) \ldots$$
$$= 60.6 \quad 30.3 \quad 7.58 \quad 1.26$$

B. Chi-Square Test

	Numbers of individuals in plots				
	0	1	2	3	4
Poisson distribution ($m = 0.5$)	60.6	30.3	7.58	1.26	0.16
An actual distribution ($m = 0.5$)	80	4	5	8	3
Difference (d) (combining 3 and 4)	+19.4	−26.3	−2.58	9.58	

$$\text{Chi-square, } \chi^2 = \Sigma(d^2/F) = \frac{376}{60.6} + \frac{692}{30.3} + \frac{6.7}{7.6} + \frac{91.8}{1.42}$$
$$= 6.2 + 22.8 + 0.9 + 64.6 = 94.5$$

Probability (with two degrees of freedom) less than 0.001

C. Variance to Mean Ratio

$$\text{Variance, } V = \frac{\Sigma(x-m)^2}{n-1} = \frac{\Sigma(x^2) - (\Sigma x)^2/n}{n-1}$$
$$= [80(0-.5)^2 + 4(1-.5)^2 + 5(2-.5)^2 + 8(3-.5)^2 + 3(4-.5)^2]/99$$
$$= (20 + 1 + 11 + 50 + 37)/99 = 1.20$$
$$\text{Variance/mean} = 1.20/0.5 = 2.4$$

location of other individuals. This is seldom the case. Various factors may cause individuals to grow close to one another, to be clumped into groups. If the individuals are clumped, the pattern of numbers of individuals in plots will be shifted: there will be more plots with larger numbers of individuals, and hence also more plots with none, for a given total number of individuals. The clumped distribution is termed *contagious*. A chi-square measurement can be used to test the significance of the difference between an actual distribution of numbers of individuals in plots, and a Poisson distribution calculated for the same number of individuals and plots. Table 3.3, part B illustrates an actual distribution of plant individuals in quadrats and a chi-square test for contagion. We judge that this distribution is contagious, for the test shows that there is less than one chance in a thousand that the actual distribution could represent only chance departure from a Poisson distribution.

If we accept that the distribution is contagious we may also want to know how strongly contagious it is—that is, its degree of contagion. This is a somewhat different question from the probability that the distribution is contagious, for which we here used chi-square. There are a number of approaches to measuring degree of contagion. One can, for example, use the ratio of the variance to the mean. Because the variance of a Poisson distribution is equal to its mean, this ratio for a random distribution is 1. A ratio significantly above 1 implies contagious or clumped distribution, a ratio less than 1 implies that individuals are more evenly spaced than a random distribution would imply. The condition in which individuals tend to be evenly spaced, rather than scattered at random, is termed *regularity* (Figure 3.3). For our example of an actual distribution in Table 3.3 the variance to mean ratio (part C of the table) is 2.4, indicating moderate contagion.

Regularity is apparently uncommon and difficult to demonstrate in natural communities. Some plants appear to have rather even distributions—among them the shrubs in some deserts in which it is reasonable to suppose that it is less likely that one shrub will be close to another shrub

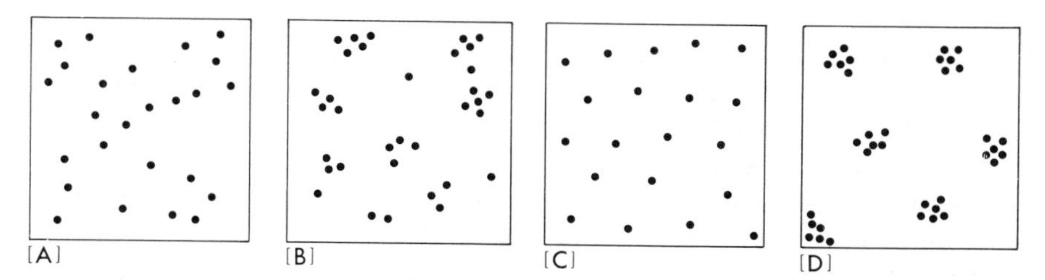

Figure 3.3. Four types of population dispersion in a community. **A:** Random dispersion (note its apparent irregularity). **B:** Clumped or contagious distribution. **C:** Regular or negatively contagious distribution. **D:** Combination of strong clumping of individuals into colonies and regular distribution of the colonies as wholes.

of the same species (hence within the area of its root effects) than farther away from it. Singing birds divide a community into territories, and each male of a species establishes and defends an area in which the pair nests and feeds; these territories may be of the same order of size for all pairs of a given species in the community. It seems clear that interaction among the birds has brought about a distribution that is more regular than random. It may be difficult, however, with either desert shrubs or singing birds to demonstrate statistically that the distribution is regular.

Departures from randomness toward contagion are very common indeed and easily established by measurement. It may be quite apparent that the plants of the forest floor are concentrated in patches, with few individuals between the patches. There may be at least three reasons for such patches:

1. Dispersal from parent plants. Seeds from a plant may fall near that plant, producing a clump of young plants when the seeds grow. Plants spreading from a parent plant by runners or rhizomes may form clumps of individuals, or of above-ground stems that are still connected.
2. Differences in environment. The forest floor is a mosaic of patches receiving more light, or less, through the canopy. Light differences among these patches may influence the development of patches of herbs. A microrelief of scarcely visible rises and depressions, or other obscure patches of soil characteristics, also may contribute to the formation of undergrowth patches.
3. Species interrelations. One species may be dependent on another (for example, an herb parasitic on the roots of a tree species), or an herb population may be denser under trees of a given species because of the effects of the trees on the soil. Patches of the herb species may then occur around or under the tree species; the herb species can be contagiously distributed whether or not the tree species is. If major species form clumps from which some minor species are excluded, then these minor species will have nonrandom distributions concentrated in the spaces between the clumps of the major species.

Clearly these three causes of contagion may be combined with one another in various ways. Effects of the first of the causes may (in the absence of the other two) tend to even out toward a random distribution, given sufficient time. The second and third, however, imply differentiation of the community in horizontal space. Such differentiation occurs as a product of microrelief in many nonforest communities. Rocky soils produce complex patterns of differing soil depths and qualities and different shelter by rocks and water drainage from rock surfaces. Different plant species respond in their distributions in the community to these microhabitat differences. Many bogs have patterns of hummocks and hollows, with different plant species occurring at the different levels of these. Frost-heaving produces distinctive patterns in arctic and some high-elevation communities. In some tundra communities (biome-type 16, Chapter 4), the repeated freezing and thawing of the soil separates rocks from finer soil materials and arranges the rocks in networks with polygonal cells. Striking patterns in plant communities develop in response to such networks (Figure 3.4). In other tundra communities there are parallel bands in the vegetation, or terraces on slopes, or hummocks (Plate 21). Plant distributions are contagious in these patterns, even if the contagion is expressed in parallel bands in some of them, rather than in separated patches.

Animal populations too show varying degrees of contagion. Marine plankton animals and fish may be strongly clumped into schools. These clumps are not easily studied, for the investigator is using equipment lowered into the ocean to take measurements on a population whose presence cannot usually be seen from the surface, whose boundaries cannot be effectively located, and which is changing its distribution in time even as it is sampled. The problem is suggestive of fishing for a cloud, but locating populations by echo sounding and other techniques aid in studying them. Patchiness in terrestrial animal communities is more easily investigated.

Figure 3.4. Patterned soils and vegetation in the tundra. Stone polygons are formed by frost action and produce a corresponding pattern in the plant communities of the arctic tundra, shown near Ny-Aalesund, Kings Bay, Spitsbergen. [Photo by F. Mattick, courtesy of Carl Troll, from *Colloq. Geogr.,* Bonn, **9:**15 (1968).]

Most of the smaller animals of the forest are likely to show contagious distributions, with the degrees of contagion ranging up to the very strong contagion of ants in their colonies. Each species may have its own degree of clumping and spacing of clumps, different from the patterns of other species.

One asks, in consequence, a second question about our forest plots: how are the distributions of different species related to one another? The undergrowth plots can be studied again to determine association among the species occurring in them. One may set up a simple table of the numbers of plots containing one, or both, or neither of a pair of species *A* and *B* (Table 3.4). A chi-square test can indicate the probability that the two

Table 3.4 Species Association

Occurrence of Two Species in 100 Sample Plots

		Species *B*, Present	Absent	
Species *A*	Present	$a = 17$	$b = 22$	$a + b = 39$
	Absent	$c = 13$	$d = 48$	$c + d = 61$
		$a + c = 30$	$b + d = 70$	$F = 100$

Chi-Square Test of Species Association

$$\chi^2 = \frac{[(ad - bc) - 0.5F]^2 \times F}{(a + b)(a + c)(b + d)(c + d)} = 4.6$$

Probability (one degree of freedom) less than 0.05.

Coefficient of association of Cole (1949)

$$C_a = \frac{ad - bc}{(a + b)(b + d)} = 0.194 \text{ (for case when } ad \geqq bc)$$

species are distributed independently, or are associated with one another. As indicated in Table 3.4, the example gives a chi-square of 4.6, which implies a probability of less than one in twenty that species *A* and *B* are independently distributed. The distributional association of the two species is significant, although it probably would not be recognized in the field (more individuals of *A* occur outside, than in, plots containing *B*). Other measurements, which range from 0 (for species distributed independently of one another) to 1.0 (complete distributional association or correspondence) and —1.0 (complete disassociation or avoidance), can be used to express degrees of distributional association. One of these measurements (Table 3.4) gives a value of 0.194 for the statistically significant but rather weak association of species *A* and *B*. The strength of distributional relationship of species in a community may also in some circumstances be measured as a species correlation. One may, for example compare numbers of individuals of the two species in the plots by rank correlation.

Associations or correlations between species are for the most part not strong, and many pairs of species in a community may show none. Among the species of a community, however, there will usually be some that are positively associated and tend to occur together, and some that are negatively associated and tend to occur separately. Parasites and insects that feed on a single plant species are expected to show association with their host or food species. Positive associations may imply either that species are linked by the dependence of one on another, or that they are responding in similar ways to the small-scale differences of environment within the community. Negative associations may imply either that one species tends to exclude another by some effect on its population, or that the two species respond in different ways to the differences in environment.

Thus both differences in environment and interactions among species may be responsible for both contagion of individual species and association between species. Contagion and species association are related phenomena. Each species in the community has its own pattern of population distribution, often correlated with the patterns of other species and yet usually not quite like the pattern of any other species. When one conceives of the forest in terms of superimposed, different distributions of dozens of plant species and hundreds of animal species, the complexity and subtlety of pattern of the community may be evident. It is most to the point that there is, in fact, marked horizontal differentiation within most communities, and that different species have different relations to this horizontal pattern.

Time Relations

We have been discussing the differentation of communities in space, but there is also differentiation in time. Environments of natural communities are rhythmic: in most environments light, temperature, and other environmental factors go through daily and yearly cycles. In some communities, especially those of ocean shores, there are also complex rhythms set by the rise and fall of tides. The physiology and behavior of organisms respond to rhythms of environment, in many cases by linkage or coupling of an intrinsic, functional rhythm of the organism with the environmental rhythm. We expect rhythms of function in natural communities in adaptation to the rhythms of environment.

For example, on a Hawaiian coral reef many species of fishes of diverse and striking colors are active in the day time. As the light fades in the evening, these fish move downward out of the water into crevices in the coral reef, or other protected places. While they are doing so some of the nocturnal fish emerge from shelter; and as the water darkens further, swarms of nighttime fish emerge and become active. Both daytime and nighttime fishes on the reef include diverse food habits; many of the daytime fish feed on algae and plankton, and some are cleaner fishes that glean parasites from the skin of other fish. More of the nighttime fish are predators on a variety of invertebrate animals, including plankton, that are active at night. The nighttime fish, in contrast to those active in the day, are mostly orange-red in color. It appears that in some sense red is the gray of the seas. Most animals active at night or in the dusk on land are gray or brown; but in the ocean fish, and many invertebrate animals, that are active in the dark or in very weak light are red. The red color appears not only in the nocturnal fishes of the reef, but in many fishes and invertebrates of the permanently dark middle depths of the ocean—below the lighted surface waters but above the deepest parts of the oceans. By the time the water above the reef is lighted, shortly before sunrise, the nighttime fish retire to shelter and are replaced by the daytime species. Both groups move into and out of shelter during the twilight of evening and morning, and during twilight both groups are most vulnerable to predation because their colors, eyes, and behaviors are adapted to either light or dark. In the twilight a third group of fishes, predators feeding on other fish, are most active. These predators are of varied color patterns but mostly, unlike the other groups, either light-colored or mottled. Evolution on the reef has thus produced three groups of fish, each rich in species, adapted to activity in the different conditions of light, dark, and dusk.

Many plankton animals, both of fresh water and the oceans, migrate up and down in daily cycles (Figure 3.5). In a common pattern of movement the plankton sink, or swim downward, as light increases in the morning so that most individuals remain below the zone of most intense light near the surface. After remaining at some depth during the daylight hours they swim upward as light decreases in late afternoon and evening, to spend the night in the water nearer the surface. A marked rhythm thus also affects the plant plankton, which grows most actively in the sunlight hours, and is harvested most effectively by animals in the nighttime hours. Patterns of movement may be quite different, however, among the species of a given plankton community. Distances of migration range from centimeters in unicellular flagellates to many meters in the larger marine plankton animals. Populations of some of these as they move up and down in the ocean are responsible for the deep scattering layer that reflects sound waves transmitted through the water by sonar or echo sounding devices to determine depth of the bottom. An animal population may produce a reflection or apparent sonar bottom that is not merely false, but mobile in daily cycles.

Plankton populations fluctuate rapidly in many water bodies, with species replacing one another in periods of days and weeks. A number of major species may occur as plankton dominants in the course of the year. Thus in fresh-water lakes species of diatoms and other yellow-green algae

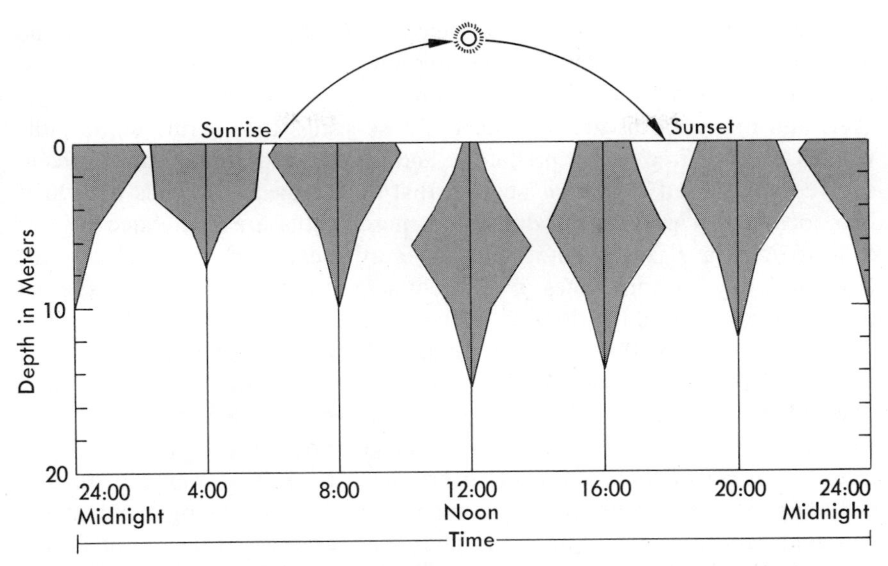

Figure 3.5. Vertical migration in a fresh-water plankton animal. Individuals move a few meters up and down in daily cycles, and the population as a whole shifts downwards below the most strongly lighted water in the daytime and upward toward the surface at night. Widths of the polygons are relative numbers of individuals at different depths.

may predominate in the winter plankton. As water temperatures warm in later spring and early summer, these are replaced by desmids and other green algae. At peak summer temperatures blue-green algae may predominate or share dominance with the green algae; as temperatures cool, dominance shifts back toward the green and yellow-green algae. Each species has an inactive stage in which it survives the season unfavorable for its activity. Each species has its own place in the annual pattern, determined by its own response to the fluctuations of temperature and other environmental factors. The plankton community shows differentiation in time: different groups of species occur at different times during the seasonal cycle. The year-round total number of plankton species is much larger than the number present at a given time.

Seasonal and daily differentiation occur also in forests. One group of insects are active in the daytime, another group at night, and a third group may be active in the twilight transitions of morning and evening. In terrestrial communities flycatchers, warblers, and other insectivorous birds are active by day, bats at night, and nighthawks in the dusk. Within one major insect order, the butterflies are adapted for daytime flight and most moths for nighttime flight. Evolution among the butterflies has featured bright colors and, in many of them, fast or erratic flight that makes their capture by birds infrequent. Some species are unpalatable to birds and have distinctive color patterns warning of their unpalatableness. Evolution in the moths has featured, in many species, obscure colors and slower flight in the protection offered by dusk or dark. This is only partial protection, however, and many moths are covered by loose scales that permit some individuals to slip free from predators and spiderwebs. Some moths have evolved specialized organs of hearing that detect the sonar calls by which bats locate their prey in the dark, and these moths are stimulated by bat calls to take a zigzag flight track that may escape an approaching bat. Night and day are thus times of different sets of predator and prey species that have evolved adaptations to each other.

The progress of the seasons is marked by the appearance of different groups of plants in flower and different groups of insects visiting these flowers. In temperate deciduous forests spring beauties, hepaticas, dogtooth violets, and other herbs develop their foliage and flower early, before the trees are in leaf. Other groups of species have their maximum growth and flowering in later spring and in summer; still others, among them asters and goldenrods, grow and flower in later summer and fall. Different flower colors predominate at different seasons, and some of these color differences represent adaptations to attract different pollinating animals. Many of the spring herb species are geophytes; most of the summer herbs are hemicryptophytes. The Sonoran desert of Arizona, which has a winter and a summer rainy season, has separate groups of herbs (with each group including hemicryptophytes, geophytes, and therophytes) blooming in the two rainy seasons.

One thinks of a tropical rain forest as having evergreen leaves and some plants in flower throughout the year, yet there are marked seasonal patterns in these forests too. Most tropical rain forest tree species have definite seasonal rhythms of flowering and fruiting, and these rhythms differ widely in different species. Some species produce abundant flowers in pulses at certain times during the year, others flower sparsely through a longer period. Different pollinating animals serve these. Some hummingbird species range widely, seeking the pulse-flowering species; other hummingbirds follow more definite search routes along which they feed on whatever flowers are available from the sparse-flowering species. Some plants have even closer adaptive fits to pollinators; among some of the epiphytic orchids each species is pollinated by a single species of bee with a seasonal timing of its life cycle matching that of the orchid.

Sparse- and pulse-flowering imply sparse- and pulse-fruiting, with different adaptations to dispersal of the seeds. In a tropical seasonal forest in Panama three major patterns of fruiting in relation to seasons were recognized. The large fruits of some species were available in a pulse from May to July, and these were dispersed primarily by agouti, a large rodent that hides fruits for later use much as squirrels do in temperate forests. A second group of smaller fruits were wind-dispersed, and these mostly fell at different times between December and April, in the driest and windiest part of the year. The seeds of a third group of smaller fruits were dispersed by passing undigested through the digestive tract of birds and mammals; many of these were sparse-fruiting species that have some fruits maturing through most or all of the annual cycle. Both temperate and tropical forest plants have evolved toward diverse ways of flowering and fruiting, differently adapted to the seasonal cycle and to animals and wind for pollination and dispersal.

Niche Difference

It appears that each species has its own time and place in the community, different from those of other species. There is much evolutionary interest in this observation.

We may choose to distinguish three aspects of the relationship of a species to environment. The *area* of a species is its geographic range, its distribution in space as this can be plotted on a map. The *habitat* of the species refers to the kind of environment the species occurs in, as this environment can be described in physical and chemical terms and often by elevation, topographic position, and so on, or by a kind of community. A species may occupy a range of different habitats, or more than one distinctive kind of habitat, in different parts of its area. Within each habitat one can describe for a species its position in the space, time and functional relationships of the natural community that occupies that habitat. The species' position in a community in relation to other species is its *niche.*

Niche is thus a term for the way a species population is specialized within a community. It is not irrelevant to recall the advantage of specialization in human societies. An individual may gain from professional specialization because of a high degree of skill or efficiency in his trade that ensures his obtaining the resources (income) he needs for his life. Two or more individuals may gain by following different specialties because they are not in competition, and each has his own assured source of income. The society at large may gain because the members' specializations serve one anothers' needs, and because the efficiency of the whole gains from the efficiency of the different specialists in their trades. It is hardly to be doubted that evolutionary advantage underlies the specializations we observe in natural communities.

For the species population a crucial advantage is a degree of protection from competition with other species. We may recall the treatment of competition in Chapter 2. Gause's experiments and the Lotka-Volterra competition equations (Table 2.3B) are in accord on an essential point. If two species are direct competitors, utilizing and limited by the same resources in the same space at the same time, then in the equilibrium condition one species will become extinct. If, however, the two species differ in their requirements or space occupied in the experimental culture, then they can coexist in a durable population balance. They can coexist if their populations are subject to different controls, whether by different resource limits or by other means of population regulation. We have said they can coexist by occupying different "spaces" in the culture, and we can extend this observation to the way species occur in communities in two different ways. Species can coexist in an area if they occur in different environments as members of different communities. This, as we illustrated with the barnacles in Chapter 2, is difference in *habitat*. Or, species can coexist within a community if they relate differently to resources and other species in that community. Such difference in species' positions within a given community, which is the concern of this chapter, is difference in *niche*. Species can coexist in a stable community if they differ in horizontal or vertical position in it, in time relation, in resources used or kind of interaction with other species, or in manner of population control—if they differ in niche.

The idea that two species cannot coexist permanently in the same niche is known as the *principle of Gause* (although a number of scientists contributed to its formulation) or the principle of competitive exclusion. We apply the idea to communities in the form of three statements, linked in a progression: (1) If two species occupy the same niche in the same stable community, one will become extinct. (2) No two species observed in a stable community are direct competitors limited by the same resources; the species differ in niche in ways that reduce competition between them. (3) The community is a system of interacting, niche-differentiated species populations that tend to complement one another, rather than directly competing, in their uses of the community's space, time, resources, and possible kinds of interactions.

Much of what was said about the variety of growth-forms and life-forms in communities takes enhanced meaning from the concept of niche differentiation. The differences of growth-form among plants in a community are visible indications of the niche differentation of these plants. Their niche differences should involve, however, a wide range of functional relationships—in physiology, life cycle and population function, and adaptation to other species—of which only a few will have any visible expression in growth-form.

Diversification of niches among the species of a community has evolved because of the selective disadvantage to one or both species of direct competition, versus the selective advantage (reliable availability of a distinct set of resources to support a given species, relatively free from competition for these resources by other species) of niche difference. For a pair of species and a single niche characteristic, one may represent the selective process as in Figure 3.6. The two populations at first overlap broadly in a

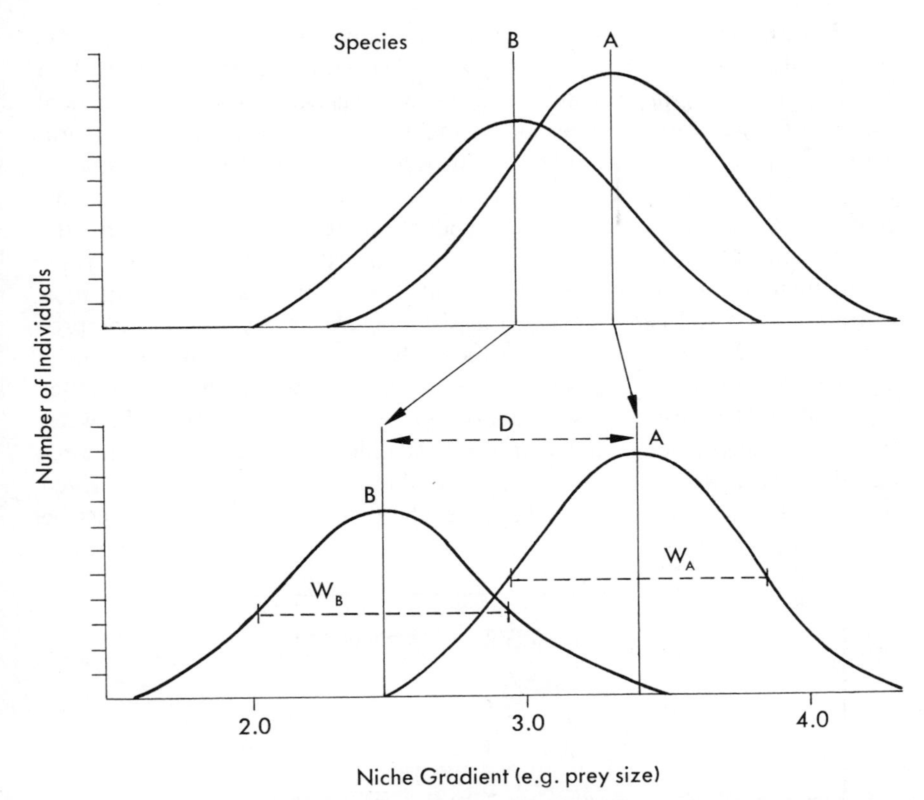

Figure 3.6. Selective niche divergence of two species. Above, two species overlap broadly in a niche characteristic, such as the mean prey-size choice by individuals. Competition reduces the survival of individuals choosing prey sizes around 3.0, and hence selects against genes determining such prey choice. Below, mean prey-size adaptations have diverged, as indicated by the arrows connecting the mean prey-size choices for the species.

niche characteristic—for example, sizes of prey taken by two species of predatory or hunting animals. Individuals of each species have a genetically determined mean food-size preference, and when individuals of *B* are competing with individuals of *A* for the same food, *A* has an advantage over *B*. The survival of individuals of *B* adapted to prey sizes where the species overlap is lower than the survival of individuals adapted to prey sizes outside this overlap and hence free of competition with *A*. The frequency of genes adapting *B* to prey sizes where it is in competition with *A* decreases relative to the frequency of genes adapting *B* to capture of smaller prey. The mean genetic composition of the population of *B* shifts (and that of *A* may also shift) until the two species, although they may still overlap in prey choice, are adapted to reliance on quite different sizes of animals for food.

If two species can thus divide a resource gradient between them, so can several. R. B. Root has termed a group of species that are closely related to one another in their niches in a given community a *guild.* If several species of a guild divide a resource gradient among them, they may form a niche sequence. For example, an Australian desert includes a number of lizard species feeding on insects. Three of the lizard species in the desert can be arranged in a size-sequence, and the insects they capture form the distributions shown in Figure 3.7. The three lizard species are closely similar in their foods (mostly termites in all three), but they are using food of different sizes. In many cases the sizes of animals in a guild are proportional to the mean sizes of the foods they consume. Figure 3.8 shows a series of fruit-eating pigeons in New Guinea for which this is true. Because different species choose fruits of different sizes, competition between them is reduced, though not completely avoided. Competition is also reduced because different species feed at different positions along a given branch. The larger the pigeon, the larger the diameter of a branch must be

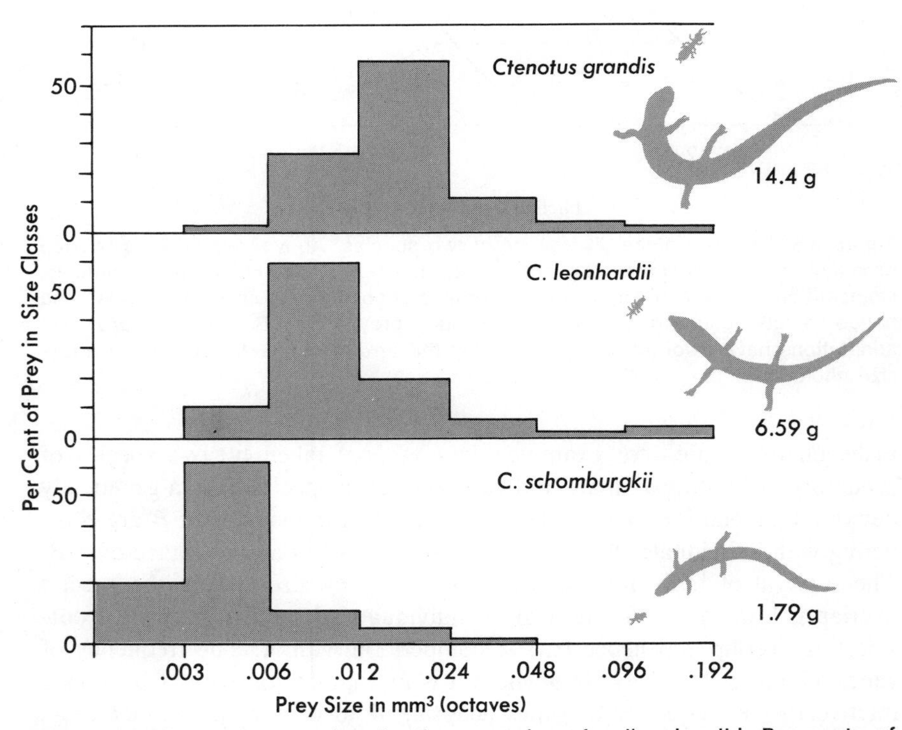

Figure 3.7. Sizes of prey taken by three species of a lizard guild. Per cents of prey are plotted for prey size classes on a logarithmic scale (by octaves or doubling units) for three species of the genus *Ctenotus* feeding mainly on termites in an Australian desert. Mean weights of the lizard species are given on the right. [Data of Pianka, *Ecology* **50:**1012 (1969).]

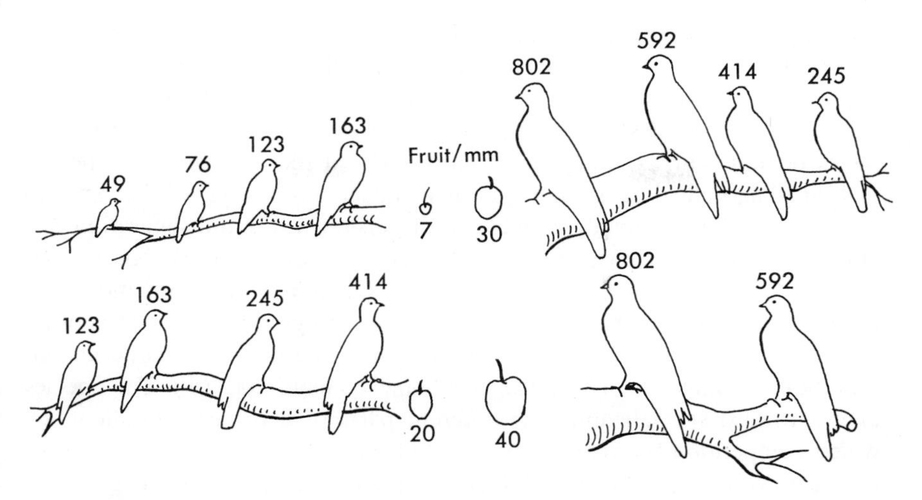

Figure 3.8. Niche relationships among eight pigeon species eating fruits in tropical rain forests, New Guinea. Four sets of birds are shown for trees bearing fruits of the four diameters in millimeters indicated. Weights in grams are indicated above the pigeons; each pigeon weighs about 1.5 times as much as the next smaller pigeon in the set. Each fruit tree attracts up to four consecutive members of this niche sequence; trees with larger fruits attract larger members of the sequence. On a given tree the smaller pigeons feed farther from the trunk where the branch is smaller in diameter. [Diamond, 1973.]

to support its weight, and the closer toward the trunk of the tree on a given branch that pigeon will feed. Different selections from the full size range of pigeons feed on trees of a given species, as illustrated. The full sequence may be the longest such size sequence within a guild reported, for it includes eight pigeon species.

Species forming niche sequences may evolve toward definite quantitative relations to one another. A number of cases of related species that occurred in the same communities and differed in size, but seemed much alike in their niches otherwise, were compiled by G. E. Hutchinson. The ratios of lengths, or other dimensions, of the larger species to the smaller ranged from 1.1 to 1.4, with a mean at about 1.26. Perhaps, however, these species should be compared not by length but by weight; the cube of the length ratio indicates a likely weight ratio. The cube of 1.26 is about 2.0. We infer that the species of a niche sequence in relation to food size may form a geometric series, with body weight approximately doubling from one species to the next. There are a number of cases among birds that reasonably approach such a doubling series—for example, hawks in given areas, and woodpeckers and other bark feeders in forests. The mean weight ratio for pairs of related birds in similar niches is 1.9 or 2.0 in New Guinea, although the pigeons in Figure 3.8 are closer together in weight than that.

There may also be some consistency in species widths and distances apart along a resource gradient. If we arrange species in a sequence, as in Figure 3.7, then a species' niche width in relation to the resource can be expressed as a standard deviation, giving a dispersion or spread for the bell-shaped curve. Distance apart can be expressed as distance between the peaks of these curves along the resource axis, in the same units as the deviation. Theoretical reasoning and field observation suggest that species evolve toward approximate equality of dispersion width and separation distance, and the lower part of Figure 3.6 has been drawn on this basis. The extent of the population overlap for species that are in competition in a niche sequence is thus limited. If, along a resource gradient, we add another species, the niche widths of the species already there will be reduced. Observations of niche sequences suggest some conclusions: (1) The species evolve toward specialization for different parts of a resource gradient, by which competition between species is reduced, and (2) given time, species evolve toward even spacing of their adaptive centers along the gradient. (3) Niche sequences can be lengthened by fitting in new species between old ones, while reducing the niche widths of the latter, but (4) there are probably limits on the number of species that can be thus "packed" into the sequence along a given resource gradient.

Niche Space

Clearly there is more to the niches of species than these relations to single resource gradients. Only a few of the many species in a community can be analyzed in such well-defined series. Many niche characteristics are very difficult to measure, or cannot be treated as linear gradients. We need, however, to proceed from the concept of niche axis to a multidimensional definition of niche that permits us to relate species by their full ranges of adaptive interrelationships.

Four scientists are principal authors of the niche concept. Joseph Grinnell originated the term niche in describing for a bird, the California thrasher, its food and nesting requirements and other adaptations that defined its place in its community, the California chaparral. Grinnell did not make clear the relationship between niche and habitat, but Charles Elton recognized the niche as the species' position in relation to other species in a community, and G. F. Gause employed the concept in this sense as a basis of the principle that bears his name. If a niche is a species' "place" or "position" in the community, we should be able to characterize it by some kind of dimensions. This G. E. Hutchinson proposed to do, considering that the niche can be defined by a number of variables of environment within the community, to which the species must be adapted. These variables include both biological ones (for example, food size), and nonbiological ones (such as vertical height above ground and seasonal time).

The variables as axes define an abstract multidimensional space that we can call the *niche space.*

Each species can adapt to or tolerate some range of each variable. The upper and lower limits of all the variables for a given species become boundaries of a part of the niche space that the species is able to occupy. Such a bounded portion of an abstract, multidimensional space, or hyperspace, is an abstract, multidimensional volume, or hypervolume. In crude terms, the species is now conceived as occupying an *n*-sided "box" in niche space, with the sides of the box being the species' tolerance limits along niche axes. Hutchinson termed this hypervolume the "fundamental niche." It should be noted that the variables of Hutchinson's fundamental niche were intensive variables, those that relate species to environment and one another within a given community. The niche concept has been subject to some confusion because other authors have applied the fundamental niche to species' relations to extensive or habitat factors, or to combinations of niche and habitat factors.

The word *abstract* was used for the concepts in the preceding paragraphs, for they are that. It is, however, just this abstract, multidimensional approach that has made possible much research on niches and interpretation of community organization. Given the multidimensional approach to community niche space, we can discuss: (1) the positions of species in the space, (2) the form of the species' response to more than one niche variable, (3) the relative sizes of niches and importances of species, and (4) some implications for the evolution of communities.

Consider first the positions of plant species in the niche space of the Sonoran desert of the lower mountain slopes in southeastern Arizona, a spectacular, tall semidesert of giant cactus (*Carnegiea gigantea*), ocotillo (*Fouquieria splendens*), palo verde (*Cercidium microphyllum*), mesquite (*Prosopis juliflora*) and a great variety of plants (Figure 3.9). A first niche axis in this community involves the relation of species to vertical height. The heights at which plant species bear the greater part of their foliage range from near zero in herbs with stems and leaves on the ground surface, to a few centimeters in other herbs, to a few decimeters in most of the semishrubs (*Encelia farinosa, Franseria deltoidea,* etc.), to 0.5 to 2.0 m in true shrubs of different species, to 2 to 5 m in arborescent shrubs (ocotillo, palo verde, mesquite), whereas the giant cactus or saguaro has a photosynthetic surface to the top of its stem, up to 6 to 9 m. As in our discussion of life-forms, average position of the buds or tissues that survive unfavorable seasons and from which the foliage develops, from the ground surface up (or in some herbs below the ground surface), provides a convenient expression of plant height.

A second set of relationships involves seasonal time. There are two rainy seasons in this desert, one in winter and one in late summer, separated by dry seasons; and two waves of plant growth correspond to the

Figure 3.9. A Sonoran semidesert of diverse plant growth-forms. Taken near Tucson, Arizona, with the Santa Catalina Mountains in the background. The giant cacti are the saguaro, *Carnegiea gigantea.* [Courtesy of W. A. Niering.]

rainy seasons. Most perennial plants use the moisture of both rainy seasons, but in different patterns of leaf and stem photosynthesis, patterns expressed in growth-forms and the seasonal behavior of foliage. The plants may again be arranged along a gradient, from those with persistent, evergreen leaves, through semideciduous species with leaves (or the leaf-bearing twigs of the semishrubs) persistent through less severe dry seasons but not more severe ones, to the deciduous mesquite, palo verde, and semishrubs. These in turn grade through forms like ocotillo, with short-lived leaves quickly produced and soon lost after rains, to the cacti, which lack leaves. In plants of the latter part of this sequence the stems and branches are green and photosynthesize to supplement the photosynthesis by the leaves; in the cacti all the photosynthesis occurs in the stem and branches. The plants form, then, a gradient of decreasing leaf persistence and increasing stem and branch photosynthesis, and we shall assume that differences along the gradient are significant in relation to plant competition. The gradient is also one of adaptive patterns in relation to water shortage—different solutions of the plant's problem of how to photosynthesize enough while also conserving water sufficiently to stay alive in a desert.

A third axis is one of horizontal pattern. Some species occur mainly underneath the larger shrubs, other species mostly in the open spaces between shrubs. We can thus arrange species along a niche axis from primarily shaded to primarily exposed. Much more than exposure to light is involved in this axis, for the shrubs modify the microclimate under them in ways other than partial shade, and the soil under the shrubs differs in higher organic content and other characteristics from that in the open. Some species show little response to this gradient; but others, including some of the larger species, are clearly affected. The seedlings of the giant cactus or saguaro survive best under shrubs, and stems of the larger of these cacti often extend up through the foliage of the shrubs that served as shelter for the seedlings.

We thus characterize plant niches in this desert by three major niche axes: vertical height, seasonal time relations, and horizontal pattern. These are not simple axes, and they by no means account for all the adaptive relationships among the plants. Differences in root depth and form are probably also important. The niche space is a simplification of, an abstraction from, the wealth of niche relationships in the community. The three-dimensional space permits us, though, to observe much that is of interest about the desert as a plant community. We can diagram species niche positions as in Figure 3.10, considering here only the seasonal and height axes (the pattern axis would be perpendicular to this page), and representing species positions only by points where the species populations seem to be centered. The plant species appear to be scattered in this space, as the principle of Gause would lead us to expect. Thus the principle of Gause

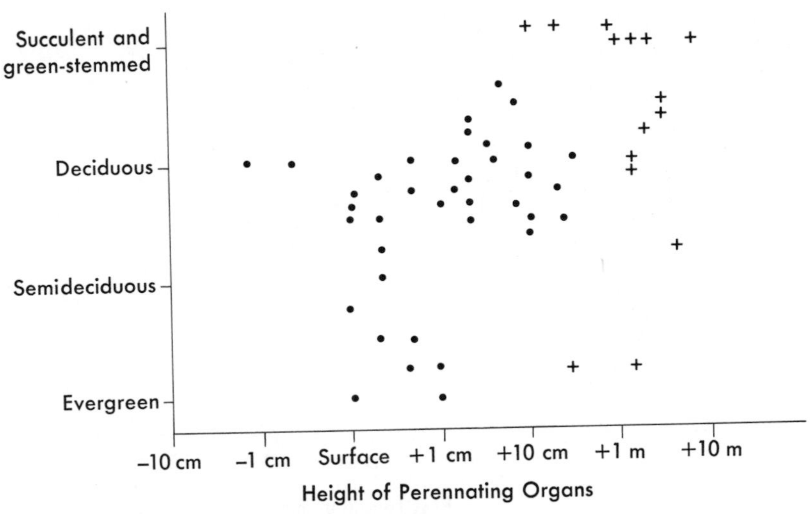

Figure 3.10. Growth-form differentiation in a Sonoran semidesert, Arizona. The plant species of the desert are plotted by the seasonal relationships of their foliage, on the vertical axis, and plant height in relation to ground surface (on a logarithmic scale) on the horizontal axis. The larger woody plants of the community are armed with spines, as indicated by the crossed points. [Whittaker and Niering, *Ecology*, **46**:429 (1965).]

provides some interpretation of the diversity of plant forms that is so striking in this desert.

The niches of singing bird species that nest in a community can also be characterized by three major axes: vertical height at which they feed and nest, food size, and food composition (whether insects, or seeds, or some combination of these). An intensive study of one bird species, the blue-gray gnatcatcher, in a California oak woodland permits a characterization of its niche as in Figure 3.11. Frequency of capture of prey of different sizes, taken at different heights above the ground, is illustrated. Since the gnatcatcher relies almost wholly on animal food, we need not plot its relation to food composition as a third axis. Figure 3.11 represents the niche as a frequency distribution in two dimensions, or a response surface that can be visualized as a hill sloping downward in all directions from its peak at *H*. The gnatcatcher's niche centers on food of 3 to 5 mm in length (leaf-

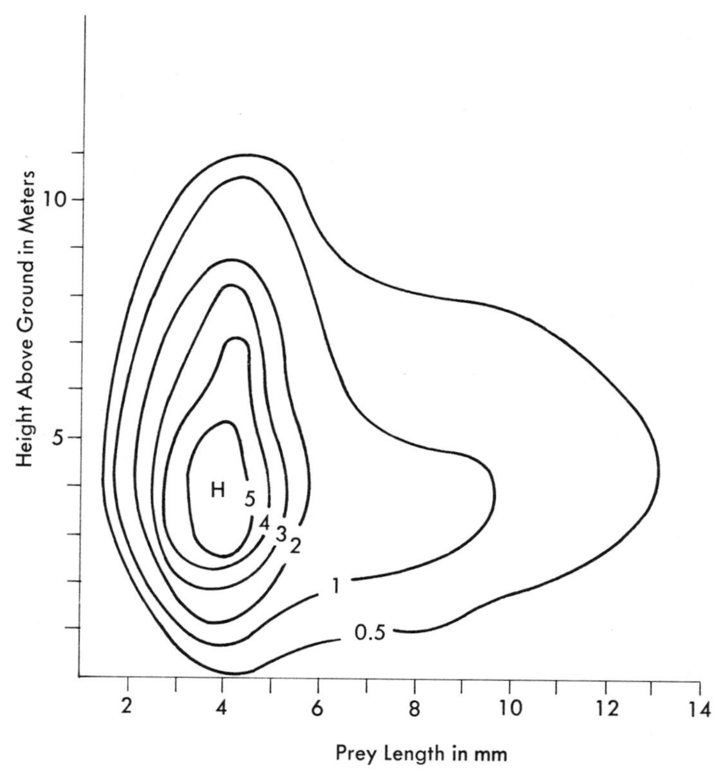

Figure 3.11. A niche response surface. Frequencies of prey capture in per cent for different prey lengths taken at different heights above the ground are outlined by contour lines. The highest capture frequencies are indicated by *H,* and frequencies decrease in all directions away from this high. The bird is the blue-gray gnatcatcher (*Polioptila caerulea*) in oak woodlands in California. [Root, 1967; Whittaker, Levin, and Root, 1973.]

hoppers, small flies, and so on), taken at 3 to 5 m above the ground. Outside this center, the gnatcatcher's niche overlaps with those of other bird species. A response pattern like the one in Figure 3.11 is a more realistic way of characterizing a species' niche than a sharply bounded, box-like hypervolume. The response surface is not so easily conceived in three or more dimensions. It can be thought of, however, as a kind of population cloud that has a dense center and becomes sparser in all directions away from that center. Inside the niche hypervolume, then, is a cloud-like population response that better characterizes the niche.

For birds, as for desert plants, the reduction of niche space to three axes is a simplification; for bird species differ in other ways, including behavior and manner of food choice and capture, that our axes do not do justice to. Our concern here, however, is with the concept of niche. Let the birds' niches be conceived of as occurring in a space of only three dimensions. Then this space, for a temperate forest, is occupied by 15 to 20 populations, each with a distribution resembling Figure 3.11 in form, but differing from it in position. The centers of the birds' populations are scattered in the space, and by this scattering competition between bird species is reduced. The populations do overlap, however, and along each axis may form a niche sequence (Figures 3.7 and 3.8).

To consider the community as a whole, including its mammals and reptiles, plants and insects, bacteria and fungi, and so on, we must consider many kinds of additional niche axes to which these many species relate. We can then generalize the niche and niche space from two or three to n dimensions. The essential basis of the conception remains the same. Our discussion of how species in communities relate to one another has led to the conclusion that: (1) The niche relationships among the species in a community can be interpreted as an n-dimensional niche space in which (2) each species has its own position or niche, with a central location that differs from those of other species because (3) evolutionary processes, including the reduction of competition, lead toward the dispersion or scattering of species' positions in the niche space.

Species Importances

The means by which a species population is controlled is a most important aspect of its niche. Control mechanisms and some other aspects of niche are not really resources. Let us assume, however, that there is some correspondence among three things: the fraction of the niche hyperspace of the community that a species occupies, the fraction of the community's resources (light, water, food, and so on) that the species uses, and the fraction of the community's productivity that the species realizes. It may be clearer for our present purposes if we set aside niche characteristics that are not resources, and thus simplify the n-dimensional niche space to an

m-dimensional resource space. We can then ask how this resource space is divided up among species, and what kinds of relative importances of species result.

Importance refers here to a group of measurements by which the species in a community can be compared. Density, the number of individual organisms per unit area, was the importance value used in the discussion of populations. Density may not be the best way to compare populations of organisms of widely different sizes. For this we sometimes use dry weights, or biomasses, per unit area for all individuals of the different species we are comparing. Plant populations can also be compared by coverage (percentage of ground surface area over which foliage of a species occurs) or frequency (percentage of small plots within the community, in which a species is recorded as present). When we can determine it, however, our preferred importance measure is likely to be species productivity (amount of dry organic matter formed, per unit area per unit time). Productivity seems most appropriate for our present purposes because it expresses the species' use of resources for population growth, and also permits comparisons of species of widely different sizes and kinds on a single scale. We now ask how the resource space of the community is divided, to produce what quantitative relations among the productivities (or other importance values) of species. We can plot the relative importances of species, from the most important to the least important, as dominance-diversity or importance-value curves (Figure 3.12).

A number of hypotheses on these curves have been suggested. When

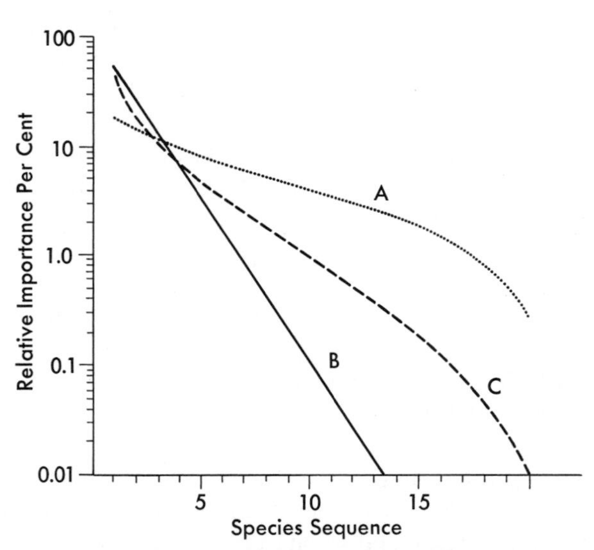

Figure 3.12. Three hypotheses on importance-value curves. The curves are all computed for a hypothetical sample of twenty species. **A:** Random niche-boundary hypothesis. **B:** Geometric series, $c = 0.5$. **C:** Lognormal distribution. (See Figure 3.14 for the manner in which the data are plotted.) [Whittaker, 1965.]

they were suggested it seemed possible to treat these as multiple working hypotheses, and to use measurements of actual importance relations of species in communities to choose among them:

1. The random niche-boundary hypothesis of Robert MacArthur. One assumes that boundaries of niche (resource) hypervolumes of species are located at random positions in the hyperspace. It is much simpler to conceive of a one-dimensional "hyperspace," that is a line, onto which the points that represent niche boundaries are cast at random. We then break the line into segments separated by these points and arrange the segments, the lengths of which represent niche sizes, in order from the longest to the shortest. The segments form a curve from the longest segment (most important species) to the shortest segment (least important species); curve A in Figure 3.12 is of this form. The lengths of the segments, and hence the importance values of species, should be distributed according to the series given in Table 3.5, part A.
2. The niche pre-emption hypothesis. Suppose that the sizes of niche hy-

Table 3.5 Importance-Value Distributions

A. MacArthur Distribution

$$n_r = \frac{N}{S} \sum_{i=1}^{r} \frac{1}{S - i + 1}$$

S is the number of species in the sample, N the total of importance values for all species in the sample, n_r is the importance value of species r in the sequence of species from least important ($i = 1$), through the species in question ($i = r$), to the most important species ($i = S$). $\sum_{i=1}^{r}$ indicates summation of values of $1/(S - i + 1)$ for all species from $i = 1$ to $i = r$.

B. Geometric Series

$$n_i = Nk(1 - k)^{i-1} = n_1 c^{i-1}$$
$$c = 1 - k, N \sim n_1/k$$

N is the total of importance values for all species in the sample and n_i the importance value for species i in the sequence from the most important to the least important species; c is the ratio of the importance value of a species to that of its predecessor in the sequence, and n_1 is the importance value of the first and most important species.

C. Lognormal Distribution

$$s_r = s_0 e^{-(aR)^2}$$
$$\Sigma s_r = S = s_0\sqrt{\pi}/a$$

Here s_r is the number of species in an octave R octaves distant from the modal octave, which contains s_0 species; a is a constant that often approximates 0.2.

pervolumes are determined primarily by the success of certain species in pre-empting part of the niche space, whereas less successful species occupy what is left. One most important species may occupy a fraction k, say 50 per cent, of niche space, using a corresponding fraction of community resources for a corresponding fraction of community productivity. This species is dominant in the community. The second species is able to occupy a similar fraction of the niche space unoccupied by the first, and the third species a similar fraction of the niche space unoccupied by the first and second, and so on. We are not implying by "first, second, and third" anything about the time when the species reached the community, only their relative success in competition. We take for granted that there will be at least some random differences in values of k for different pairs of species in the sequence. The importance values of these species will, however, approach a geometric series, as suggested by I. Motomura and given in part B of Table 3.5, and when plotted in the manner of Figure 3.12 will form a straight line, curve B.

3. The lognormal distribution of F. W. Preston. The extent of the resource space occupied by a species may be determined by a large number of factors affecting the relative success of one species in competition with other species. If the relative importances of species populations are determined by a number of independent variables that differently affect different species, than a bell-shaped or normal distribution of importance values should result. If we again divide a line into segments with lengths representing importance values for a species, we can group the segments by ranges of importance values to form a frequency distribution. A central or modal range of importance values will have the largest number of segments (that is, species) in it, and smaller numbers of segments (species) will occur in importance-value ranges on each side of it. There will then be few very important species, few very rare species, and many species of intermediate importance values in the community.

Figure 3.13 represents such a treatment—a frequency distribution of species by importance-value classes—from the same Sonoran semidesert as Figures 3.9 and 3.10. A bell-shaped normal curve has been fitted to the numbers of species in the ranges. The ranges, however, are related by units not of a linear but of a geometric scale; they are octaves whose limits are set by doubling of importance values from one octave to the next. The horizontal axis of the figure is consequently a logarithmic scale. A frequency distribution that becomes a symmetrical normal curve on a logarithmic scale is lognormal; our hypothesis on importance values of species is consequently that they form lognormal distributions.

The logarithmic scale is appropriate because the response of a population to environmental factors and combinations of factors is geometric, not linear. If the environment becomes more favorable, the population may

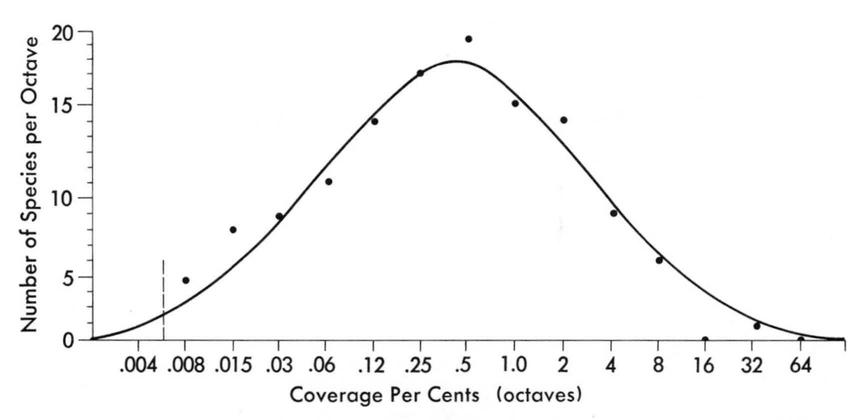

Figure 3.13. A lognormal distribution for plant species in a Sonoran semidesert, Arizona. Percentage of the ground surface covered by the foliage of a species is used as the importance value on the horizontal axis. The scale is logarithmic, with the species grouped by doubling units, or octaves, of coverage per cents. The largest numbers of species occur in the middle octaves of 0.12 to 2.0 per cent, and there are smaller numbers of species with very high or very low coverages. The dashed line on the left is a "veil line" below which no measurements are available. (The fitted curve is $S_r = 17.5e^{-(0.245R)^2}$, where S_r is the number of species in an octave, R octaves distant from the modal octave, which contains $S_0 = 17.5$ species.) [Whittaker, 1965.]

increase not by a given number of individuals, but by a given fraction of the population present at that time. Favorable environment might add 20 per cent to the population, say, hence 20 individuals to a population of 100 but 200 individuals to a population of 1,000. The effect of two factors on the populations is likely to be multiplicative rather than additive. Hence two factors that would each (with the other held constant) produce a doubling of the population might together produce effects approaching a fourfold increase. In comparing the importance values of species in a community it is more appropriate to compare them by ratios of importance values—hence a logarithmic scale—than by absolute differences of importance values on a linear scale. Formulas for the lognormal distribution are given in Table 3.5, part C. In figure 3.12 a lognormal distribution gives the sigmoid curve C, steeper than curve A.

Figure 3.12 shows together the forms of the three curves applied to hypothetical communities. There are now many sets of data from communities that can be compared with these curves. The results are not quite what the proponents of any of the three hypotheses would have expected. Figure 3.14 illustrates some actual samples: (A) a small sample of nesting bird populations, which approaches the random niche-boundary curve, (B) a high-elevation fir forest in the Great Smoky Mountains, which approaches the geometric series, and (C) a cove forest rich in species from the Great Smoky Mountains, which approaches lognormal distribution. Different

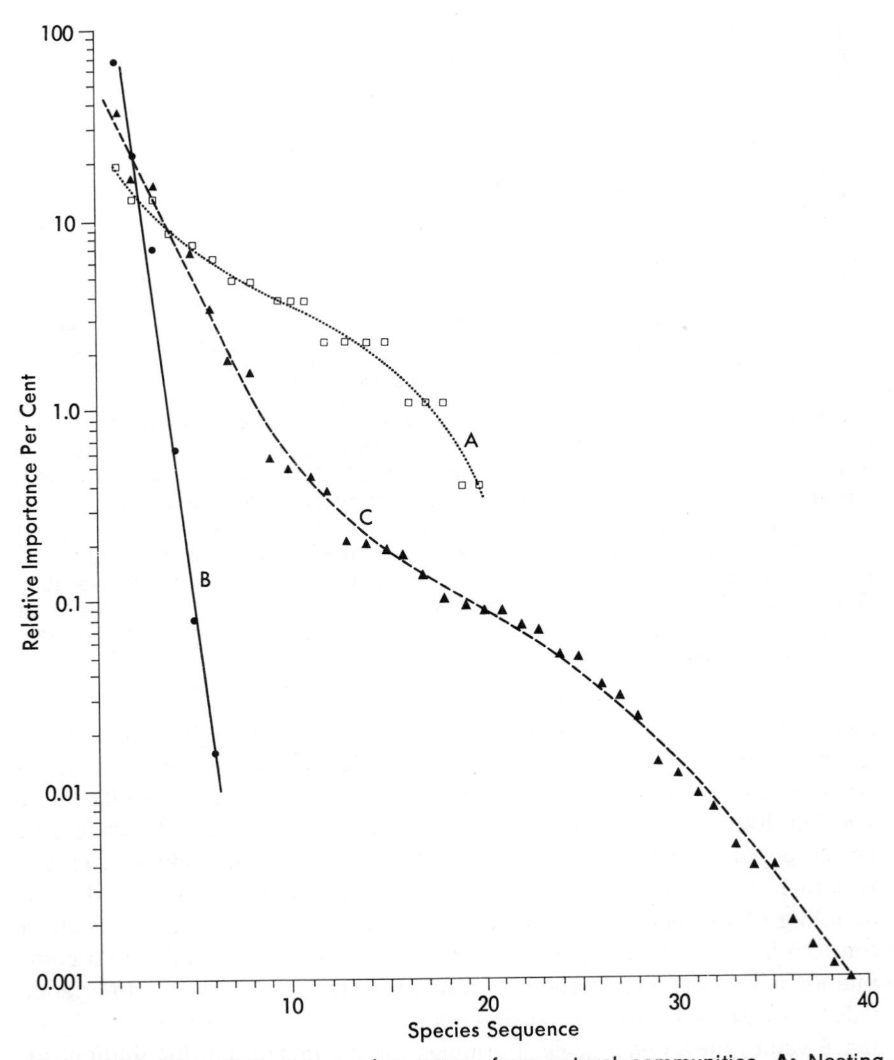

Figure 3.14. Three importance-value curves from natural communities. **A:** Nesting bird pairs (densities) in a deciduous forest, West Virginia; the data fit the random niche-boundary hypothesis as shown in Figure 3.12A. **B:** Vascular plant species by net production in a subalpine fir forest in the Great Smoky Mountains, Tennessee. The data fit the geometric series and niche pre-emption hypothesis as shown in Figure 3.12B. **C:** Vascular plant species by net production in a deciduous cove forest in the Great Smoky Mountains of Tennessee, a community of much higher species diversity. The data approach the lognormal distribution, as shown in Figure 3.12C and (in a different kind of plot) Figure 3.13. Each species is represented by a point located by that species' relative importance (the percentage that that species represents of the total net production, or total density, of all species in the community, on a logarithmic scale) on the vertical axis, and its position in the sequence of species from highest to lowest importance values, on the horizontal axis. Thus in curve *B,* there are points for six species in a sequence of decreasing importance values, from the most important species at the top with 69 per cent, through other species with 23, 7.0, 0.62, and 0.08 per cent, to the least important species sampled, at the bottom, with 0.016 per cent of net primary production for the forest.

samples from communities appear to fit all three models, together with intergradations among them. Instead of choosing among the three, we must ask when and why importance curves for species in communities approach one of the three models.

It appears that:

1. The curve we have given for the random niche boundary hypothesis is approached by some small samples of taxonomically related animals from narrowly defined, homogeneous communities, for example, the nesting birds of a limited area in a forest. Fits are obtained only for some such samples, and primarily for higher animals with contest competition, stable populations, and relatively long lives. The random niche-boundary hypothesis thus may describe a limiting case that some animal samples approach.
2. Some plant communities, especially those of severe environments and small numbers of species, approach geometric series. In such communities the phenomenon of dominance is strongly developed (in contrast to the animal groups just discussed, in which niche space is divided without strong dominance of any species). Also, the importances of species in a given stratum in a plant community may approach a geometric series, even though those of the community as a whole do not. The geometric series may thus express the outcome of scramble competition among a limited number of species dependent on some of the same resources. The resulting steep curves are an opposite limiting case from the rather flat random niche-boundary curves.
3. In communities that are rich (in numbers of species), importance values in homogeneous samples approach lognormal curves. Such is the case with the Sonoran semidesert (Figure 3.13), the cove forest (Figure 3.14) and tropical rain forests, which are even richer than these. Samples that are not homogeneous, but combine species from a range of environments and communities (such as the collection of insects caught in a light trap), also approach lognormal distributions. The lognormal form may appear for a mixed sample even though the curves for individual communities (among those being combined) might approach the forms for either the random niche-boundary hypothesis or the geometric series.

We conclude that: (1) A common theme—division of resource (and niche) space with reduction of competition among species—may underlie the varied forms of importance-value curves, but that (2) a variety of forms for these curves, ranging from geometric through lognormal to random-boundary, appear when importance values are plotted for different groups of organisms and different communities; among these (3) samples including a limited number of species related to one another by competition in the same community may approach the random niche or geometric form, whereas (4) samples including larger numbers of species, whether or not they are competitors and whether or not they are from the same community, will approach the lognormal.

Study of importance-value distributions has not produced the single mathematical form and choice among the three hypotheses that the early work suggested might be possible. The study has led not only to a range of curves, but also to increasing recognition of the limitations of these curves for interpreting community relationships. There is no doubt that forms as different as the curves for birds and spruce forest plants in Figure 3.14 express something significant about the groups of species the curves represent, very likely the differences we have suggested in the way competition occurs and resources are divided among species. But we cannot be sure what the curves express, for more than one reasonable hypothesis can be suggested as interpretations for each of the types of curves illustrated. The three hypotheses given above are plausible but not necessary interpretations. We cannot draw strong conclusions from these curves, and they have not made possible more penetrating analysis of competitive relationships. In this respect importance-value curves are like other approaches in this chapter, such as life-form composition and the measurement of horizontal pattern. These are useful quantitative descriptions of certain relationships of species in communities; but they have not led as far beyond description, toward further understanding of communities, as we would wish them to.

Species Diversity

We have found that these importance-value curves are related to another community characteristic of interest—relative richness in numbers of species. The fir forest and the cove forest of Figure 3.14 grow in the same mountain range within a few kilometers of each other. The fir forest included in a tenth hectare sample (20 × 50 m) 8 vascular plant species, the cove forests 46 to 68 species. These communities are in striking contrast in their richness in species or their *species diversities.*

Many indices of species diversity have been proposed; Table 3.6 gives some of the most useful ones. The simplest and most basic measure is the number of species, S, in a sample of standard size; S can also be called richness in species, or "species density" in distinction from population density. The numbers of species in different-sized samples from a given community are approximately proportional to the logarithms of the areas of the samples. Because of this relation the measure d can be used to compare diversities of samples that differ in size, but are not too widely different in size. The fir and cove forests are in contrast in other ways, notably in the slopes of the importance-value curves in Figure 3.14. The cove forest curve's slope is more gentle, and adjacent species consequently have im-

Table 3.6 Diversity and Related Measurements

1. Species Diversity or Richness

S = the number of species in a sample (of standard size)
$d = S/\log A$, or $S/\log N$
A is the sample area (usually in square meters), N is the total number of individuals in the sample.

2. Dominance Concentration

$$C = \sum_{i=1}^{S} p_i^2 = \sum_{i=1}^{S} \left[\frac{n_i}{N}\right]^2 \qquad \text{(Simpson index)}$$

N is the total of importance values for all species in the sample, n_i the importance values of S individual species, and p_i the relative importance values (as decimal fractions) for these same species.

3. Equitability

$$H' = -\sum_{i=1}^{S} p_i \log p_i, \text{ or antilog } H' \qquad \text{(Shannon-Wiener or information index)}$$

$$E_c = S/(\log n_1 - \log n_s) \qquad \text{(species per log cycle index)}$$

S is the number of species in the sample, n_1 the importance value of the most and n_s of the least important species.

4. Sequential Comparison Index, for Class Use Without Species Determinations

$$D_S = N_r/N$$

N_r is the number of "runs" of different kinds of organisms observed while counting in sequence, or one plus the number of changes observed while counting in sequence. For example: 10 individuals of species, *A, B,* and *C* are counted in the sequence, *AA B AAA B C BB,* six runs and five changes, $D_s = 0.6$.

(Diversity and related indices should not be "scaled" in the forms, $J = H/H_{max}$ and $H^* = (H' - H'_{min})/(H'_{max} - H'_{min})$, for this increases their fluctuation with sample size.)

portance values that are closer together than those of adjacent species in the fir forest curve. This quality of relative gentleness of the importance-value slope, and relative similarity of adjacent importance-values, is *equitability*.

One simple way of expressing equitability is the mean number of species per log cycle of the importance-value curve, because the more species occur in a given log cycle of importances, the less steep the curve connecting them will be. The measure E_c in Table 3.6 gives values of 1.7 for the fir forest, 8.4 for the cove forest, and 11.9 for the bird community plotted in Figure 3.14. In dealing with plant communities it is appropriate to ask also how strongly the importance values are concentrated in the first, or the first two or three, species—the dominants. The Simpson index, C, in Table 3.6 expresses concentration of dominance, and it gives values of 0.52, 0.18, and 0.10 for the three communities. Equitability is positively correlated with diversity, and dominance concentration is inversely correlated with both. Because diversity and equitability are correlated, equitability indices can be used to express relative diversity, and the index H' has been widely used in this way. (Because H' is logarithmically related to diversity, the antilogarithm of H' may be preferred.) The correlations are sometimes weak, and it is often desirable to measure both species richness (S) and either dominance concentration (especially in land plant communities) or equitability (in other kinds of communities).

We use these measures to express the relative diversities of communities, so that we can try to interpret diversity differences. The fir forest grows in cool climates of high elevations and the cove forests in warmer climates at lower elevations. We judge from this pair (and can support the judgment with other data) that species-diversities of plant communities increase toward lower elevations and warmer climates. We can extend the relation to the still warmer climates of the tropical rain forests, in which a sample may contain more than one hundred species of trees, plus many other plant species. This trend of increasing species-diversities from communities of the high Arctic (and Alpine and Antarctic) that are poor in species, toward the tropical rain forest and coral reef where profusion of species overwhelms the naturalist, is one of the major generalizations of biogeography.

It appears that few species are able to survive in the rigorous and widely fluctuating environments of the Arctic. A large share of the adaptation of these species must concern means of gaining the necessary food for growth and reproduction during the short favorable season, and surviving the unfavorable season. Evolution is affected more strongly by selection for survival in relation to problems of physical environment, less strongly by selection involving interaction and competition with other species. In the tropical rain forest the climate is warm and moist throughout the year and has been so for long periods of geological history. Very many species have been able to accommodate to a favorable and undemanding climate and to survive there. They can survive, however, only by solving the problems of interaction with many other species also able to survive there—that is, by niche differentiation. Their evolution is affected less strongly by selection for the solution of problems in the physical environment, and more strongly by selection for success in solving, by more finely drawn and more varied niche differentiation, the problems of surviving in relation to many other species in the same community. It is not assumed that this is the whole explanation of the diversity gradient in relation to latitude and altitude (which differently affects different groups of organisms). The result, however, is a biogeographic gradient connecting contrasting extremes: (a) the tropical rain forests with their rich life, relatively stable populations, wealth of strikingly adapted forms, and occurrence of surrvivors of some ancient and primitive groups, versus (b) arctic communities with relatively poor and monotonous composition, more narrowly modern in the evolutionary relations of species, and with striking year-to-year fluctuations in some of their populations.

We can call the decrease of species diversity from the lowland tropics to high latitudes and high altitudes the master gradient of diversity. It applies to many different groups of organisms, both terrestrial and aquatic, but it does not apply to all. Beyond this major gradient, diversity relations seem to branch into different kinds of trends in different major groups of organisms. We cannot review many of these, but shall illustrate some of them in four groups—birds, lizards, the marine benthos, and vascular plants.

We have already laid the basis for discussing the diversities of bird communities. Let us assume that the "lengths" of two of the three major axes of the niche space we defined—the ranges of food sizes and food kinds—do not differ too much among temperate-zone communities. These communities differ much, however, in the third axis—vertical height and complexity of vegetation structure. A grassland has a single major stratum of plants in which birds live, within which niche differences in food habits have evolved. A woodland may have two principal strata—small trees, and grass—, and a forest has three—canopy trees, shrubs (and small trees), and herbs. In the forest the birds can occupy parallel niches (as regards food habits) in the three strata; it is as if three sets of food-differentiated bird species were superimposed on the different vertical levels of the forest vegetation. MacArthur and others have shown strong correlation between species diversity in bird communities and the structural complexity of the corresponding plant communities, as shown in Figure 3.15. Note especially the rapid increase in bird diversity as the coverage increases from 80 to 120 per cent and the structure of the vegetation from one to three plant strata. On the other hand, species diversities of bird communities mostly bear little relationship to the species diversities of plant communities. (In some deserts, however, numbers of bird species are related to numbers of plant species and growth-forms.)

In communities of comparable structure, bird species diversity increases from temperate to tropical lowland climates. MacArthur found that much of the increase in numbers of bird species toward the tropics was made possible by narrower division of habitat gradients between bird species (this we shall call beta or between-habitat diversity in Chapter 4), and greater geographic differentiation of tropical bird communities. Apart from these, however, individual samples of bird communities are richer in tropical forests than in temperate forests. The additional species seem mainly to be specialists on kinds of food that are scarcely abundant enough to support bird species in temperate climates. These foods include very large

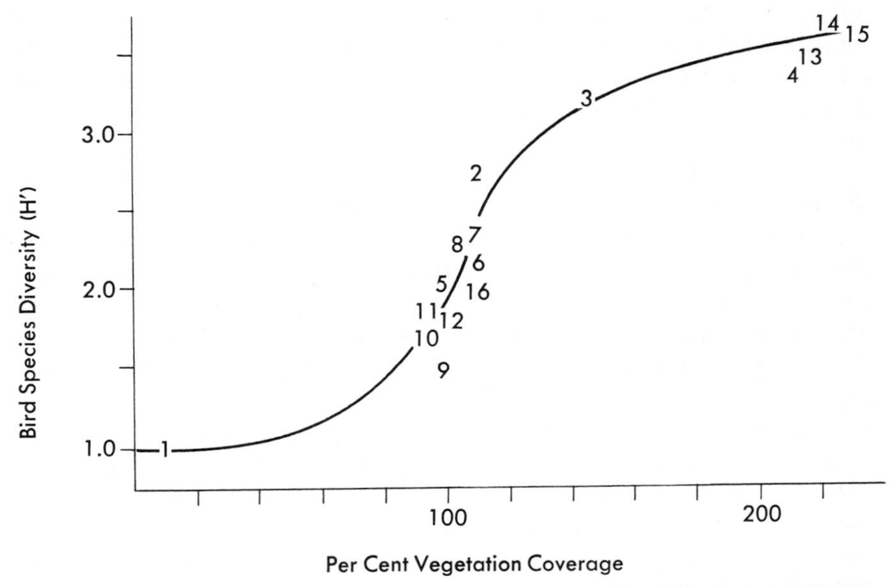

Figure 3.15. Bird species diversity and vegetation coverage. The Shannon-Wiener Index (Table 3.6, H′) for bird samples, on the vertical axis, is plotted against total per cent coverage of vegetation (herb plus shrub plus tree strata) on the horizontal. Samples were taken in Illinois (1–4), Texas (5–8), Panama (9–15), and the Bahamas (16). [Karr and Roth, *Amer. Nat.*, **105**:423 (1971).]

insects, fruits that extend through a range of sizes much larger than those of temperate-zone berries and that help support pigeons, parrots, and other birds, and the nectar of flowers used by hummingbirds and others. (There are of course some migratory hummingbirds that feed on nectar in temperate American communities, but there are many more flower-feeding birds in the Tropics.) Thus the niche space for birds becomes "larger" in the Tropics, with longer food-size axes and an additional axis formed by flowers as food resources for birds. The "larger" niche space supports a larger number of bird species. The work on birds has produced one most interesting observation: Within temperate climates, or within tropical climates, the diversities of bird communities are similar in communities of similar vegetation structure. This relationship implies that it is possible to *predict* the likely diversity of a bird community from the structure of the corresponding plant community. It further suggests that there is a kind of ceiling on the number of bird species that can be packed into the niche space of a community with a given kind of vegetation structure.

Numbers of birds and other species tend to be lower on islands than in corresponding areas on the continent. The number of species on an island must depend on the rate at which species immigrate to the island from the continent or other islands, and on how long these species survive on the island. The farther the island is from the continent, the lower the immigration rate, and the smaller the number of species to be expected on the island. The smaller the island, the fewer the habitats it is likely to offer, and the higher the extinction rate is likely to be because of the small sizes of the populations it can support. Given time, the extinction rate will come into balance with the immigration rate, and the number of species on the island will then be in a steady state. For this steady state we can, if we know something of immigration rates (as affected by distance from the continent) and extinction rates (as related to island size and other factors), predict species diversity, *S*. There is more to the interpretation of island diversities than this; but field observations and experiments with small mangrove islands support these concepts: diversity increase with island size and decrease with island distance; and the steady state within which there is continuing species turnover, with some species becoming extinct and others immigrating and replacing them.

The marine benthos is the community of the ocean bottom (biome types 35 and 36 in Chapter 4). The animals of the benthos—crabs and other crustaceans, worms, clams, brittle stars and so on—can be collected by dredges or other devices that bring up samples from the ocean bottom. Food is more abundant for the shallow-water benthos than for the deep-water benthos. We might suppose that diversity is higher in environments that are in some sense "favorable" for the group of organisms whose diversity trends we are observing. We should be mistaken if we supposed that benthic diversity is higher in shallow waters where food is more abundant. Studies of the benthos by H. L. Sanders show that diversities increase from the continental shelf into the deeper ocean (but diversities are lower in the deepest waters than in middle depths). We may well be surprised by the richness in animal species of the deeper benthos with its limited food, intense pressure, frigid temperatures (near 1°C), and slow metabolism. The key to the diversity of the benthos may be environmental stability, or stability and evolutionary time. Environmental conditions fluctuate in the shallower waters, but they are very stable in deep waters. Environmental fluctuations apparently limit the numbers of species that can occur in shallow waters. In the stable conditions of the deeper waters, more of the benthic species are able to survive and, given evolutionary time, develop the niche differentiation that we assume permits their occurrence together. For the benthos, the trend into the cold and stable depths is suggestive of the trend toward the warm Tropics we observed for terrestrial organisms.

As we have indicated, the latter trend is conspicuous among land plants. Apart from this trend we might expect that diversity of terrestrial plant communities would be correlated with their productivities, since productivity expresses the resources used. We should be mistaken; diversity and productivity are not simply correlated, and in some cases the relation is inverse. Redwood forests are highly productive but of low diversity; saltmarshes can be even more productive and even lower in diversity. We might also expect diversity to decrease from wet to dry environments, because soil moisture has so profound an influence on land plants. We should again be mistaken if we assumed a simple correlation. Maximum diversities, in the Temperate Zone at least, are in the middle part of the moisture gradient, as shown in Figure 3.16. Grasslands and semideserts are often richer in species than forests of more humid environments. Dry woodlands, communities of small trees in open growth, generally have more plant species than closed forests of more humid environments. The open structure of woodlands admits more light to the undergrowth, and also offers that undergrowth a mosiac of light and shade and different soil effects of the trees. A greater range of resource conditions, less strongly affected by the canopy trees, may be available in the woodland. Thus the wood-

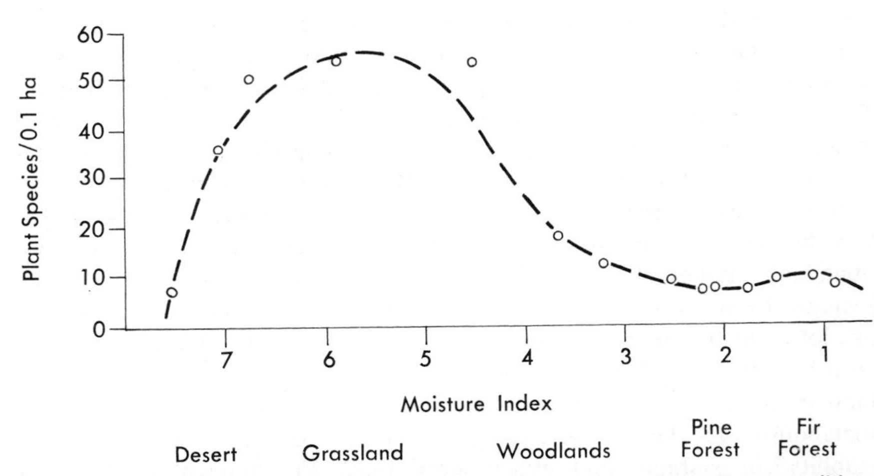

Figure 3.16. Plant species diversity in relation to elevation and moisture conditions in the Santa Catalina Mountains, Arizona. Diversity is measured as *S*, the number of species in 20×50 m^2 quadrats. Precipitation and humidity increase toward higher elevations, so that the moisture gradient indicated is also a gradient from low elevations on the left to high elevations on the right. Species diversity increases from forests (above 2200 m, moisture index 0.9 to 2.5), through woodlands (1810 to 2040 m, index 3.2 to 3.7), to open oak woodland and desert grassland (1310 and 1220 m, index 4.5 and 5.8); it is high in the Sonoran desert of the mountain slopes (1020 m, index 6.8) and decreases down the desert plain below the mountains to the creosote bush desert (760 m, index 7.5) near the city of Tucson. At high elevations diversity decreases somewhat from fir forests (above 2500 m, moisture index 0.9 to 1.8) to pine forests (moisture index 2.0 to 2.5). For the moisture index species are first classified into groups by their distributional relations to the moisture gradient, and the numbers for these groups are then used to calculate weighted averages for the species compositions of samples. [R. H. Whittaker and W. A. Niering, unpublished.]

land can be more diverse in its undergrowth—and consequently in its total vascular plant species—than a nearby forest.

Our observations on the benthos suggest that plant communities might be more diverse in stable environments. The climate of the Sonoran desert is of marked instability, with alternations of two wet and two dry periods per year, and with high year-to-year variability in the amounts of precipitation in the rainy seasons. Some parts of the Sonoran desert are also notably rich in species (Figures 3.9 and 3.13). It appears that in this desert environmental instability, rather than limiting diversity, has become an aspect of environment to which plants respond with niche differentiation (Figure 3.10) and consequent diversity. The vegetation of Israel and surrounding countries has been gravely disturbed by man, subjected to fire and cutting and heavy and varied grazing pressures of sheep, goats, cattle, and camels. The grazed, structurally altered woodlands and shrublands are very rich in species adapted to these disturbances, especially species of annual plants and bulb plants. The fact that the Sonoran, Israeli, and other warm-climate plant communities are so rich despite drought and environmental instability suggests that temperature, rather than moisture or stability, is a key influence on vascular plant diversity. One other observation on terrestrial plant communities may be mentioned—broadleaf deciduous forests average markedly richer in species than evergreen coniferous forests of similar environments. The kind of dominant species, as this affects the character of the leaf litter and the organic chemistry of the soil, is a significant influence on the diversity of a terrestrial plant community.

Plant diversity patterns seem less orderly and predictable than those of birds. We may wonder whether predictable patterns are more typical, and as a kind of test consider another animal group—lizards. The diversity of desert lizards has been studied by E. R. Pianka in three areas: the southwestern United States, the Kalahari desert of southwestern Africa, and Australia. These desert areas have always been widely separated geographically, and their plants and animals have evolved independently of one another. It is of interest to see whether evolution has produced convergent diversities in their lizard communities. In the United States, lizard diversity is not correlated with plant species diversity, but is correlated with the volumes of the desert shrubs and with summer temperature and length of the growing season. In the Kalahari lizard diversity is correlated with plant species diversity, but it is negatively correlated with precipitation and plant coverage. In Australia these correlations were not established, but it is evident that lizard evolution has featured a much more highly developed division of habitat gradients among species (beta diversity). Also, in Australia some lizards are nocturnal, whereas lizards are almost wholly diurnal in the other areas. The Australian lizards may, in a sense, be occupying niche space that would be filled by nocturnal small mammals in other deserts. The Australian deserts are far richer in species ($S = 15$

to 40 species) than the Kalahari (12 to 16 species), which is in turn richer than the American deserts (5 to 9 species).

Lizards may to some extent compete with birds as insectivores. Figure 3.17 plots bird and lizard species diversities against each other for samples from the three areas. What the figure shows is that diversities are *different* in the three areas, not only in richness in species, but in the way bird and lizard diversities are related to one another in Australia, compared with the other areas. The research indicates that lizard diversities have not converged in the three areas; they have instead diverged in their separate evolutionary histories.

We have been looking, without much success, for generalizations on diversity that might fit all groups of organisms. We find instead different major influences on diversity in different groups of organisms: vegetation structure for birds, environmental stability in the benthos, temperature and intermediate moisture conditions for plants, and varied relationships in different areas for lizards. We find still other influences when we look

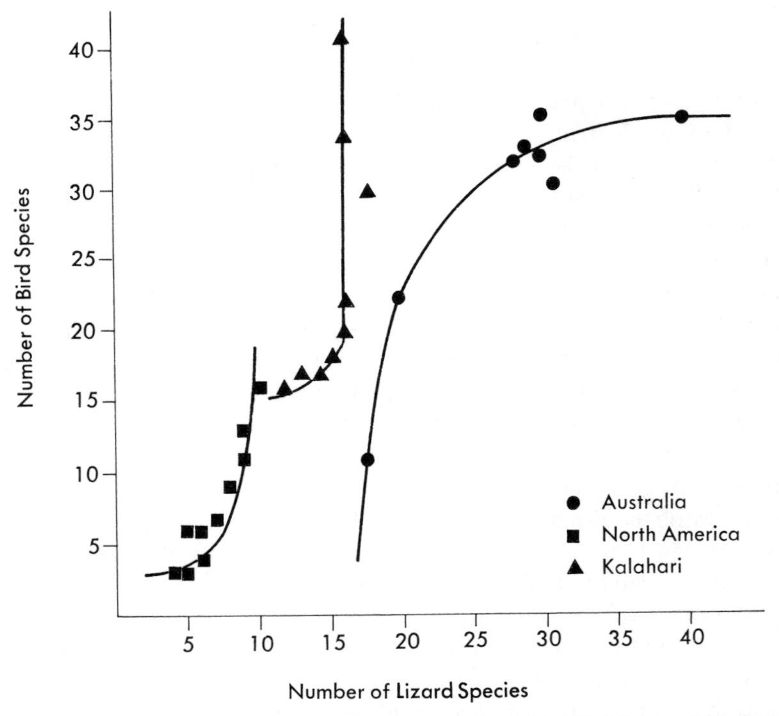

Figure 3.17. Desert bird versus lizard species diversities. Numbers of bird and of desert species in samples are plotted against each other for samples from desert areas on three continents. [Pianka, 1971.]

at other groups of organisms. How then is diversity in general to be interpreted?

Consider again the niche space for a group of organisms in a community. Along each axis of that space the number of species tends to increase in evolutionary time as additional species enter the community, fit themselves in between other species along the axes, and increase the packing of species along the axes. Species can also be added as specialists on marginal resources, and they can be added by the evolution of new resource gradients and species adapted to utilizing them. Thus the evolution of modern flowering plants in the Tropics has made available new resources of nectar and fruits, and bird species have evolved that rely on these. Considered for a given group of organisms, diversity increases through evolutionary time by the "lengthening" of niche axes and the packing of more species with narrower niches along niche axes, and by the addition of new axes—by the "expansion" and complication of the niche space.

Consider next two interacting groups of organisms, such as plants and grazing animals. The greater the diversity of plant species, the greater the variety of resources for the grazers. Evolution of diversity in the first group should be expected to make possible evolution of diversity in the second group. But the influence works in the opposite direction also. The grazing animals may increase diversity of the plant community by preventing any plant species from becoming too strongly dominant. By acting as controls on dominant plant species the grazers can increase equitability and permit a larger number of minor plant species to survive in the community. Furthermore, a species that controls the population of another species is an important part of the latter's niche. Increased numbers of grazing species may imply increased numbers of niches distinguished by these as population controls, for the grazed species. Interacting groups of species can thus each facilitate the other's increase in diversity. We should expect the same effect between predators and animal prey, and between plants and symbiotic fungi.

We can thus say that diversity begets diversity. Species diversity is a self-augmenting evolutionary phenomenon; evolution of diversity makes possible further evolution of diversity. The diversities we observe are to be interpreted as products of this evolution, through time, under the influence of different environmental factors that affect the survival, niche differentiation, and accumulation of species in communities. Rates of evolution and kinds of niche axes differ between different groups of organisms. Even if evolutionary time were equivalent for different communities and groups of organisms, we might expect what we observe: that different groups of organisms have evolved different diversity patterns in response to different environmental influences.

In some cases either biological characteristics (of birds) or environmental circumstances (of islands) imply that diversity increase is determinate, limited by some kind of ceiling or steady-state value. In other cases no such ceiling can be recognized. The tropical rain forest plants and insects in particular give the impression of an evolution, based on refinement of niche differences and elaboration of niche space, that has no well-defined limit. In some special cases in which diversity evolution seems determinate, as with singing birds and islands, we can make some reasonable predictions about diversity. In other cases we may regard the species diversity of communities as a product of evolution that is interesting to observe and interpretable, but not really predictable.

SUMMARY

When we study the structure of natural communities we find them to be mixtures of plants and animals with different ways of life. Communities show vertical differentiation based on different plant growth-forms and animal species on different levels in the community. Communities show horizontal differentiation, expressed in the patchy or mosaic-like occurrence of species, and in correlations of species that tend to occur together or to occur separately in relation to the patches. Communities show temporal differentiation, with different species carrying on similar functions at different times in seasonal and daily cycles. Each species has its own place in vertical and horizontal space, time, and way of relating to other species in a given community; the species' place in the community in relation to other species is its niche. Species evolve toward difference in niche, by which competition between them is reduced. In general no two species will occupy the same niche, using the same resources at the same time and place and subject to the same limiting factor, in the same stable community.

The gradients of niche characteristics by which species of the community can be compared may be treated as axes of an abstract niche space, or hyperspace. Species evolve toward the dispersion of their adaptive centers in niche space. Species differ widely in relative importance in the community. We may conceive importance as dependent on the fraction of niche space the species occupies and the fraction of community resources it uses, and measure importance as the species' fraction of the productivity (or other dimension) for the community as a whole. Importance values for species in a community can be arranged in progressions from most to least important species. The resulting curves are of at least three intergrading forms that can be interpreted by three different hypotheses on niche division and determination of relative importances: random niche boundaries, niche pre-emption and the geometric series, and the lognormal distribution.

Communities differ also in species diversity, or richness in numbers of species. Species diversities show a broad trend of increase from arctic, antarctic, and alpine environments into the lowland tropics. Apart from this, different groups of organisms show different correlations of diversity with vegetation structure, environmental stability, moisture conditions, and so on. Two interacting groups of organisms (such as predator and prey species) can each make possible increase in the other's diversity; and their diversities can increase in parallel through evolutionary time. Evolution of diversity is consequently a self-facilitating process; evolution of diversity makes possible further evolution of diversity. The diversities we observe are in many cases not predictable, but can be interpreted as products of evolution. The communities that result from this evolution are functional systems of niche-differentiated species; and community structure, differentiation in time and space, importance-value curves, and species diversities are interrelated expressions of the organization of species in communities.

References

Growth-forms and Life-forms

BRAUN-BLANQUET, J. 1932. *Plant Sociology*. New York: McGraw-Hill. xviii + 439 pp.

CAIN, S. A. 1950. Life-forms and phytoclimate. *Botanical Review* **16**:1–32.

CAIN, STANLEY A. and G. M. DE OLIVEIRA CASTRO. 1959. *Manual of Vegetation Analysis*. New York: Harper. xvii + 325 pp.

DANSEREAU, PIERRE. 1957. *Biogeography: An Ecological Perspective*. New York: Ronald. xiii + 394 pp.

RAUNKIAER, CHRISTEN. 1934. *The Life Forms of Plants and Statistical Plant Geography*. Oxford: Clarendon. xvi + 632 pp.

SHIMWELL, DAVID W. 1971. *The Description and Classification of Vegetation*. London: Sidgwick & Jackson. xiv + 322 pp.

Vertical Structure and Light

*ALLEE, W. C., O. PARK, A. E. EMERSON, T. PARK, and K. P. SCHMIDT. 1949. *Principles of Animal Ecology*. Philadelphia: Saunders. xii + 837 pp. (Ch. 26, pp. 441–495).

ANDERSON, M. C. 1964. Studies of the woodland light climate. *Journal of Ecology* **52:**27–41,643–663.

†BAINBRIDGE, RICHARD, G. C. EVANS, and O. RACKHAM, editors, 1966. *Light as an Ecological Factor* (British Ecological Society Symposium no. **6**). New York: Wiley. xi + 452 pp.

FEDERER, C. A. and C. B. TANNER. 1966. Spectral distribution of light in the forest. *Ecology* **47:**555–560.

FENCHEL, T. and B. J. STRAARUP. 1971. Vertical distribution of photosynthetic pigments and the penetration of light in marine sediments. *Oikos* **22**:172–182.

HOLMES, R. W. 1957. Solar radiation, submarine daylight, and photosynthesis. *Geological Society of America Memoir* **67**, Vol. 1:109–128.

HORN, HENRY S. 1971. *The Adaptive Geometry of Trees.* Princeton Univ. xi + 144 pp.

HUTCHINSON, G. E. 1957. *A Treatise on Limnology,* Vol. **1**. New York: Wiley. Ch. 6, pp. 366–425.

MONSI, M. 1968. Mathematical models of plant communities, pp. 131–149 in *Functioning of Terrestrial Ecosystems at the Primary Production Level,* ed. F. E. Eckardt. Paris: Unesco.

Horizontal Pattern

BARBOUR, M. G. 1969. Age and space distribution of the desert shrub *Larrea divaricata. Ecology* **50**:679–685.

CASSIE, R. M. 1963. Microdistribution of plankton. *Oceanography and Marine Biology, an Annual Review* **1**:223–252.

COLE, L. C. 1946. A study of the cryptozoa of an Illinois woodland. *Ecological Monographs* **16**:49–86.

COLE, L. C. 1949. The measurement of interspecific association. *Ecology* **30**:411–424.

†GREIG-SMITH, P. 1964. *Quantitative Plant Ecology*. 2nd ed. London: Butterworths. xii + 256 pp.

HUTCHINSON, G. E. 1953. The concept of pattern in ecology. *Proceedings of the Academy of Natural Sciences,* Philadelphia **105:**1–12.

KERSHAW, KENNETH A. 1973. *Quantitative and Dynamic Plant Ecology*. 2nd ed. New York: Elsevier. x + 308 pp.

LEVIN, S. A. and R. T. PAINE. 1974. Disturbance, patch formation, and community structure. *Proceedings of the National Academy of Sciences, USA* **71**:2744–2747.

STEWART, G. and W. KELLER. 1936. A correlation method for ecology as exemplified by studies of native desert vegetation. *Ecology* **17**:500–514.

WASHBURN, A. L. 1956. Classification of patterned ground and review of suggested origins. *Geological Society of America Bulletin* **67**:823–865.

WATT, A. S. 1947. Pattern and process in the plant community. *Journal of Ecology* **35**:1–22.

*WHITFORD, P. B. 1949. Distribution of woodland plants in relation to succession and clonal growth. *Ecology* **30**:199–208.

WIEBE, P. H. 1970. Small-scale spatial distribution in oceanic zooplankton. *Limnology and Oceanography* **15**:205–217.

*WOODELL, S. R., JR., H. A. MOONEY, and A. J. HILL. 1969. The behaviour of *Larrea divaricata* (creosote bush) in response to rainfall in California. *Journal of Ecology* **57**:37–44.

Time Relations

ALLEE et al. 1949, Chapter 28, pp. 528–562.

BECK, STANLEY D. 1968. *Insect Photoperiodism*. New York: Academic. viii + 288 pp.

CROAT, T. B. 1969. Seasonal flowering behavior in central Panama. *Annals of the Missouri Botanical Garden,* St. Louis **56**:295–307.

HOBSON, E. S. 1972. Activity of Hawaiian reef fishes during the evening and morning transitions between daylight and darkness. *Fishery Bulletin* **70**:715–740.

HUTCHINSON, G. E. 1957, 1967. *A Treatise on Limnology*. Wiley, New York. Vol. 1, ch. 7, vol. 2, chs. 23 and 25.

KORRINGA, P. 1957. Lunar periodicity. *Geological Society of America Memoir* **67**, Vol. **1**:917–934.

*MOSQUIN, T. 1971. Competition for pollinators as a stimulus for the evolution of flowering time. *Oikos* **22**:398–402.

MUNZ, F. W. and W. N. MCFARLAND. 1973. The significance of spectral position in the rhodopsins of tropical marine fishes. *Vision Research* **13**:1829–1874.

PARK, O. 1940. Nocturnalism—the development of a problem. *Ecological Monographs* **10**:485–536.

PEARRE, S., JR. 1973. Vertical migration and feeding in *Sagitta elegans* Verrill. *Ecology* **54**:300–314.

*SMYTHE, N. 1970. Relationships between fruiting seasons and seed dispersal methods in a neotropical forest. *American Naturalist* **104**:25–35.

SWEENEY, BEATRICE M. 1969. *Rhythmic Phenomena in Plants*. New York: Academic. ix + 147 pp.

Niche

BROADHEAD, E. and A. J. WAPSHERE. 1966. *Mesopsocus* populations on larch in England—the distribution and dynamics of two closely-related coexisting species of Psocoptera sharing the same food resource. *Ecological Monographs* **36**:327–388.

GAUSE, G. F. 1934. *The Struggle for Existence*. Reprint, 1964, New York: Hafner. ix + 163 pp.

DEBACH, P. 1966. The competitive displacement and coexistence principles. *Annual Review of Entomology* **11**:183–212.

Diamond, J. M. 1973. Distributional ecology of New Guinea birds. *Science* **179**:759–769.

Elton, Charles. 1927. *Animal Ecology*. London: Sidgwick & Jackson. xxi + 207 pp.

Grinnell, J. 1917. The niche-relationships of the California thrasher. *Auk* **34**:427–433.

*Hardin, G. 1960. The competitive exclusion principle. *Science* **131**:1292–1297.

Hutchinson, G. E. 1958. Concluding remarks. *Cold Spring Harbor Symposia on Quantitative Biology* **22**:415–427.

Lack, David. 1947. *Darwin's Finches, an Essay on the General Biological Theory of Evolution*. Reprint 1968, Gloucester, Mass.: Peter Smith, x + 204 pp.

Levin, S. A. 1970. Community equilibria and stability, and an extension of the competitive exclusion principle. *American Naturalist* **104**:413–423.

MacArthur, R. 1968. The theory of the niche, pp. 159–176 in *Population Biology and Evolution,* ed. R. C. Lewontin. Syracuse Univ.

May, R. M. and R. H. MacArthur. 1972. Niche overlap as a function of environmental variability. *Proceedings of the National Academy of Sciences,* U.S.A. **69**:1109–1113.

Pianka, E. R. 1974. Niche overlap and diffuse competition. *Proceedings of the National Academy of Sciences,* U.S.A. **71**:2141–2145.

Root, R. B. 1967. The niche exploitation pattern of the blue-gray gnatcatcher. *Ecological Monographs* **37**:317–350.

Schoener, T. W. 1974. Resource partitioning in ecological communities. *Science* **185**:27–39.

†Whittaker, R. H., S. A. Levin, and R. B. Root. 1973. Niche, habitat, and ecotope. *American Naturalist* **107**:321–338.

Williamson, M. H. 1957. An elementary theory of interspecific competition. *Nature* **180**:422–425.

Species Importances and Diversity Measurement

*Cairns, J., Jr., D. W. Albaugh, F. Busey, and M. D. Chanay. 1968. The sequential comparison index—a simplified method for non-biologists to estimate relative differences in biological diversity in stream pollution studies. *Journal of the Water Pollution Control Federation* **40**:1607–1613.

Cohen, J. E. 1968. Alternative derivations of a species-abundance relation. *American Naturalist* **102**:165–172.

Fisher, R. A., A. S. Corbet, and C. B. Williams. 1943. The relation between the number of species and the number of individuals in a random sample of an animal population. *Journal of Animal Ecology* **12**:42–58.

Hill, M. O. 1973. Diversity and evenness: a unifying notation and its consesequences. *Ecology* **54**:427–432.

KING, C. E. 1964. Relative abundance of species and MacArthur's model. *Ecology* **45**:716–727.

MACARTHUR, R. 1960. On the relative abundance of species. *American Naturalist* **94**:25–36.

MAY, R. M. 1974. Ecosystem patterns in randomly fluctuating environments. *Progress in Theoretical Biology,* ed. R. Rosen and F. Snell **3**:1–50.

MAY, R. M. 1975. Patterns of species' abundance and diversity, in *The Ecology and Evolution of Communities,* eds. M. L. Cody and J. M. Diamond. Cambridge: Harvard Univ. (in press).

MCNAUGHTON, S. J. and LARRY L. WOLF. 1973. *General Ecology*. New York: Holt, Rinehart & Winston. x + 710 pp.

LLOYD, M. and R. J. GHELARDI. 1964. A table for calculating the 'equitability' component of species diversity. *Journal of Animal Ecology* **33**:217–225.

PEET, R. K. 1974. The measurement of species diversity. *Annual Review of Ecology and Systematics* **5**:285–307.

PRESTON, F. W. 1948. The commonness, and rarity, of species. *Ecology* **29**:254–283.

SIMPSON, E. H. 1949. Measurement of diversity. *Nature* **163**:688.

WHITTAKER, R. H. 1965. Dominance and diversity in land plant communities. *Science* **147**:250–260.

†WHITTAKER, R. H. 1972. Evolution and measurement of species diversity. *Taxon* **21**:213–251.

WILLIAMS, C. B. 1964. *Patterns in the Balance of Nature, and Related Problems in Quantitative Ecology*. New York: Academic. vii + 324 pp.

Species Diversities

DIAMOND, J. M. 1969. Avifaunal equilibria and species turnover rates on the Channel Islands of California. *Proceedings of the National Academy of Sciences,* U.S.A. **64**:57–63.

JOHNSON, R. G. 1970. Variations in diversity within benthic marine communities. *American Naturalist* **104**:285–300.

HUTCHINSON, G. E. 1959. Homage to Santa Rosalia, *or* why are there so many kinds of animals? *American Naturalist* **93**:145–159.

KARR, J. R. 1971. Structure of avian communities in selected Panama and Illinois habitats. *Ecological Monographs* **41**:207–233.

MACARTHUR, R. H. 1965. Patterns of species diversity. *Biological Reviews* **40**:510–533.

MACARTHUR, R. H. 1969. Patterns of communities in the tropics. *Biological Journal of the Linnean Society* **1**:19–30.

MACARTHUR, ROBERT H. and E. O. WILSON. 1967. *The Theory of Island Biogeography*. Princeton Univ. xi + 203 pp.

MARGALEF, R. 1967. Some comments relative to the organization of plankton. *Oceanography and Marine Biology, an Annual Review* **5**:257–289.

ORIANS, G. H. 1969. The number of bird species in some tropical forests. *Ecology* **50**:783–801.

*PAINE, R. T. 1966. Food web complexity and species diversity: *American Naturalist* **100**:65–75.

PIANKA, E. R. 1966. Latitudinal gradients in species diversity: a review of concepts. *American Naturalist* **100**:33–46.

*PIANKA, E. R. 1971. Lizard species density in the Kalahari desert. *Ecology* **52**:1024–1029.

PORTER, J. W. 1972. Patterns of species diversity in Caribbean reef corals. *Ecology* **53**:745–748.

SANDERS, H. L. 1968. Marine benthic diversity: a comparative study. *American Naturalist* **102**:243–282.

SIMBERLOFF, D. S. and E. O. WILSON. Experimental zoogeography of islands. *Ecology* **50**:267–314.

STEHLI, F. G., R. G. DOUGLAS and N. D. NEWELL. 1969. Generation and maintenance of gradients in taxonomic diversity. *Science* **164**:947–949.

TOMOFF, C. W. 1974. Avian species diversity in desert scrub. *Ecology* **55**:396–403.

TRAMER, E. J. 1974. On latitudinal gradients in avian diversity. *Condor* **76:**123–130.

†WHITTAKER 1972. Evolution and measurement of species diversity. *Taxon* **21:**213–251.

†WOODWELL, G. M. and H. H. SMITH, editors. 1969. Diversity and Stability in Ecological Systems. *Brookhaven Symposia in Biology* **22,** vii + 264 pp.

4

Communities and Environments

One travels toward the mountains across the lowlands, somewhere west of the Rocky Mountains, through miles of semidesert scrub. The sagebrush scrub extends up onto lower foothills, but as one climbs scattered small junipers appear. Somewhat higher in the mountains, large and more numerous junipers are joined by piñon pines to form an open woodland of small trees with an undergrowth of grasses and shrubs. As one climbs further the woodland is denser, and scattered large trees of western yellow or ponderosa pine appear; then the piñon and juniper decrease in numbers while the pines increase until one climbs through a forest of ponderosa pines. Still higher these pines give way to forests of Douglas fir and white fir; still higher these in turn give way to forests of Engelmann spruce and alpine fir. One climbs now through the uppermost forests of the mountain, until the trees become small and scattered or reduced to shrubby patches in meadows. Beyond this forest edge are the alpine meadows of the high country. The meadows extend upward, but become sparser toward higher altitudes and finally are reduced to lichens and a few herbs among the rocks of the mountain summit.

In this ascent we have seen change not only in the occurrence of plant (and animal) species, but also in the structure or physiognomy of vegetation—from scrub through woodland and forest to meadow. Changes of vegetation with altitude in mountains are observed throughout the world; description of these changes in the form of "life-zones" by C. H. Merriam was one of the early developments in ecology. In Merriam's treatment the vegetation forms a series of zones in relation to temperature gradients, each zone being characterized by major plant and animal species. We are observing, however, a phenomenon broader than elevation zones—the response of species populations and communities to environmental gradients. We therefore need to ask more general questions than Merriam's:

1. How are species populations distributed in relation to one another and communities along an environmental gradient?
2. How are the kinds of communities in an area related to patterns of more than one environmental gradient?
3. How can we best classify these communities?
4. How are we to interpret world-wide relationships of community structure to environments?

Species Along Environmental Gradients

Consider first a single environmental gradient, which could be a long, even, uninterrupted slope of a mountain. The slope is occupied by communities that include many species of plants, animals, and saprobes. We have seen that these species have evolved in relation to one another, that they influence one another's populations, that some are competing, and that these will have evolved in such ways that competition is reduced by niche differentiation. What kind of distributional relations among species will result from this evolution? It was assumed by early ecologists that species formed groupings that characterized distinct, clearly bounded types of communities that were often termed *associations*. This view—of communities as consisting of well-defined units—may be called the *community-unit theory* and contrasted with a different view advanced independently in Russia in 1924 by L. G. Ramensky and in the United States in 1926 by H. A. Gleason as the *individualistic hypothesis.* This hypothesis states that species are variously, "individualistically" distributed and do not form groupings that characterize clearly bounded types of communities. As a basis for choosing between the community-unit and individualistic concepts we can suggest multiple working hypotheses on ways species populations might be distributed, and then turn to evidence from the field for decision. Species might be distributed in relation to one another and communities in one of these ways:

1. Competing species, including dominant plants, exclude one another along sharp boundaries. Other species evolve toward close association with the dominants and toward adaptations for living with one another. There thus develop distinct zones along the gradient, each zone having its own assemblage of species adapted to one another, and giving way at a sharp boundary to another assemblage of species adapted to one another (Figure 4.1A). The zones then represent well-defined, relatively discontinuous kinds of communities.
2. Competing species exclude one another along sharp boundaries, but do not become organized into groups with parallel distributions (Figure 4.1B).
3. Competition does not, for the most part, result in sharp boundaries between species populations. Evolution of species toward adaptation to one another will, however, result in the appearance of groups of species with similar distributions (Figure 4.1C). These groups characterize different kinds of communities, but the communities intergrade continuously.
4. Competition does not usually produce sharp boundaries between species populations, and evolution of species in relation to one another does not produce well-defined groups of species with similar distributions. Centers and boundaries of species populations are scattered along the environmental gradient (Figure 4.1D).

Studies of how species are actually distributed along environmental gradients were carried out independently by the author in his research and thesis on the Great Smoky Mountains, 1947–8, and by J. T. Curtis and his students in Wisconsin; first results from both studies were published in 1951. The same kind of study could be applied to the mountain vegetation described in the introduction to this chapter. Vegetation samples, with suitable measurements of importances of plant populations, are taken along the elevation gradient from the tops of the mountains down to the bases. To study the distributions of populations effectively we should (1) have samples from comparable environments—open, south-facing slopes, say—for all elevations, and (2) have a large enough number of samples, so that they can be grouped and averaged at elevation intervals. We may want, say,

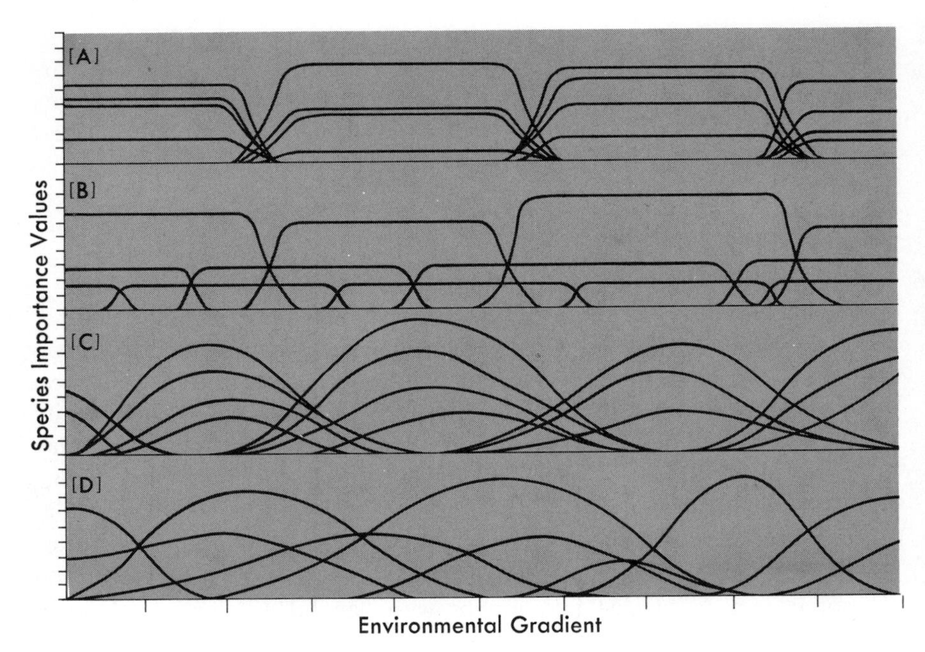

Figure 4.1. Four hypotheses on species distributions along environmental gradients. Each curve in each part of the figure represents one species population and the way it might be distributed along the environmental gradient.

to average five or more samples for every 100-m elevation change to smooth out irregularities in population distribution. The result will be a *transect* with tables showing the way each species population is distributed in relation to the elevation gradient, other species, and the kinds of communities we have observed. We can do the same along the topographic moisture gradient. This is the gradient of soil moisture and atmospheric humidity from moist canyon bottoms through environments that are increasingly dry—lower slopes of the canyon, open north-facing slopes, intermediate (east- and west-facing) slopes, to the driest open south- and southwest-facing slopes. To use the moisture gradient as the variable for such a topographic transect, we should hold elevation constant. Samples from the same, or closely similar, elevations are then averaged for different positions along the topographic moisture gradient.

When we study the manner in which plant populations rise and fall along environmental gradients in this way, the results support hypothesis 4 (Figure 4.1D, and Figure 4.2). (Cases of sharp boundaries between competing species and of close distributional association of species are known, but populations are usually distributed as in Figure 4.2.) Observations agree with Ramensky and Gleason's principle of species individuality:

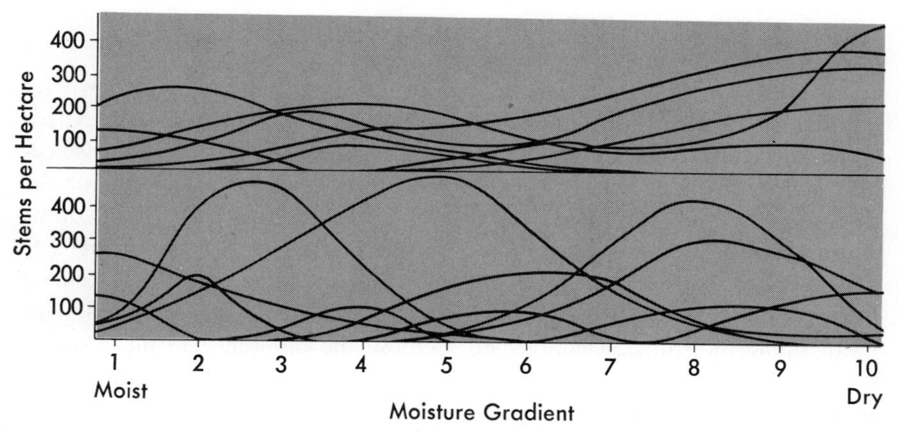

Figure 4.2. Actual distributions of species populations along environmental gradients. Species populations are plotted by densities: numbers of tree stems 2 cm or more in diameter per hectare (10^4 m^2), on the vertical axis, in samples taken along a topographic moisture gradient from moist environments of ravines to dry environments of southwest-facing slopes, on the horizontal axis. The data are from the Siskiyou Mountains, Oregon, 760–1070 m elevation, above, and the Santa Catalina Mountains, Arizona, 1830–2140 m elevation, below. [Whittaker, 1967.]

> Each species is distributed in its own way, according to its own genetic, physiological, and life-cycle characteristics and its way of relating to both physical environment and interactions with other species; hence no two species are alike in distribution.

The observations further agree with Ramensky and Gleason's principle of community continuity:

> The broad overlap and scattered centers of species populations along a gradient imply that most communities intergrade continuously along environmental gradients, rather than forming distinct, clearly separated zones. (Either environmental discontinuity or disturbance by fire, logging, and so on, can of course produce discontinuities between communities.)

It is useful to recognize life-zones, as Merriam did, as major kinds of communities in relation to temperature. But the zones are continuous with one another: distributions of the major plant species by which we recognize the zones overlap broadly, and other plants and animals do not form groups with distributions closely similar to those of the dominants. The zones are kinds of communities man recognizes, mainly by their dominant plants, within the continuous change of plant populations and communities along the elevation gradient. The zones can be compared to the colors man recognizes, and accepts as useful concepts, within the spectrum of wave lengths of light which are known to be continuous.

It is of interest to ask *why* species do not evolve to form groups with parallel distributions. We have shown in Figure 3.6 how two species that come into competition in relation to a niche gradient tend to diverge: selection increases the difference between the mean adaptative positions of the two species populations along the niche gradient. For niche gradient, now, we may substitute a habitat gradient as the horizontal axis of Figure 3.6. The two species are in close competition (in the same or closely related niches) within the same range of a habitat gradient, such as elevation or topographic moisture. Selection will increase the difference in mean adaptive positions along the habitat gradient. As competing species evolve toward difference in niche, so they evolve also toward difference in habitat.

In a simplified example of this process, a number of species occur along an environmental gradient (Figure 4.3). A new species, number 4, enters the area. Its potential distribution along the gradient (on the basis of genetic characteristics evolved in another area) is indicated by the dashed line in the upper part of the figure. Because species 4 is strongly in competition with species 3, however, its population shifts position toward 5 and occupies a position between the other two species, in the lower part of the figure.

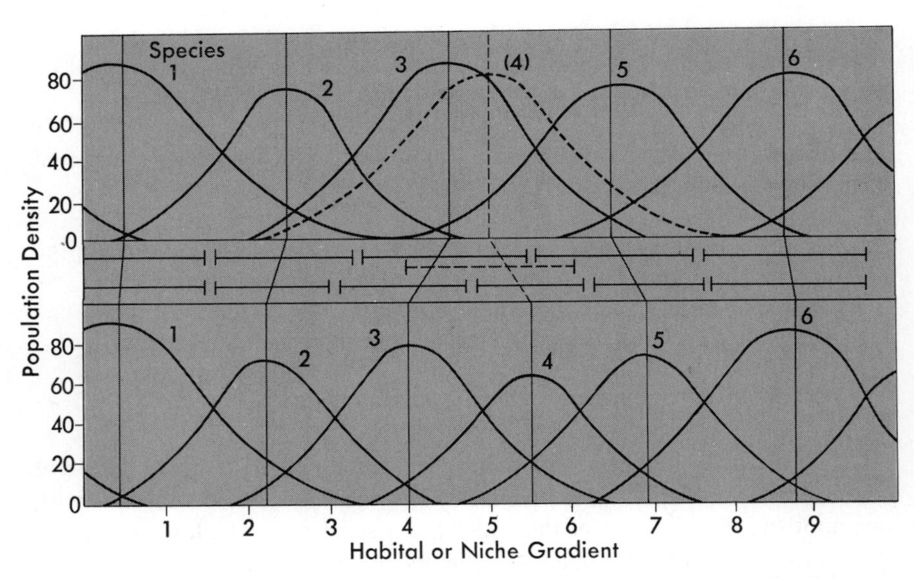

Figure 4.3. Establishment of a new species in a community gradient. The new species, number 4, has a potential distribution along the habitat gradient as represented in the dashed line of the upper figure. In competition with species 3 and 5 it fits in between these, as indicated in the lower figure. The bars between the figures represent dispersions, the degrees of deviation or spread of the populations on each side of their mean positions along the gradient.

Compensating shifts occur in distributions of other species, and widths of their distributions along the gradient, indicated by the bars between the figures, are reduced. Species 4 survives and takes its own position along the gradient by virtue of both a sufficient niche difference from species 3 and 5 to survive in their presence, and a new range of habitat preference that reduces the competition, which does result from partial niche overlap.

The evolution of populations in directions that reduce competition thus leads both toward niche difference and toward habitat difference and the scattering of population centers along environmental gradients. Selection toward a distinctive niche and a distinctive habitat preference will normally occur at the same time, for niche and habitat are closely related aspects of the species' total adaptation to environment. However, the consequences of selection lead primarily away from, not toward, formation of groups of species with parallel distributions.

Figure 4.2 illustrates a further expression of this evolutionary process. The two community gradients shown occur along similar environmental gradients (the topographic moisture gradient from canyons to south-facing slopes of the same elevations in mountains). Individual samples of the forests in the two areas are of similar species diversity. Yet the number of species encountered along the gradient in the figure below is higher, for the species' distributions are narrower. There is a more extensive floristic turnover of species replacing one another along the gradient. We may thus distinguish *alpha* or within-habitat diversity, in the sense of richness in species of particular community samples, and *beta* or between-habitat diversity, or the degree of change in species composition of communities along a gradient. The lower transect of Figure 4.2 has conspicuously higher beta diversity.

Alpha diversity can be measured as S or one of the other expressions in Table 3.6. Beta diversity, in contrast, should be measured as the extent of change in, or degree of difference in composition among, the samples of a set. One simple measurement of beta diversity is $BD = S_c/\bar{S}$, in which S_c is the total number of species occurring in a transect or set of samples, counting each species only once whether or not it occurs more than once, and $\bar{S}$ is the average number of species per individual sample. For a single sample $BD = 1$, for two samples that have no species in common $BD = 2$. For three samples sharing no species, or a larger set of samples with the same total number and mean number of species as those three, $BD = 3$, implying the extent of difference in composition among the samples of the larger set is equivalent to that in three samples with no species in common. For tree strata of the two transects in Figure 4.2, $BD = 2.6$ above, and 3.7 below.

We can also approach measurement of beta diversity through the degree of change in composition of samples along the gradient, or the extent of difference in samples from opposite ends of the gradient. There are a number of ways of measuring the relative similarity of samples from communities. The two simplest are coefficient of community and percentage similarity. The coefficient of community is the ratio of two times the number of species that occur in both samples, to the total number of species occurrences in both samples (Table 4.1, 1). Coefficient of community thus expresses floristic similarity of the samples in terms of presence and absence of species only. Percentage similarity is based on the sum of the signless differences in importance values (for each species, in the two samples).

Table 4.1 Similarity and Distance for Community Samples

A. Sample Similarities

		Sample A	Sample B	
Species	1	.40	0	Importance values of species
	2	.30	.30	
	3	.20	.40	
	4	.10	.25	
	5	0	.05	
Total		1.0	1.0	

1. Coefficient of Community

$CC = 2S_{ab}/(S_a + S_b) = 2(3 \text{ species})/(4 + 4 \text{ species}) = 6/8 = 0.75$
S_a is the number of species in sample A, S_b the number in sample B, and S_{ab} the number in both samples.

2. Percentage Similarity

$$PS = 1 - 0.5\ \Sigma|p_a - p_b| = \Sigma \min\ (p_a \text{ or } p_b)$$
$$= 1 - 0.5\ (.40 + 0 + .20 + .15 + .05) = 0 + .30 + .20 + .10 + 0$$
$$= 1 - 0.5\ (.80) = 0.60$$

Here, p_a is a decimal importance value for a given species in sample A, p_b the decimal value for the same species in sample B.
(If the importance values in a given sample do not sum to 1.0, then percentage similarity can be calculated as, $PS = 2\Sigma\min\ (n_a \text{ or } n_b)/(N_a + N_b)$. In this, n_a and n_b are the importance values of a given species in samples A and B, and N_a and N_b are the totaled importance values of all species in samples A and B.)

3. Euclidean Distance

$$ED = \sqrt{\Sigma(n_a - n_b)^2}, \text{ or}$$
$$ED_r = \sqrt{\Sigma(p_a - p_b)^2/2} \qquad \text{(Relative Euclidean distance)}$$
$$= \sqrt{(.16 + 0 + .04 + .0225 + .0025)/2} = \sqrt{.1125} = .335$$

Table 4.1—Continued

B. Pythagorean Distance Along a Polar Ordination Axis

$x = (L^2 + D_1^2 - D_2^2)/2L$

L, D_1, and D_2 are all expressed as decimal (or per cent) dissimilarities or distance values, usually either $(1 - PS)$ or $(1 - CC)$. L is the distance value between the two end points, D_1 is the distance of a sample being located on the axis from the first end point, and D_2 is that sample's distance from the second end point. The location of the sample on the axis is then x. The height of the sample above the axis, e, may be desired when end points for another axis are chosen, but e is not used for ordination.

$e = \sqrt{D_1^2 - x^2}$

It can be calculated more simply by summing the smaller of the two relative importance values (for each species in the two samples), including zeros for species that are absent from one of the two samples (Table 4.1, 2). Percentage similarity, unlike coefficient of community, expresses the degree to which the two samples are alike in quantitative representation of species. Euclidean distance, the third measure in Table 4.1, is used in principal components analysis and some other techniques. Apart from these, it is less suitable for community samples than the other two, for it weights heavily the importance values of dominant species and sample errors in measuring these. Percentage similarity and coefficient of community may be subtracted from 1.0 to obtain sample dissimilarity, or distance measures.

These measurements express a kind of *ecological distance*—the degree to which samples differ from one another in species composition because of their separation along environmental gradients, or other factors. The higher beta diversity of the lower community gradient of Figure 4.2 is expressed in lower percentage similarities of the samples at the extremes of the gradient (0.9 per cent versus 18 per cent in the upper gradient), hence greater ecological distance between those extremes.

Alpha and beta diversity will be recognized as consequences of niche diversification and habitat diversification of species, respectively. The two aspects of diversity may vary in parallel along some climatic gradients; both increase from the coastal redwood belt inland in the area of mountains from which the upper transect of Figure 4.2 was taken. In studying bird communities along the diversity gradient from cold climates into the Tropics MacArthur found the two somewhat independent. As we have indicated, alpha diversity for birds, which is closely related to vegetation structure, was not much higher in tropical than in temperate communities of similar structure. Beta diversity of birds, however, increased into the Tropics in a striking way. It appears that there is a point of saturation, or maximum feasible division of niche space among bird species for a given vegetation structure. Evolution in the Tropics acts not to increase alpha diversity beyond this saturation but to fit additional species in along environmental gradients by habitat differentiation and narrowed habitat distributions—it acts to increase beta diversity.

Community Patterns

We have been discussing the population structure of communities along environmental gradients. Three concepts are implicit in this treatment. First, there is the concept of *community gradient* (*coenocline*), represented in terms of populations in Figure 4.2. Second, there is the conception of environmental factors that change through space together. Thus the "elevation gradient" includes decreasing mean temperatures, decreasing lengths of growing seasons, increasing rainfall, increasing wind speeds, and so on, toward higher elevations. All these factors act together on plants and animals, and it may be difficult without experiment to judge what factors are most important for a given population. The assemblage of environmental factors that change together through the space along which a community gradient occurs and that influence its populations is termed a *complex-gradient*. Third, the complex-gradient and coenocline together are a gradient of communities-and-environments, or a gradient of ecosystems, an *ecocline*.

The approach to communities through ecoclines is important because it permits us to study the way gradients of species populations and community characteristics change in response to, or in concurrence with, gradients of environment. Research approaches that relate gradients to one another on these three levels—environmental factors, species populations, and community characteristics—are termed *gradient analysis* and are the major alternative to approaches to communities through classification. (In the latter, types of communities are recognized, and these community-types may then be characterized in terms of environmental measurements, species composition, and other community characteristics.) The preceding section dealt with results from gradient analysis applied to questions of how species populations relate to one another and communities along particular environmental gradients. We need also to consider applications of gradient analysis to patterns of communities in relation to more than one environmental gradient.

In mountains, for example, communities show striking changes along both elevation and topographic moisture gradients, and both gradients should be studied. It is quite possible to use transects both of elevation and of topographic moisture to study mountain vegetation. The two kinds of transects cross one another to form a grid covering the vegetation pattern by which the pattern can be analyzed. For some purposes, however, a different approach to a pattern of mountain communities is more effective. The two kinds of complex-gradients may be used as vertical and horizontal axes of a chart as in Figure 4.4. Vegetation samples taken essentially at random from many positions on the mountain topography can be located at points in relation to these axes, and at each of these points importance values for species and classification of the sample into a community type can be entered. Species populations and community types can then be outlined on the chart to show their relations to one another and to mountain environments.

Figure 4.4 shows the relations of vegetation types to elevation and topographic moisture in the Santa Catalina Mountains of southeastern Arizona, a mountain range with strong Mexican influence. These points of interest may be observed from the chart:

1. The pattern is dominated by effects of increasing aridity toward lower elevations. Apart from the canyon forests this pattern (but not those of some more humid mountains) forms a series of zones in response to moisture conditions related to elevation. At a given elevation north-facing slopes and canyons are cooler and more humid, south-facing slopes warmer and more arid. The zones are consequently tilted: a given zone occurs at higher elevations on south slopes than on north slopes and in canyons.
2. The zones are defined by growth-forms and dominant species. A sequence of growth-forms replace one another as community dominants in a continuous, flowing fashion; from high elevations to low these are needle-leaved trees, sclerophyll trees (evergreen oaks), small needle-leaved trees (juniper and piñon) together with evergreen-sclerophyll and rosette shrubs, grasses, and the spinose shrubs and the semishrubs of the desert.
3. A species population has a bell-shaped distribution along each environmental gradient. Its population distribution in relation to two gradients may be conceived as a bell-shaped or hill-shaped solid (Figure 4.5). In some species different ecotypes are indicated by separate peaks, with different central points of adaptation. In two dimensions as in one, species populations overlap broadly, and the locations of their population centers are scattered in the pattern. Because a community pattern like that of Figure 4.4 includes hundreds of vascular plant species, it can be conceived as a complex population continuum formed by these many, overlapping species populations.
4. Community and soil characteristics too show patterns in relation to the pattern of environments. Species diversity increases from high elevations to low, hence from moist forests to the woodland, grassland, and desert of lower mountain slopes in this area. The amount of organic matter produced, on the other hand, in this area decreases from high elevations to low in response to the moisture gradient. Soil pH and

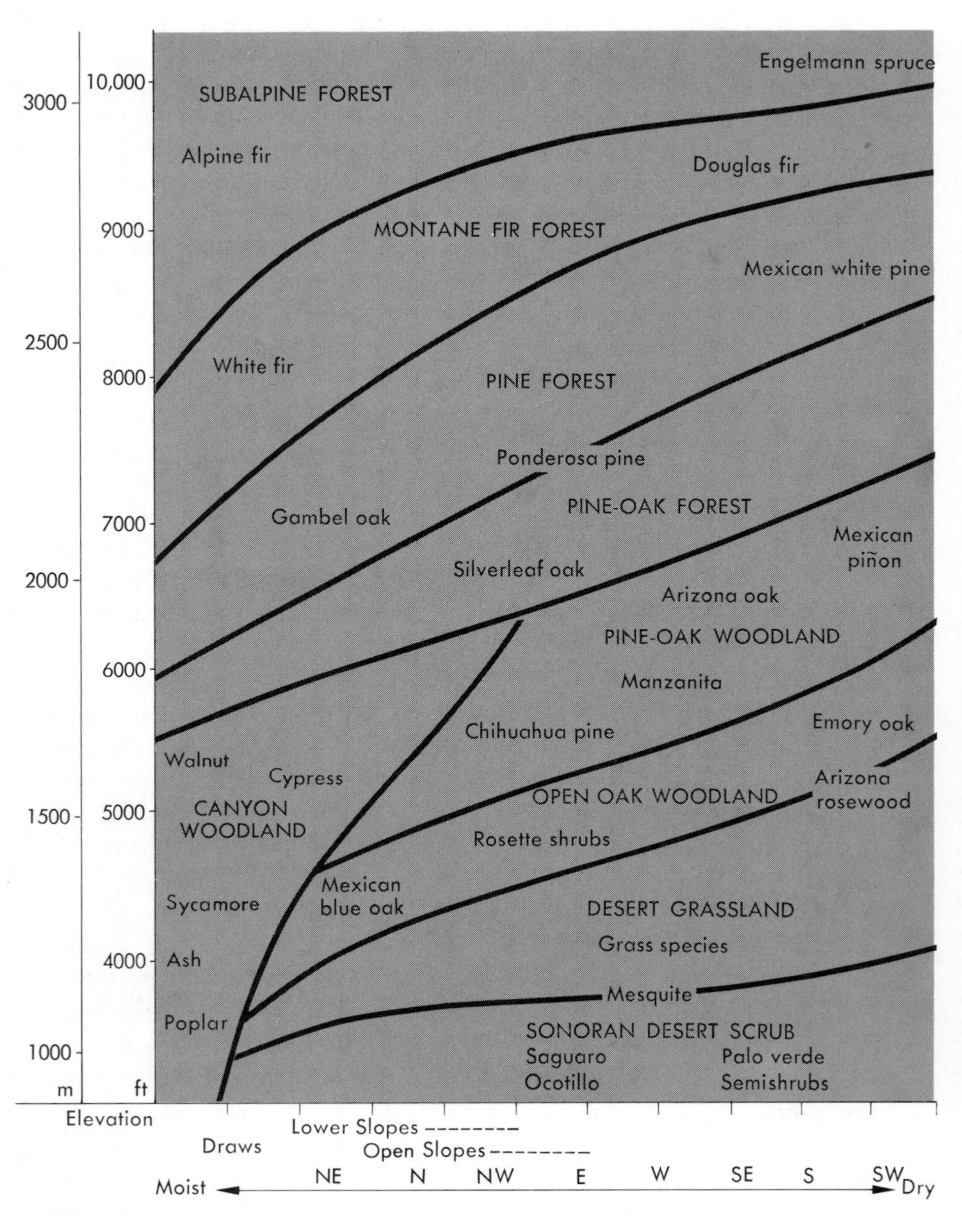

Figure 4.4. Vegetation chart for the Santa Catalina Mountains, southeastern Arizona. (The pattern above 9000 ft is for the nearby Pinaleno Mountains.) Four hundred vegetation samples were plotted on the chart by their positions in relation to the elevation gradient, on the left, and the topographic moisture gradient, on the bottom. Boundary lines were drawn to connect the mean positions at which one community-type, as these had been defined in this study, gave way to another. Dominant species are indicated in the parts of the pattern where they are most important. [Whittaker and Niering, 1965.]

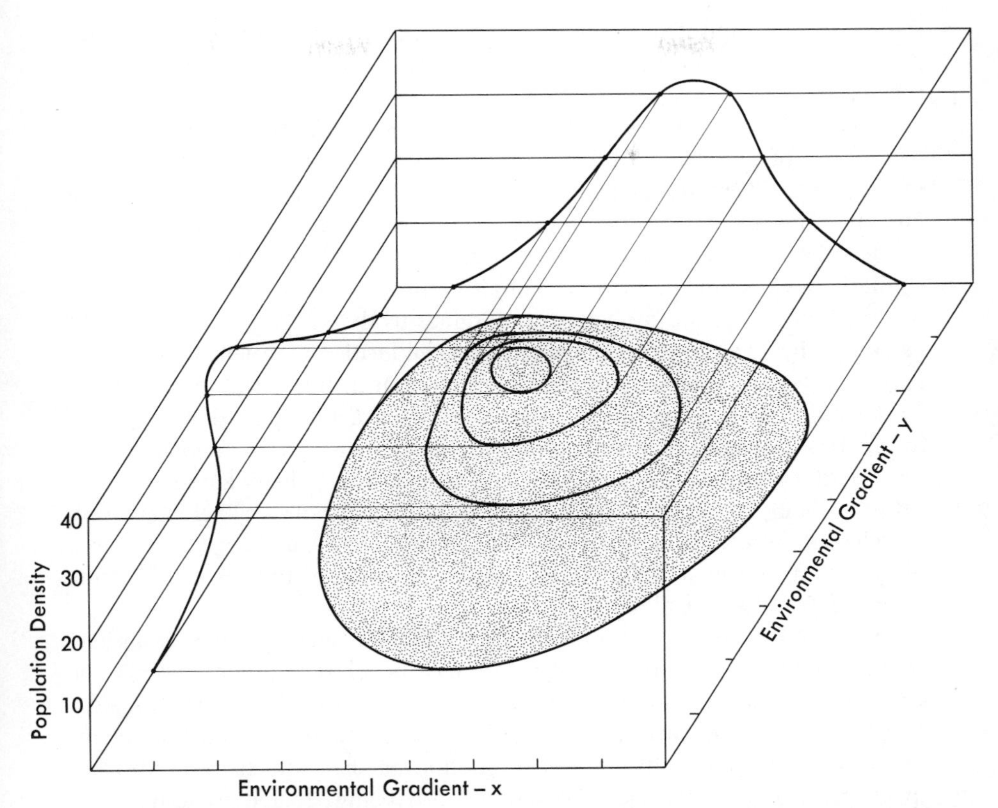

Figure 4.5. Population response to two environmental gradients. The distribution forms a bell-shaped or hill-shaped figure, with population density decreasing in all directions away from the population center or peak. In any transect of communities along a single environmental gradient that cuts through this population solid, a bell-shaped curve of population density will be obtained.

nutrient content increase and soil organic matter content decreases from high elevations to low.

5. The chart thus permits us to represent many of the relationships of environmental factors, species populations, and communities to one another; it makes possible a conception of the vegetation as a pattern of populations and communities corresponding to a pattern of environments.

In many areas three or more environmental gradients have significant effects on communities. Various kinds of soil factors and disturbance factors must often be considered, and it may not be appropriate to base the study on assumptions about which of these are most important. In this case it is possible to approach the recognition of major environmental gradients indirectly, causing them to emerge from an analysis of community samples, rather than accepting them in advance as was done in the study just described.

One of the techniques by which this is possible is called *polar ordination.* A set of samples are taken to represent a range of communities to be investigated. These samples are compared with one another in all possible combinations by coefficient of community, or percentage similarity (see Table 4.1A), or other related measurements. A table results that contains similarity measurements for each sample compared with every other sample. Some of the samples, however, are marginal to the set as a whole, as indicated by the fact that the sums of their similarities to other samples are particularly low. Two samples that are marginal in this sense, and that are least similar to one another, can be chosen as a first pair of end points for an axis. Each of the other samples in the set may now be compared with these end point samples and may be located, somewhere along the axis between them, by relative similarity to the two end points. The calculation by which they are located is based on the Pythagorean theorem and uses the equation given in Table 4.1B. All the samples of the set can be thus arranged (ordinated) along the axis.

Some of the samples near the middle of the axis may be quite dissimilar. A pair of these that are least similar may be chosen as end points for a second axis, and all other samples may be arranged along this axis also. The process can be carried on to a third axis, and if necessary a fourth. The product is an arrangement, or ordination, of the samples in an abstract space defined by the axes. The samples are being treated in terms of a hyperspace of axes that are not niche gradients, but gradients of change in the species composition of samples. These gradients presumably correspond to ecoclines, but they are not yet identified as such.

Figure 4.6 illustrates such an arrangement in a simple case, limited to two dimensions and ten samples representing different types of communities. The ten types are from a forested landscape in Poland; they range from wet ash-alder woods (carr) and bog with scattered pines to dry pine-bilberry forest and hazelnut brush. With the samples arranged in the chart, values for environmental variables can be plotted at the points for samples. It is then apparent that soil moisture decreases from upper left to lower right across the pattern, whereas soil fertility increases from lower left to upper right. The arrangement thus represents a pattern of communities in relation to a pattern of soil characteristics. The axes are in fact ecoclines, but their directions in the pattern are oblique in relation to the conventional gradients of soil moisture and soil fertility. It is usually possible to interpret such a pattern in terms of known environmental factors. In some cases axes represent effects of environmental variables that had not been recognized before the sample arrangement was prepared and studied. The axes are in some cases gradients of disturbance effects, or of community

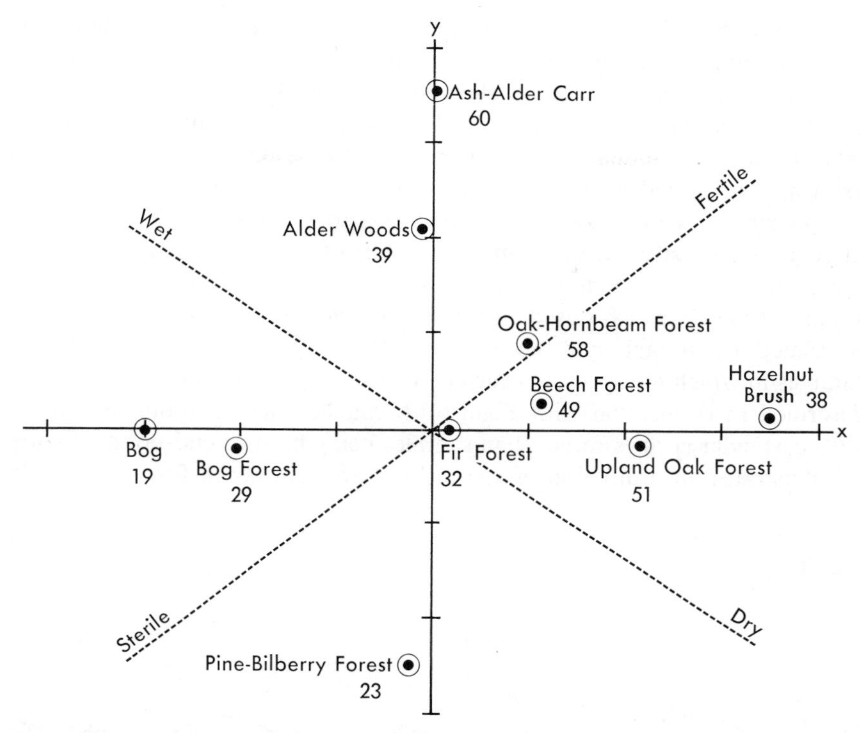

Figure 4.6. An arrangement of ten Polish forest communities in relation to two axes by polar ordination. Each sample is an average of a number of field samples representing a community-type. [Data of Frydman, *Ecology,* **49:**896 (1968).] The bog and hazelnut brush served as end-point samples for the first, *x,* axis; pine bilberry forest and ash-alder carr for the second, *y,* axis. Samples were located by relative similarity (coefficients of community) to these end-point samples; units of ecological distance marked on the axes are 10 per cent coefficient of community. The dashed lines are gradients of soil characteristics that are oblique in relation to the axes; the numbers are mean numbers of plant species in the field samples.

development, rather than of stable communities in relation to habitat gradients.

Community characteristics form gradients through the pattern also. Diversities (numbers of plant species in samples) increase, as illustrated, along the fertility gradient (but are lower in the dry hazelnut brush). In the central place in the pattern are the fir forests, which grow in habitats intermediate for the area—neither especially wet, nor dry, nor of especially low fertility, nor on soils rich in lime. For the uplands of the area the fir forests form the largest part of the vegetation pattern where it is not disturbed by man; they are a prevailing community type for the landscape. It may be observed that at a given elevation one type of community may prevail in the vegetation of Figure 4.4 also; at 8,000 feet, for example, pine forests are most extensive but fir forests occur in moist, pine-oak forests in dry habitats. The undisturbed community that is most extensive in the vegetation of a given area, or elevation belt, is thought to express most effectively the climate of that area. The treatment of community types in relation to climates that follows is consequently based on relating the most extensive, or prevalent, communities to climatic gradients.

As samples can be arranged in patterns, so can species. Measurements of relative similarity of distribution in a set of samples can be computed for each species, compared with every other species. These distributional similarity values can be used to arrange species in an abstract space that is related to, though not the same as, that obtained by ordinating the samples in which the species occurred. The same polar ordination technique described in connection with Figure 4.6 can be used to ordinate species, with particular species rather than samples being used as end-points. Figure 4.7 illustrates an ordination of species in European beech forests, using in

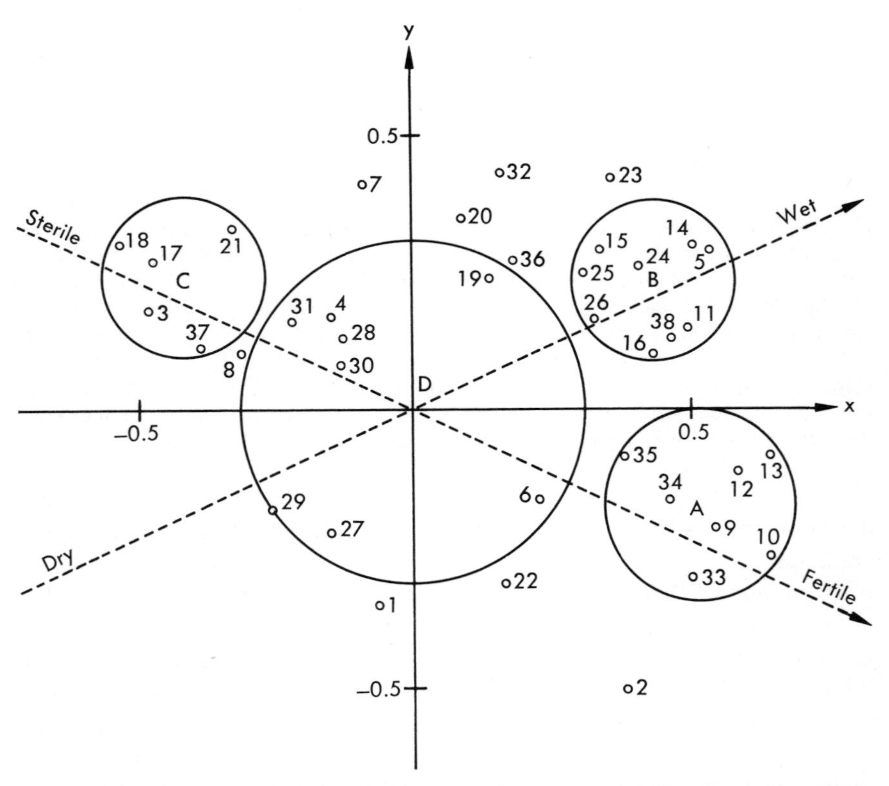

Figure 4.7. An arrangement of species for French beech forests, using factor analysis. Species (indicated by numbers) are located in relation to two extracted factors; ecologically related groups of species are enclosed in circles. The dashed lines indicate soil moisture and fertility gradients that are oblique in relation to the extracted factors. [Dagnelie, *Bull. Serv. Carte phytogéogr.*, Paris, *B.* **5**:7 (1960).]

this case not polar ordination but the more formal mathematical technique of factor analysis. This ordination, like Figure 4.6, can be interpreted in terms of gradients of soil fertility and soil moisture. Circles have been drawn around groups of species with different distributional relations to these gradients. The species in circle *A* occur mainly in forests on moist and fertile soils, those in *B* on moist but less fertile soils, and those in *C* on drier and less fertile soils. Species in circle *D* are "companion" species that have wide distributions, occurring in all these types of beech forest. Circles *A, B,* and *C* enclose *ecological groups,* sets of species that are similar in their distributional response to environmental gradients, and may have their centers of distribution in relation to these gradients close together. It may be noted that the grouping is rather arbitrary, for the species are scattered in the ordination space according to the principle of species individuality.

More formal mathematical techniques such as factor analysis and principal components analysis are often experimented with as means of ordination. In some cases, when the range of communities sampled is not too large, they produce useful ordinations as in Figure 4.7. In general, however, these techniques give disappointing results in ordinating community samples, because of the complex and curvilinear relationships of species distributions illustrated in Figure 4.2. When samples representing a wide range of environments and communities are ordinated by principal components analysis or factor analysis, the ordination axes may be uninterpretable. Formal ordination techniques that are more appropriate to community data have been developed, but in many applications the simpler techniques give effective ordinations. The approaches described—treatment in terms of recognized environmental gradients, ordination by quantitative comparison of samples, and ordination based on distributional correlation of species—usually produce convergent results. When these approaches can be applied to the same set of samples, they show similar relationships of species and communities to environment and one another.

Community Classification

The concept of the vegetation pattern in Figs. 4.4 and 4.6 is a complex population continuum, but the diagrams break the continuum into types. Classifications of communities are often needed. There is no real conflict between the principle that communities are generally (but not universally) continuous with one another, and the practice of classifying these communities as a means of communication about them. No one argues that words for colors should not be used because colors are subjectively distinguished fractions of a continuous spectrum. The results of gradient analysis do, however, have major significance for understanding the practical problems of classification.

The kinds of communities we recognize are abstract *classes,* each grouping together a number of particular communities by some characteristics they share. Communities can be classified by a number of different kinds of characteristics—growth-form dominance, species dominance, stratal structure, species composition, and so on. We can term a class or abstract grouping of communities by any kind of shared characteristic a *community-type.* The different characteristics of communities by which they may be classified do not simply change in parallel with one another. Use of different characteristics will lead to different classifications, including different community-types, when applied to the same communities. In any classification the boundaries between community-types will be more or less arbitrary, for these boundaries are determined by the characteristics chosen for classification and the ecologist's choice of where to place the boundaries. There is no single correct way to classify communities. A number of different systems of classification have developed:

1. The physiognomic approach classifies communities by structure—generally by the dominant growth-form of the uppermost stratum or the stratum of highest coverage in the community. A major kind of community characterized by physiognomy (and environment) is a *formation* or *biome.* Structural or physiognomic classification is the usual approach to description of the communities of a continent, or of the world, and it is widely used by geographers, climatologists, and soil scientists as well as ecologists.

2. Classification by dominant species is a natural and widely used approach. Community-types defined by their dominant species can be termed *dominance-types,* but often they are called simply "types." Although dominance-types are one of the easiest ways of classifying communities, they are not always the most satisfactory way. In tropical rain forests the many tree species make definition of dominance-types difficult. In contrast to this, many forests in the western United States are dominated by tree species of such wide geographic distribution that types defined by them are broad collections of different kinds of communities. Most communities are dominated by more than one species, and subjective decisions on what combinations of major species ought to be recognized as dominance-types are necessary to make the classification work. Dominance-types can provide a quite workable way of classifying communities, however; and physiognomy and dominance can be used together, with dominance-types subordinate units within formations. The community-types in both Figure 4.4 and Figure 4.6 are dominance-types (but those of Fig. 4.6 are also characterized as associations in the sense of paragraph 5).

Some American authors have followed F. E. Clements in using as units very broadly defined dominance-types, or groups of dominance-types. Thus E. Lucy Braun, in her classification of the eastern forests of North America, recognizes several types or "associations" defined by species or genera of dominant trees: the oak-hickory, beech-maple, maple-basswood, oak-chestnut, and oak-pine "associations," and the rich mixed mesophytic forest, dominated by a larger number of tree species (Plate 3).

3. Community-types can be defined also by considering the different strata of communities. Types can be characterized by the dominant species of all the strata, not only the uppermost stratum. In this case one tree species (red spruce) two shrub species (hobblebush, and mountain cranberry), and two herb species (spinulose fern, and oxalis) could define two community-types: red spruce-hobblebush-spinulose fern, and red spruce-cranberry-oxalis. These types defined by stratal dominance are termed *sociations*. Forests can also be classified into *site-types* defined by the undergrowth. A site-type may group together forests with different dominant tree species, but with similar undergrowth composition indicating that these forests grow in similar environments, or sites. In the *synusial* approach to classification each stratum or life-form grouping is classified separately from others. These three classifications developed in northern Europe and seem most appropriate to northern vegetation that is poor in species and has well-defined strata. The synusial approach is also used for communities of epiphytes and of aquatic plants.

4. Quantitative or numerical approaches to classification employ measurements of the relative similarity of samples to group these, or measurements of the relative distributional similarity of species to group these. The measurements can be the same as those used for ordination. Community-types derived by numerical classification may be termed *noda*. Numerical classifications seek to escape the subjectivity of other approaches; but they do not really escape arbitrary choices and have not, on the whole, been as useful as other approaches to classification.

5. Communities can be classified also by considering all their plant species, their whole floristic composition. The basic unit for such classification is the *association,* but this is a different unit from the American association of Clements and Braun. The floristic approach developed by J. Braun-Blanquet has been little used by English-speaking ecologists, but very widely used in continental Europe. We shall describe it further as an illustration of some principles and possibilities of community classification.

For this approach we shall make three assumptions about how to classify communities. First, classification and interpretation of communities should be based on their floristic composition. The full species composition of communities should better express their relationships to one another and environment than dominance or any other community characteristic. Second, some species in a community give more sensitive expression of relationships than others. For practical classification the approach will emphasize certain diagnostic species whose distributions make them effective indicators of particular relationships. Third, these diagnostic species will be used to organize communities into a hierarchical classification paralleling in form the one used for the taxonomic classification of individual organisms.

A plant ecologist using this approach takes a number of samples from the vegetation in which he is interested. The samples should each include a full listing of species present, together with some indication of their relative importances. For the latter, a five-point scale of coverage and abundance is used. The samples of a set are compiled into a first community-table in which the rows are species and the columns samples. The ecologist studies the distribution of species in the table, with particular attention to species that are present neither in almost all the samples, nor in very few samples. Species and groups of species that can be used to characterize different kinds of communities represented in the table are sought. The table is rearranged to group together species that are most alike in their distributions in the samples, and to group together samples that are most alike in their species compositions. Repeated rearrangement may be needed to produce a "differentiated table" that is satisfactory. Table 4.2 is an example of such a table, reduced for illustration from the original.

In this table groups of samples are characterized by sets of diagnostic species, those outlined by boxes. Diagnostic species distinguish different community-types by their presence in samples of a given community-type, contrasted with their absence in samples of other community-types. Two sorts of diagnostic species are used. *Character-species* are centered in a particular kind of community; they are present in the samples of one community-type, but absent or of less importance in the samples of all other community-types. *Differential-species* are present in the samples of one community-type, but absent from most or all of those of another community-type when just these two types are compared. The differential-species need not be centered in the first of these types; and it does not matter how they are distributed apart from the border or transition between the two types being compared. Character-species characterize community-types by their distributional centers, differential-species by their distributional limits. Figure 4.8 illustrates the way these may be related to the community-types along an environmental gradient.

Three community-types can be distinguished by diagnostic species groups in Table 4.2, but the table does not tell us whether these species are character-species or differential-species. Knowledge of the way they are distributed in a wider range of communities is necessary to determine which kind of diagnostic species these are. In the case of Table 4.2, such broader knowledge indicates that the species of groups *B* and *D* include character-species of two associations already ordinated in Figure 4.6. The species of group *B* are character-species for the fir forest association, and the first two species of group *D* are character-species for the pine-bilberry forest association. The last three species of group *D* are more widely distributed,

Table 4.2 A Differentiated Community Table. Samples are arranged to show diagnostic species groups, Polish fir and pine forests. Numbers are coverage-abundance scale values; + indicates presence at very low coverage. [Rearranged and shortened from data of Frydman, 1968.]

		Fir Forests				*Pine-Bilberry Forests*							
						Moist				*Dry*			
	Sample number	1	2	3	4	5	6	7	8	9	10	11	12
Group	Species												
A:	*Abies alba*	4	2	2	2	+	+	+	+	+	+		+
	Pinus silvestris	+	+	+	+	4	3	2	4	4	1	2	3
	Picea excelsa	+	+	2	+		2	+	+	+	+		+
	Vaccinium myrtillus	+	2	+	+	5	4	2	+	1	+	+	2
	Vaccinium vitis-idaea	+		+	+	+	+		+	1	3	3	2
B:	*Lycopodium selago*	+		+	+								
	Circaea alpina	+	+		+								
	Pyrola secunda		1	+	+								
	Pyrola minor		+	+	+	+							
C:	*Lycopodium annotinum*	+		+	+	+		+	+	+			
	Ptilium crista-castrensis	2		4	+	2	+	3	3				
	Dicranum undulatum	4	+	2	2	+	+	+	+				+
	Entodon schreberi			+		5	1	5	2		+		
D:	*Pyrola chlorantha*					+			+	+	+		
	Melampyrum vulgatum					1	+	1	2		+		
	Calluna vulgaris						+	+		2	+	+	
	Cladonia silvatica					2	+			3	+	3	+
	Cladonia rangifera					1			1	2	+	4	+
E:	*Quercus sessilis*		+						+	+	+	+	+
	Betula verrucosa									+	+	+	+
	Thymus ovatus							+		+	2	+	+
	Lycopodium clavatum									+	+	+	1
Number of species		35	37	38	37	20	17	24	25	39	41	32	34

but in this table they serve as differential-species that help to distinguish the two associations. The species of group *C* are shared between the fir forests and moister stands of the pine-bilberry forests, whereas the species of group *E* are limited to the drier pine-bilberry forests or are shared between these and dry oak forests. Groups *C* and *E* are thus differential-species groups, that distinguish moist and dry subassociations or variants of

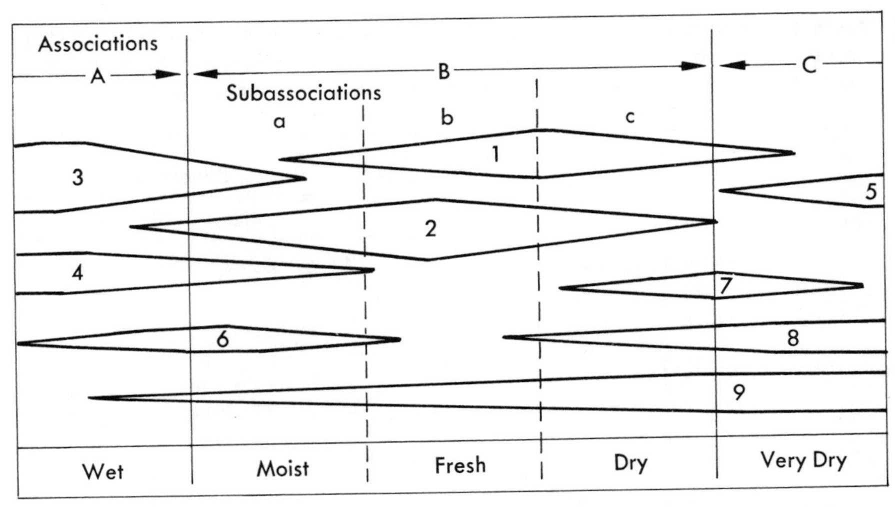

Figure 4.8. Diagnostic species along a moisture gradient. Species 1 and 2 are character-species for association B and have their populations centered in (or largely confined to) that association. Species 3 and 4 are character-species for association A, and species 5 is a character-species for association C. Species 4 and 6 are differential-species for subassociation a of association B, and species 7 and 8 are differential-species for subassociation c. In each case presence of the differential-species distinguishes the moist or the dry subassociation from the intermediate or typical subassociation b. Species 9 is a more widely distributed species that might be used as a character-species for a higher unit, such as an alliance uniting associations B and C with other associations. [Westhoff and van der Maarel, *Handbook of Vegetation Science,* **5**:617 (1973).]

the pine-bilberry association. The species of group *A* are dominant trees and shrubs: *Abies* = fir, *Pinus* = pine, *Picea* = spruce, and *Vaccinium* = bilberry or blueberry. These dominants do not distinguish the three community-types by presence and absence, but they do differ between types in coverage values.

Character-species are used to define the basic unit of the floristic classification, the *association.* Other species than the character-species are considered in recognizing a characteristic species combination for an association, but the character-species are usually emphasized as part of the formal definition of the association. Ideally, in fact, the association is the lowest unit that can be characterized by one or more character-species. Differential-species are used to define subassociations and other units below the association. To produce a hierarchy associations must be grouped into higher units; and for consistency of the approach these higher units should, like associations, be defined by character-species. Species whose distributions are too wide to permit their use as character-species for associations can become character-species for higher units. The first five species in Table 4.2, for example, can be used as character-species for higher units that include both the fir and pine-bilberry associations. The increasingly broad units of the hierarchy, defined by character-species of increasingly broad distribution, are: association, alliance, order, class, and division. Formal names can be created for each of these, using standard suffixes applied to the names of one or two defining species, as in Table 4.3. The table gives in a formal hierarchy the ten associations ordinated in Figure 4.6.

Table 4.3 Formal Classification of Associations. Braun-Blanquet classification for an area of forests in southeast Lublin Province, Poland [Frydman & Whittaker, 1968].

Class Oxycocco-Sphagnetea
  Order Ericeto-Ledetalia
    Alliance Oxycocco-Ericion
      1. Association Sphagnetum medii, raised bog
Class Vaccinio-Piceetea
  Order Vaccinio-Piceetalia
    Alliance Vaccinio-Piceion
      2. Association Pineto-Vaccinietum uliginosi, bog pinewood
      3. Association Pineto-Vaccinietum myrtilli, pine-bilberry forest
      4. Association Abietum polonicum, Polish fir forest
Class Alnetea glutinosae
  Order Alnetalia glutinosae
    Alliance Alnion glutinosae
      5. Association Cariceto elongatae-Alnetum, wet alder wood
Class Querceto-Fagetea
  Order Fagetalia silvatica
    Alliance Alno-Padion
      6. Association Circaeo-Alnetum, ash-alder carr
    Alliance Carpinion
      7. Association Querceto-Carpinetum medioeuropaeum, oak-hornbeam forest
    Alliance Fagion
      8. Association Fagetum carpaticum, Carpathian beech forest
  Order Quercetalia pubescentis
    Alliance Quercion pubescentis-sessiliflorae
      9. Association Querceto-Potentilletum albae, upland oak forest
      10. Association Coryleto-Peucedanetum cervariae, hazelnut brush

When ordinated on the axes of Figure 4.6, the twelve samples of Table 4.2 lie along a gradient of soil moisture (and fertility), with the moist pine-bilberry samples in positions between those indicated for the fir forests and the dry pine-bilberry forests. The community-types, and the groups of species, are thus diagnostic for soil conditions as well as kinds of communities. This use of species as *indicators* of environment is one of the important features of the floristic approach to classification. The vegetation of an area can be mapped by associations and lower-level units, and the units can indicate soil fertility, moisture conditions, and other characteristics of environment. Such maps can then serve as aids for land use planning.

Thus the approach realizes some of its major objectives. Detailed knowledge of local vegetation is embodied in a formal classification. Species distributions and the species composition of communities both determine the units of classification, and express their relations to environment. Relating information on the three levels of environment, species populations, and communities is an objective of classification as it is of gradient analysis. Only the means of relating these kinds of information differs, being based on units when they are classified, and on gradients when they are ordinated. Both Figures 4.4 and 4.6 involve both classification and ordination. It is possible either to subject the vegetation to a gradient analysis, and use the results as a basis of classification as in the study of which Figure 4.4 is a part, or to classify the vegetation and ordinate the resulting units to clarify their environmental relationships, as in Figure 4.6. Gradient analysis and classification are not antagonistic but can be aids to one another; they are complementary ways of studying communities.

Table 4.3 reflects another major objective of the floristic approach of Braun-Blanquet: coordination of the work of many ecologists into a general, standard scheme of vegetation classification. Because many ecologists use this same classification, their vegetation studies can be effectively related to one another; and as new kinds of communities are described, they can be fitted into the classification by their floristic relations to other community-types already known. The floristic approach has made possible a master-hierarchy classifying the vegetation of Europe, from which Table 4.3 represents a few units of forest vegetation.

Clearly, such a classification is a human construction. The form of the hierarchy is not inherent in vegetation, but is determined by ecologists' choice of assumptions about classification and of ways of applying these assumptions. Because of species individuality and community continuity, the particular ways species are grouped and community-types distinguished must be to some extent arbitrary. Community classification is justified not by theory, but by usefulness.

A final question is the relationship of animals to a community classification based on plants. It is possible to describe animal distributions in terms of any of the approaches to classification given here; all these have, in fact, been applied to animal communities. Because of the species individuality of both plants and animals, there seems to be no simple fit of animal communities to any one classification of plant communities. Many animal species have distributions that are wider than the associations of the floristic approach. This seems particularly to be true of vertebrate animals, whose distributions are not determined by particular species of plants. Many students of animal communities have felt that the best units for animal communities are broader than associations and dominance-types, and are best defined by vegetation structure rather than by plant species. Many animal ecologists consequently prefer the physiognomic approach to classification that we describe next.

Biome-Types

A major kind of community on a given continent, as recognized by physiognomy, is a *biome* or *formation.* (Formation is used when the concern is with plant communities only, biome when the concern is with both plants and animals.) Thus the broad-leaved deciduous forests of the temperate eastern United States are a biome, and the prairies and other grasslands of temperate, moderately dry climates of the Middle West and West are a biome. A biome is a grouping of terrestrial ecosystems on a given continent that are similar in vegetation structure or physiognomy, in the major features of environment to which this structure is a response, and in some characteristics of their animal communities.

Similar deciduous forests of temperate continental climates occur in North America, Europe, and eastern Asia. Temperate grasslands occur in similar climates in North America, Eurasia, and the Southern Hemisphere. The still broader grouping of convergent biomes or formations of different continents is a *biome-type* or *formation-type.* These very broad units are few enough in number that we can describe them here. The concepts of biome and biome-type are defined for land communities. They can be applied to aquatic communities as well, but this application seems less natural; for aquatic communities intergrade with one another in different ways and are less clearly characterized by structure in response to climate.

In principle, formations and biomes, formation-types and biome-types, should be defined by physiognomy; but in practice they have to be defined by combinations of physiognomy and environment. There are six major physiognomic types on land: forest, grassland, woodland (of small trees, generally in open spacing and with well-developed undergrowth), shrubland (dominated by shrubs, with coverage of these and other plants generally above 50 per cent), semidesert scrub (semiarid communities with shrubs and other plants forming a sparser cover), and desert (plant cover very sparse, often much below 10 per cent). These intergrade in various directions with one another and certain other types. Each of the six types occurs in so wide a range of environments that more than one biome-type is defined within it on the basis of major differences in climate. Thus alpine meadow, temperate grassland, and tropical savanna are separate biome-types although all are dominated by grasses or grass-like plants; and rain forests are in biome-types separate from other forests. On the other hand some communities, particularly the tundra, are dominated by more than one growth-form; and importance of these growth-forms varies with environment to form patterns of physiognomically different communities in a given area. The tundra biome-type is defined by environment and a range of physiognomic types.

In principle, formations and biomes should be very similar units, since both are defined by physiognomy and environment. However, definitions with vertebrate animals in mind tend to lead to a smaller number of more broadly defined biomes; definitions based more directly on vegetation structure tend to lead to a larger number of more narrowly defined formations. An effort has been made here to bring these units into accord by giving each type both defining characteristics, and a range of accepted variation in structure. Some of the biome-types thus include kinds of communities that in other classifications become different formations and formation-types. On the other hand, more biome-types are recognized here than in some other classifications. The statements on biome-types that follow are limited to major communities and are very brief; further accounts and illustrations of other classifications can be found in the reference list and other books on plant geography. Geographic distributions of some of the biome-types are shown on the map inside the back cover.

1. Tropical rain forests (Plate 1) occur in the humid Tropics where rainfall is abundant and well distributed through the year, in South and Central America, Africa, and the largest rain forest area extending from Southeast Asia through Indonesia to northeast Australia. Trees are often tall, of numerous species, some with buttressed bases; and many of the tree species are remarkably like one another in appearance, with smooth bark and very similar oval, medium-sized, evergreen leaves. Tree-ferns are sometimes present, large woody climbers or lianas extend from the forest floor to the canopy, and the canopy is adorned with many orchids and other epiphytic plants. These forests thus have their plant species concentrated in the canopy, unlike temperate forests in which the greater number of species is usually in the undergrowth. Lower levels of mature rain forests are not generally dense and jungle-like, but are fairly open and not difficult to walk through. Most mature rain forests have heights in the range normal for many temperate and tropical forests, 30 to 40 m, but climax rain forests both taller and lower than this are reported. The invertebrate animal life is exceedingly rich, and striking in the colors, sizes, and special adaptations for defense by concealment and mimicry among its members. Vertebrate animal life is also rich although not so conspicuously so beyond that in forests of the Temperate Zone. In contrast to temperate forests, a high proportion of the mammals and reptiles of tropical rain forests are arboreal.

2. Tropical seasonal forests include monsoon forests and a range of other communities, semi-evergreen, or largely or wholly deciduous. They occur in humid tropical climates with a pronounced dry season during

which some, or many, or all of the trees lose their leaves. Canopy heights decrease and canopy coverage tends to decrease as the climate dries until the forests are reduced to woodlands. In some areas there is a parallel change in seasonal conditions, with increasing proportions of deciduous species toward drier climates. Tropical seasonal forests are extensive in India and southeast Asia and occur also in West Africa, South and Central America and the West Indies, and Northern Australia.

3. Temperate rain forests occur as giant forests along the Pacific Coast of North America in a belt from the mixed coniferous rain forests of the Olympic Peninsula of Washington to the coast redwood forests of California and Oregon (Plate 2). The climate is cool and maritime, with abundant winter rainfall and much summer cloudiness and fog. The coast redwood forests are "rain forests" (implying year-around rainfall) only because of the fogs from which their foliage collects water in the summer. These Pacific Coast forests are (along with Australian temperate rain forests of *Eucalyptus regnans*) the tallest in the world; heights of these forests (60 to 90 m in some stands, with some trees over 100 m) are of a different order from those of most forests. The low species diversities of the Coast forests contrast with the richness of tropical rain forests. Temperate rain-forest types range from the giant forests, and smaller coastal Sitka spruce forests north to Alaska, through forests of southern beech (*Nothofagus*) and southern-hemisphere conifers in New Zealand and Chile, to montane forests and mossy forests dominated by smaller broad-leaved evergreen trees in tropical mountains. Bamboos are conspicuous in the undergrowth of some tropical mountain forests, and in some areas bamboos become dominants of a belt of mountain vegetation.

4. Temperate deciduous forests grow in moderately humid, continental climates, with summer rainfall and severe winters. Broad-leaved deciduous trees—oak, beech, maple, ash, basswood, and others—are dominant in forests that in their best development are of heights comparable to tropical rain forests. American deciduous forests include the species-rich mixed mesophytic forests of the southern Appalachians (Plate 3), beech-maple and maple-basswood forests in the north, and a range of oak-hickory, oak-pine, and oak or oak-chestnut forests in the western, southern, and eastern parts of the biome. The deciduous forests of eastern Asia also are rich and varied, but those of western Europe are poor in tree species because of extinctions during glaciation. Mammals are primarily ground-dwelling. In the United States the white-tailed deer and black bear are characteristic large mammals; the red-eyed vireo, ovenbird, and ruffed grouse are characteristic birds. Temperate deciduous forests occur also in limited areas of the Middle American Highlands and southern Chile.

5. Temperate evergreen forests occur in varied climates and with varied structure. Forests of less humid, maritime (summer-dry) climates in California, the Mediterranean area, and southern Australia are dominated by

trees with sclerophyll leaves (tough, evergreen, relatively small, broad leaves). Extensive needle-leaved evergreen forests occur in continental climates of the western United States and include Douglas fir and white fir forests in more humid habitats, various pine forests in drier habitats, and the mixed forest of Sierra redwood with pine, fir, and incense cedar in the Sierra Nevada. Many needle-leaf forests are of more open structure than deciduous forests, and some pine forests have quite open canopies (Plate 4). Other evergreen forests occur in eastern Asia and in places where soil conditions or fire frequency, or both, favor pines over broadleaved trees in the eastern United States and western Europe. Temperate evergreen forests are dominated by southern-hemisphere conifers and southern beeches in Chile and New Zealand, and by *Eucalyptus* in Australia (Plate 5).

6. The taiga or subarctic-subalpine needle-leaved forests are usually separated from the temperate evergreen forests as a biome-type of the cold edge of the climatic range of forests. Evergreen spruces, firs, or pines are dominant in most such forests, but deciduous larches are dominant in some; in many, dominance is shared by a spruce and a fir species (Plate 6). Because there are few tree species, such forests give an impression of monotony, but differences in undergrowth composition express differences in environment. In arid mountains open pine woodlands occur in the subalpine zone; in humid subarctic areas extensive bog or muskeg may occur. In the Far North the taiga opens northward to a woodland of scattered trees with tundra-like undergrowth. At its high-elevation margin the taiga may be either opened to a woodland or reduced at timberline to a wind-sheared shrub community (krummholz) dominated by the tree species of the taiga. Characteristic animals in North America include the moose, wolverine, lynx, and varying hare, red-breasted nuthatch, slate-colored junco, and warbler species. The taiga extends around the world in the northern part of North America and Eurasia and extends to the south at higher elevations in mountains.

7. Elfinwoods occur in the subalpine zone on tropical mountains, in cold but nonseasonal climates of rather constant rain, fog, and wind. Elfinwoods are dominated by large shrubs or small trees in dense, thicket-like growth with contorted branches and a low canopy of broad evergreen leaves (Plate 7). Lichens and mosses coat and festoon the branches and hang from them like draperies, making the denser of these communities visually, as well as physically, impenetrable. The elfinwoods are rain forests in miniature, rainshrub communities; they give a distinctive impression of luxuriance combined with stunting. Elfinwoods occur above the upper temperate rain forests or mossy forests on mountains of Africa, South America, and New Guinea.

8. Tropical broadleaf woodlands of small trees replace tropical seasonal forests toward drier climates (and less favorable soils) in some areas. Ex-

tensive areas of central Brazil are cerrado (Plate 8), which in the broad sense includes a range of communities from woodlands with canopy typically at 4 to 7 m, but sometimes taller, to a thicket or scrub less than 3 m tall; and from a closed but not dense canopy to open scrub or scrub with scattered trees. The trees and shrubs mostly have large, rigid, semievergreen leaves, grotesquely twisted branches, and thick, fire-adapted bark. Palms and leguminous trees with small leaflets are usually present, but spinose plants and succulents are almost absent. The cerrado occurs between seasonal forest and thornwood on deep, acid, sandy soils. The physiognomy of these communities appears to be governed by relative infertility of the soils. Evergreen or semievergreen woodlands and thickets occur also in northern South America and the West Indies; woodlands occupy a large area of interior southern Africa (miombo) and occur in Burma and probably elsewhere in southeast Asia.

9. Thornwoods occupy tropical climates more arid than those of seasonal forests (and of broadleaf woodlands where these occur). Spiny species of *Acacia* and other genera of the pea family are widespread as dominants. These plants mostly have small drought-deciduous leaflets, but succulent plants also occur and increase toward more arid environments. A wide range of communities are included here, from thorn woodlands approaching forest in density and structure, through more typical woodlands of small trees in open growth, to thorn thicket or scrub of large shrubs in dense growth (Plate 9). The latter in turn with increasing drought becomes lower and more open until it gradates into semidesert scrub (Type 17). Thornwoods are widespread in South America as the Brazilian caatinga and other types; they occur also on limestone in the West Indies and Middle America, in Burma and areas of India and Thailand, and in Africa and Madagascar.

10. Temperate woodlands are by definition communities of small trees, but the communities included range from nearly full canopy cover to open woodlands with scattered trees, transitional to grasslands without trees. The dominant trees may be of any of the three growth-forms—needle-leaved trees, sclerophylls, and deciduous broad-leaved trees—or any combination of these. Pygmy conifer woodlands (dominated by piñon and juniper) are a widespread and distinctive community-type in the western United States (Plate 12), oak woodlands occur in the Great Valley of California (Plate 14) and the Cross Timbers of Texas, and evergreen oak and oak-pine woodlands are extensive in the southwestern states and Mexico. Structurally related communities occur around the Mediterranean and in the southern continents (Plate 13). Woodlands are most extensive in climates too dry for true forests; toward still drier climates woodlands give way to grassland, shrubland, and semidesert.

11. Temperate shrublands include the sclerophyll shrublands that develop in climates of the Mediterranean type—moderately dry, warm temperate, maritime climates with little or no summer rain. Sclerophyll leaves

Plate 1. Lowland tropical rain forest in Panama (biome-type 1). Such forests are very difficult to photograph, but the density of the vegetation, the characteristic leaf shape, and the prominence of woody vines or lianas may be seen in this view. [Courtesy of W. J. Smith, from R. E. Ricklefs, *Ecology,* Chiron Press, Newton, Mass. (1973).]

Plate 2. Temperate giant rain forest (biome-type 3). Coastal redwoods (*Sequoia sempervirens*) in dense stand with undergrowth of sword-fern (*Polystichum munitum*), Jedediah Smith Redwoods State Park, California. The coastal redwoods in California are the southernmost extension of the giant rain forests of the Northwest Coast of the United States, adapted to a less humid climate than that of the rain forests of the Olympic Peninsula. The redwoods are limited to a coastal belt in which fog moderates the effect of summers with little or no rainfall. [Reproduced by permission of Philip Hyde, Taylorsville, Calif.]

Plate 3. Temperate deciduous forest (biome-type 4). Southern Appalachian cove forest or mixed mesophytic forest with tulip poplar (*Liriodendron tulipifera*) dominant, Chattahoochee National Forest, Georgia. Such forests include far more tree and undergrowth species than the coastal rain forests. They are probably (along with comparable forests of eastern Asia) richest in species of temperate forests of the world and most like temperate forests of wide extent in the Northern Hemisphere in Tertiary time. [Courtesy of U.S. Forest Service, photo by Philip Archibald.]

Plate 4. Temperate evergreen (needle-leaved) forest (biome-type 5). An old-growth stand of Jeffrey pine (*Pinus jeffreyi*) in a flat near Mammoth Lakes Recreation Area, California. [Courtesy of U.S. Forest Service, photo by L. J. Prater.]

Plate 5. Temperate evergreen forest (biome-type 5). *Eucalyptus obliqua* (stringybark) and *E. globulus* (blue gum) are canopy dominants in this dry sclerophyll forest on the east coast of Tasmania at 43°30′ south latitude; *Lomandra longifolia* and *Olearia viscosa* are the principal undergrowth species. Precipitation here is about 90 cm/yr; ground fires are frequent. The heights of the trees in this forest are up to 45 m, whereas northern hemisphere sclerophyll forests are mostly dominated by small trees, generally below 20 m. [Courtesy of the Forestry Commission, Hobart, Tasmania.]

Plate 6. Taiga, a subalpine spruce-fir forest in the Great Smoky Mountains of Tennessee (biome-type 6). The forest is near the summit of Mt. LeConte, Great Smoky Mountains National Park, Tennessee, at an elevation above 1830 m. Fir *(Abies fraseri)* is dominant, with spruce (*Picea rubens*) also present. In the humid climate of this stand (precipitation around 200 cm/yr) the forest floor has an almost complete cover of mosses and a low herb (*Oxalis montana*); a fern (*Dryopteris spinulosa* var. *americana*) is also present. [Reproduced by permission of Thompsons, Inc., Knoxville, Tennessee, from Whittaker, *Ecol. Monogr.,* **26**:1, Fig. 17 (1956).]

Plate 7. Timberline in the Uluguru Mountains, Tanzania, at the upper limit of the elfinwood (biome-type 7), 2600 m, looking from alpine grassland into the dense woodland. Trees are *Podocarpus, Myrica,* and *Pygeum* with lichens (*Usnea*) hanging from their branches; the rosette-shrubs at the edge of the woodland are *Lobelia ulurensis.* [Courtesy of Carl Troll, from "Zur Physiognomik der Tropengewächse," (1958).]

Plate 8. Tropical broadleaf woodland (biome-type 8). This is Brazilian cerrado in an open woodland phase, taken in Brasília at the beginning of the dry season, when it is subject to ground fires. Fire effects may contribute to the openness and sparse shrub cover of this phase of the cerrado. [Courtesy of George Eiten, from *Bot. Rev.,* **38:**228, Fig. 7 (1972).]

Plate 9. Tropical thorn scrub or thornwood (biome-type 9). The community is on limestone on a nearly level site at about 150 m elevation in Pedernales Province, southwestern Dominican Republic. The rather dense mixture of tall spiny shrubs, or small trees, with succulents behind the clearing in the foreground is typical of many thornwoods. [Courtesy of John Terborgh.]

Plate 10. Temperate shrubland (biome-type 11). Hard chaparral, fire-adapted communities of chamise (*Adenostoma fasciculatum*), manzanita (*Arctostaphylos* spp.), and other evergreen shrubs in the mediterranean (dry-summer) climate of southern California, Angeles National Forest. [Courtesy of U.S. Forest Service, photo by C. Miller, from Gleason and Cronquist, 1964.]

Plate 11. Temperate shrubland in South Africa (biome-type 11). This Cape scrub or fynbos, which resembles the chaparral in physiognomy, is far richer in species than any shrubland in the Northern Hemisphere. The grasslike plants are members of the Restionaceae, a southern-hemisphere family that may be ancestral to the true grasses. [Courtesy of H. C. Taylor.]

Plate 12. Temperate woodland (biome-type 10). An extensive stand of pygmy conifer woodland dominated by juniper and piñon pine with shrub and grass undergrowth, near Flagstaff, Arizona. Climate too dry for true forest is expressed in the small size and open growth of the trees. [Courtesy of U.S. Forest Service, photo by Clime, from Gleason and Cronquist, 1964.]

Plate 13. Temperate woodland in western Australia (biome-type 10). *Eucalyptus wandoo* is dominant in this evergreen wandoo woodland, east of Mundaring. The trees are tall for a woodland (15–20 m, with some trees to 30 m); rosette shrubs of the lily family (*Xanthorrhoea preissii* and *X. reflexa*) give the community an appearance unlike northern hemisphere woodlands. [Courtesy of the Forests Department, Perth, Western Australia.]

Plate 14. Temperate woodland (biome-type 10). Open woodland of evergreen oaks with a grassy undergrowth on foothills of the western slope of the Sierra Nevada, part of the extensive woodland belt surrounding the Central Valley of California. [Courtesy of U.S. Forest Service, photo by A. Gaskill.]

Plate 15. Savanna (biome-type 12) of widely scattered trees in tropical grassland. East African game plains, Olduvai, Tanzania. [Reproduced by permission of E. S. Ross, Calif. Acad. Sci., San Francisco.]

Plate 16. Temperate grassland (biome-type 13), a mixed-grass prairie in central Nebraska. [Courtesy of U.S. Forest Service, photo by B. W. Muir, from Gleason and Cronquist, 1964.]

Plate 17. Temperate grassland in the Southern Hemisphere (biome-type 13). This is the pampas, *The Purple Land* of W. H. Hudson, a tussock-grassland in the province of Buenos Aires dominated by *Stipa brachychaeta.* [Courtesy of Heinrich Walter, from *Die Vegetation der Erde,* Vol. II, Fig. 503 (1968).]

Plate 18. Alpine shrubland (biome-type 14) and alpine desert (biome-type 21). In this view of Mount Kenya, Africa, at the head of Telaki Valley at about 4200 m, tropical alpine shrubland convergent with South American paramo appears in the foreground and gives way to rock and snow desert on the higher slopes of the mountain. The rosette-shrub in the lower right corner is *Senecio keniodendron,* of the sunflower family. [Photo by A. Holm, reproduced by courtesy of O. Hedberg from *Acta Phytogeogr. Suecica,* 49 (1964).]

Plate 19. Detail of alpine shrubland (biome-type 14) resembling paramo, above the tree line at 4250 m on Mount Kenya, Africa. The rosette-shrubs in the foreground are *Lobelia telekii; Senecio keniodendron* and tussock grasses may also be seen. [Courtesy of Carl Troll, from "Zur Physiognomik der Tropengewächse" (1958).]

Plate 20. Alpine grassland (biome-type 15). A dry grassland phase of the puna, dominated by *Festuca orthophylla,* near Orcotuncu Pass, Andes Mountains of northern Chile. Small annual forbs and grasses grow between the tussock grasses in the rainy season and are grazed by flocks of wild vicuñas. [Courtesy of Carl Troll, from *Colloq. Geogr.,* Bonn, **9:**15 (1968).]

Plate 21. Tundra (biome-type 16), a distinctive patterned ground of earth hummocks in the High Arctic. Dwarf willow (*Salix arctica*) with sedges and grasses in the foreground, hummocks with *Cassiope tetragona* and *Dryas integrifolia* in the background, on a south southwest-facing slope of mixed glacial drift and alluvium, Alex Heiberg Island, Northwest Territories, Canada. [Courtesy of R. E. Beschel, from *U.S. Nat. Acad. Sci. Publ.,* **1287:**13–20 (1963).]

Plate 22. Warm semidesert scrub (biome-type 17) in the Mareb Valley near Teramni, Eritrea. This African vegetation is in many ways convergent with the warm-temperate desert scrub shown in Figure 3.9. The tall succulent is *Euphorbia abyssinica;* it is not a cactus but is convergent in form with the giant cactus in Figure 3.9. The flat-topped tree is *Acacia etbaica;* mesquite (*Prosopis juliflora*) is similar in form in the Arizona desert. The rosette-shrub in the foreground is *Aloe abessinica,* convergent in form with *Agave palmeri* and other species in the Arizona deserts. [Courtesy of Carl Troll, from "Zur Physiognomik der Tropengewächse" (1958).]

Plate 23. Semidesert scrub in Australia (biome-type 17). White mallee (*Eucalyptus dumosa*) with *Triodia irritans* (spinifex, or porcupine grass) undergrowth south of Broken Hill, South Australia. Rainfall in this area is about 25 cm and irregular, comparable to that of the American semidesert in Figure 3.9. The Australian community is very different from equivalents in America and South Africa (Plate 22). The malees are dwarf species of *Eucalyptus* that form giant shrubs with a massive underground stem or lignotuber from which a number of above-ground shoots or branches grow; the grass is a spiny tussock-grass. [Courtesy of the Forestry Commission of New South Wales, Sydney.]

Plate 24. Cool-temperate semidesert scrub (biome-type 18). Sagebrush semidesert dominated by big sagebrush (*Artemisia tridentata*) and bluebunch wheatgrass (*Agropyron spicatum*), near Big Butte, Idaho. [Courtesy of U.S. Forestry Service, photo by J. F. Pechanec, from Gleason and Cronquist, 1964.]

Plate 25. Alpine semidesert (biome-type 19). Dry puna vegetation dominated by hard cushions of *Azorella diapensioides,* on the Chilean-Bolivian border west of Sajama at 4400 m elevation in a climate that is both arid and tropical-alpine (without a distinct winter, but with nighttime frosts in all months). The cushions are formed by very dense branching of the thin, resin-containing branches; the shrubs are collected and used as fuel in the treeless puna. [Courtesy of Carl Troll, from *Colloq. Geogr.,* Bonn **9**:15 (1968).]

Plate 26. Alpine semidesert (biome-type 19) and alpine grassland (biome-type 15). A patterned community dominated by the low shrub *Epacris petrophila* is shown in the foreground on an almost level, wind-exposed feldmark or alpine rockland on Mount Kosciusko, southeastern Australia. Such communities go through a cyclical pattern of erosion and regeneration (Barrow et al., *J. Ecol.* **56**:89, 1968), but may eventually develop into the alpine grassland of *Poa caespitosa* and *Celmisia longifolia* seen in the background. [Courtesy of A. B. Costin, from *Vegetatio,* **18**:273 (1969).]

Plate 27. Subtropical true desert (biome-type 20). Extreme desert with dunes and few plants, Namib Desert, South-West Africa. [Reproduced by permission of E. S. Ross, Calif. Acad. Sci., San Francisco.]

(see paragraph 5) prevail. Size and coverage of the shrubs range from aborescent (2 to 5 m tall) with a closed canopy, to below 1 m and quite open, in response to moisture and other factors. Mediterranean maquis and California chaparral (Plate 10) are best known among such communities; equivalent biomes occur in South Africa as the cape scrub or fynbos (Plate 11), and in Chile and West and South Australia. The dominant plants are affected by frequent fires, which burn off the above-ground stems, regrowth then occurs from root systems that survive the fire. Other temperate-zone shrublands often treated as separate formations include the chaparral-like deciduous shrublands of some inland mountain areas, the aromatic shrubland of drier mediterranean climates (garrigue, soft chaparral), heaths mostly of cooler maritime climates, and distinctive shrublands in the Southern Hemisphere (Figure 8.1).

12. Savannas are tropical grasslands, with or without scattered trees or shrubs. Savannas are most extensive in Africa (Plate 15), where they support the richest fauna of grazing animals in the world; equivalent, less extensive, and less rich communities occur in Australia, South America, and southern Asia. Some of the African and Australian savannas occur in climates too dry for forest; but soil conditions or fire, or both, rather than climate probably cause the appearance of savannas in less arid climates, particularly in South America. Both savannas and temperate grasslands are subject to fires, which affect the structure of the communities and their extent into climates that might otherwise support forest.

13. Temperate grasslands occur in great areas of moderately dry, continental climates in North America and in Eurasia (the steppes). Major divisions of the North American grassland are the tall-grass and mixed-grass prairies (Plate 16) and short-grass plains of the Middle West, the palouse (bunch-grass) prairie of Washington, the California grasslands, and the desert grasslands of the Southwest. Many of the latter include scattered rosette shrubs or small trees, and some have been heavily invaded by woody plants because of overgrazing. There are also temperate grasslands in Africa (veldt) and South America (pampas) (Plate 17), but grassy woodlands with eucalypts occur in most of the comparable climates of Australia. Despite the limitation of vegetation structure to a single major stratum, plant species-diversity in grasslands can be high compared with most forests. Bird life is more limited than in forests because there is only one major vegetation layer; the meadowlarks, dickcissel, and grasshopper sparrow are North American grassland birds. Grasshoppers are abundant in late summer. Mammalian faunas are characterized by the smaller burrowing and larger running herbivores. Ecological equivalents on different continents among the latter include the bison and pronghorn antelope in North America, the wild horse and ass and saiga antelope of Eurasia, the larger kangaroos of Australia, and the zebras, antelopes, and others of the rich warm-temperate grasslands of Africa.

14. Alpine shrublands are part of the mountain lands above timberline (the upper limit of tree growth) that are termed "alpine." Although grasslands are thought typical of the alpine zone, shrublands also occur there. In humid mountain climates of South America, the elfinwood borders on the paramo, distinctive communities of rosette shrubs, tussock grasses, and lower shrubs. Part of the paramo is an alpine grassland; but in extensive areas the rosette shrubs are dominant, and in some cases they are arborescent (3–5 m tall), forming a kind of small alpine woodland. The heights of the shrubs decrease toward drier environments and higher elevations but in some areas paramo extends up to close to the snow line. Communities of strikingly similar physiognomy occur in the alpine zone of African mountains, as illustrated in Plates 18 and 19. Some of the rosette shrubs in Africa are composites of the genus *Senecio,* and thus are unlikely-looking relatives of small herbs of the North Temperate Zone that resemble asters but have yellow flowers. Other paramo rosette shrubs belong to other genera of composites or to other families. These shrubs are thus a striking case of evolutionary divergence (within *Senecio*) and convergence (by independent evolution to similar forms in different genera and families). The African alpine zone also includes heaths, and heaths are dominant in the alpine shrublands of the Himalaya Mountains. Alpine and subantarctic shrublands in New Zealand are dominated by composites (*Senecio* and *Olearia*) that in this case resemble large heaths (such as *Rhododendron*) rather than the rosette shrubs.

15. Alpine grasslands are the principal communities above timberline. Alpine meadows dominated by sedges are extensive above timberline in north temperate mountains, but miniature shrub communities of dwarfed willows, heaths, or other woody plants also occur. Cushion plants are present in many alpine meadows and increase relative to sedges and grasses, as community coverage decreases toward higher elevations and drier sites. Characteristic animals of the North American alpine meadows are the pica, marmot, mountain goat and sheep, and grizzly bear; birds include ptarmigan, rosy finch, and American pipit. Alpine animals and plants must be adapted to a brief summer of activity. Many of the larger animals migrate to lower elevations in winter, while smaller mammals hibernate or remain in shelter; animals that are active in the winter (mountain goat and sheep, ptarmigan) are protectively colored white. In tropical mountains the alpine environment is very different, for the cold is by night throughout the year rather than seasonal. In South America, in alpine climates drier than those of paramo, the grassland phases of the puna occur (Plate 20). Alpine grasslands occur also in African mountains and in New Guinea. In New Zealand tussock grasslands occur in the alpine zone (and on subantarctic islands) and are continuous with the tussock grasslands of lower elevations with temperate climates.

16. Tundras are the treeless arctic plains, the vegetation of which may

form varied and often complex patterns of dominance by dwarf-shrubs, sedges and grasses, mosses, and lichens. The tundras of North America and Eurasia are quite similar; their principal herbivorous mammals include the musk ox, caribou (and reindeer of Eurasia), arctic hare, and lemming. Longspurs, plovers, snow bunting, and horned larks are characteristic birds; reptiles and amphibians are few or absent, as in alpine grasslands. In many tundras and some alpine communities the deeper layers of the soil are permanently frozen, and only the surface soil is thawed and becomes biologically active during the summer. In many of these communities striking internal patterning of the plant communities results from frost effects (Chapter 3, Figure 3.4, Plate 21).

17. Warm semidesert scrubs occur in dry warm-temperate and subtropical climates. The most widespread type in North America is an open scrub of creosote bush (*Larrea divaricata*) that occurs, usually with smaller semishrubs, on valley plains of both major parts of the North American warm deserts—the Sonoran from Arizona to California and south through Sonora, and the Chihuahuan from eastern Arizona to Texas and south through Chihuahua—and also in the Mojave desert of California and Nevada. Quite different, rich and diverse shrub communities occur on mountain slopes and the upper margins of desert plains—saguaro-paloverde desert in Arizona (Figure 3.9), a lower spinose scrub of prickly pear, acacia, agave, and so on in the Chihuahuan area, Joshua tree "woodland" in the Mojave, and other types elsewhere. In the warm semideserts there is no such convergent dominant form as in the cool semideserts and other biomes; evolution here has produced the divergence of plant forms referred to in Chapter 3. North American mammals include the collared peccary, gray fox, white-throated wood rat, kangaroo rats and pocket mice; birds the roadrunner, cactus wren, and thrasher and dove species; and reptiles a relatively rich lizard and snake fauna. Warm semideserts very similar to these occur in South America, from Argentina through Chile to Peru. Great areas of semidesert surround the Sahara in northern Africa and extend through the Arabian Peninsula to Iran and the Thar Desert of India. More limited areas with distinctive semidesert occur in parts of the arid areas in East Africa (Somali-Chalbi) and Southwest Africa (Kalahari and Karoo) (Plate 22). The Australian communities, evolved in long isolation, are most unlike the other areas; succulents are lacking and spiny tussock grasses (spinifex, *Triodia*) are major plants (Plate 23).

18. Cool semideserts occupy extensive landscapes of the Great Basin, between the Rocky Mountains and the Cascades in the western United States, as a gray and rather drab, cool-temperate semidesert scrub. The most widespread communities are dominated by sagebrush (*Artemisia* species) with perennial grasses (Plate 24). Communities of the biome range, however, from dry grassland with sagebrush to a sparse cover of shrubs, from dominance of sagebrush to dominance of shadscale (*Atriplex*)

in drier areas and on more alkaline soils, and from strong dominance of one of these to mixtures of several shrub species. As a natural climax the semidesert scrub occurred in climates drier than those of grasslands; but grazing has converted extensive dry grasslands into sagebrush and spread a Eurasian annual, cheatgrass (*Bromus tectorum*), through both these and natural sagebrush communities. Animals include the black-tailed jack-rabbit, pronghorn antelope, sage grouse, sage sparrow, and sage thrasher. Small jumping mammals (pocket mice and kangaroo rats) are common in both cool and warm desert scrubs in North America and are convergent with other such mammals, independently evolved in desert and dry grass-land areas of other continents. Closely parallel communities occur in central Asia and Iran; cool temperate semideserts of different character occur in South America (Patagonia and the Andes) and Australia.

19. Arctic-alpine semideserts occur in arid climates above timberline. Arid phases of the puna are dominated by spiny cushion plants and succulents in open growth, physiognomically a semidesert (Plate 25). Semidesert with plants of the tragacanth form—low, spiny cushion plants formed by *Astragalus* and other genera—occur in arid mountains from the Mediterranean region into Central Asia. Small sagebrushes dominate communities suggesting miniature cool semidesert scrub in the alpine zone of the White Mountains of California. Sparse communities with low shrubs or cushion plants occur in other mountains (Plate 26) and in the Arctic. The semiarid areas of the Arctic are generally treated as part of the tundra; but in physiognomy, plant adaptations, and soil characteristics, and in the sparseness of their animal life, they are desert-like.

20. True deserts are primarily subtropical; they are in most cases continuous with areas of warm semidesert scrub, but in climates of more severe aridity. In areas with precipitation less than 2 cm per year vegetation is almost wholly lacking; in areas with 2 to 5 cm the vegetation is very sparse; but fog and dew are significant moisture supplements for some of these deserts. Because of their sparse plant cover, these landscapes are dominated not by vegetation but by the ground surface—sand, stony desert pavements, salt crust, or naked rock. Plant forms range from rather typical low desert shrubs, to some very distinctive plants, and lichens in areas with fog. Whereas some of the semideserts are floristically rich and have relatively abundant animal life, the true deserts are poor in plant and animal species. Subtropical and warm temperate true deserts include a vast Saharan area in northern Africa from Mauretania to Egypt and the Arabian Peninsula, the smaller Namib desert of coastal southwest Africa (Plate 27), the west coast of South America from the Atacama desert of Chile to Peru, and extreme desert areas of Australia and the southwestern United States. True deserts of cooler temperate climates occur in central Asia (Takla-Makan), the Andes and other arid mountains, and locally in other areas of semidesert.

21. Arctic-alpine deserts, at the cold extremity of climate, are deserts in the sense of landscapes of very low plant cover, dominated in this case by snow, ice, or rock (Plate 18). Some algae occur on ice surfaces, lichens on rocks, and scattered vascular plants in less extreme environments. Arctic-alpine deserts include most of Greenland, Antarctica, and mountains well above timberline throughout the world; and alpine deserts occur also in the lower alpine zone in very arid mountain ranges.

22–26. Hydric communities are adapted to very wet soils. Cool temperate bog (22) occurs as a local type in some areas, but can also become prevailing vegetation in such cool, humid, maritime climates as those of the blanket bogs of Scotland and Ireland. Other hydric types determined by local excess of soil moisture are not climatic types, but some are sufficiently extensive to be recognized as formations: tropical (23) and temperate (24) fresh-water swamp forests, the mangrove swamps (25) of tropical coasts and estuaries, and the saltmarshes (26) of temperate coasts.

27. Fresh-water lentic communities (lakes and ponds) include a vast range of water-body sizes and nutrient conditions in diverse climates. Lake communities virtually defy subdivision into biomes because of the number of directions of relationship and the fact that shore, planktonic, and bottom communities merge in ponds. There is major biological significance, however, in the three developmental directions (bogs, saline lakes, and the oligotrophic-eutrophic series) discussed under lake production (Chapter 7), and the large, ancient lakes that have become biologically very rich (Lakes Tanganyika, Nyassa, and Victoria, Lake Baikal, and Lake Ohrid are notable among these).

28. Fresh-water lotic communities (streams) have biotas quite different from lakes, although with some overlap. Plankton is sparse in most streams, and the bottom communities differ according to water speed and kind of substrate from rock through gravel and sand to mud. Where the bottom is mud, clams, burrowing mayfly nymphs, and midge larvae occur; the contrasting communities of rapids include animals adapted to life on or among rocks in swift current—darters, blackfly larvae, flattened mayfly and stonefly nymphs, caddisfly larvae attached to rocks, water pennies, and so on.

29. The next three communities occur where the land meets the sea and are termed *littoral*—part of the coastal belt washed by waves and alternately exposed and submerged by tides where these occur. The first of these types, marine rocky shores, will be described below as an ecocline.

30. Marine sandy beaches also occur around the world in the littoral belt. Sandy beaches are difficult environments, for they combine alternate exposure and submergence with instability and constant movement of the surface sand when it is submerged. Diatoms and other algae occur in the surface layers of the sand, but the beach animals are largely dependent on

plankton and dead particles or detritus brought in by the water for their food. The barren appearance of the beach is deceptive; crustaceans (amphipods, isopods, and highly specialized crabs) and worms burrow in the sand, and microscopic forms (copepods, rotifers, and protozoans) live in the water between the sand grains. Fish and shrimps move up and down with the water, and at lower tide levels clams and snails occur.

31. Marine mud flats occur where the water is quieter, in bays and estuaries. Microscopic algae occur in the mud surface; and the community may have a microstratification, with photosynthetic bacteria beneath the algae. Apart from the snails that may be present on the surface, the mud often shelters an abundant animal life of clams, distinctive worms, specialized crustaceans, and microscopic forms. Like beaches, mud flats receive most of their food from outside. A large part of this food may come as dead remains of saltmarsh plants, but plankton and particles also are brought in by river drainage and the tides. Estuaries, where the fresh waters of rivers are mixed with tidal water from the ocean, are thus complex ecosystems that often combine saltmarsh and mudflat with deeper water, and that link fresh-water communities with the seas.

32. Coral reefs are communities of tropical oceans, formed as fringes around islands, or forming hollow circlets of islands (atolls), or great offshore barrier reefs (along the eastern coast of Australia and northwest of Fiji). A coral is a coelenterate animal, related to the jellyfish and sea anemones, that forms a colonial structure embedded in a skeleton or mass of calcium carbonate. Within the cells of the coral are numerous cells of symbiotic algae (dinoflagellates). The coral is thus a partnership, with the algae contributing food from photosynthesis and the coral providing the algae with structural support and nutrients, including some of those obtained by the coral's feeding as an animal on the plankton. The collaboration goes further, for filamentous green algae are generally abundant in the coral skeleton, although they are separate from the tissues of the coral animal, and contribute to the secretion of calcium carbonate; some coral reefs are formed primarily by red algae rather than corals. Coral reefs are highly productive communities, for in conditions with water abundant and temperatures high, they combine a stable, depth-differentiated photosynthetic system with means of capturing, conserving, and cycling nutrients from the ocean water with its plankton and particles that moves past. Coral reefs are thus islands of biological abundance in the tropical seas with their impoverished nutrient economies. As in tropical rain forests, the species diversity and richness in form and color and striking adaptation among animals of Pacific coral reefs are legendary. Calcareous byrozoans, worms, and molluscs occur in the coral mass; sea anemones, sponges, and gorgonians (sea plumes) may compete with the corals for space on the coral reef surface; crabs, snails, and echinoderms are abundant. The corals them-

selves are of varied colors, and the daytime fish of the reefs suggest tropical butterflies in their multicolored brilliance and occurrence of warning coloration and mimicry.

33. Marine surface pelagic. The communities of open ocean waters are referred to as *pelagic;* they include the plankton and the larger swimming animals. We shall separate here the surface pelagic communities, in which light supports photosynthesis, from those of deeper waters. In the surface plankton diatoms, small flagellates, and dinoflagellates are major plant groups, and copepods and arrowworms are major animal plankton; but the marine plankton includes a diverse mixture of other animals—shrimps, jellyfish and ctenophores, tunicates, planktonic winged snails, protozoans and so on (Figure 1.1). In coastal waters the plankton is more abundant and includes larval stages of many coastal and bottom animals.

34. Marine deep pelagic communities are those below the lighted surface waters; and these communities are heterotrophic—wholly dependent for food on outside sources, particularly the settling of plankton and particles from the surface waters. Many of the fish and crustaceans swim between the surface and deeper waters and contribute organic matter to the latter by egestion, excretion, and death. The dark pelagic is thus, in its nutritive sources, the shadow of the surface plankton; but it is also distinctive as a range of communities. Carnivorous copepods and other crustaceans are abundant, along with colonial protozoans—foraminifera and radiolaria. Toward greater depths animals confined to deep water are encountered; many of these are of reduced vision or without functional eyes, and some are luminescent or have light-producing organs. Among the latter are bizarre deep-water fish. The predominant colors of animals change, from the blue, green, or transparent forms occurring in surface waters, to red species in middle depths, to prevalence of red and violet invertebrates and black fishes in deep waters.

35. Continental shelf (neritic) benthos. Below the littoral zone, benthic or bottom communities extend with changing composition and character into the depths of the oceans. The influences bearing on the shallow and deep-water benthos are different, for in the shallower waters light penetrates to the bottom, the bottom is in some areas sand, silt, or clay brought by rivers (or rock), and temperatures change with seasons. Where light penetrates to the bottom, extensive algal communities occur; these range from giant kelp "forests," through smaller communities of green, brown, or red algae, to a film of unicellular algae on the bottom. An abundant animal life of fish, crustaceans, and nudibranchs lives among, and in part feeds on, the larger algae where they occur. Clams, snails, crustaceans, echinoderms, worms, and other forms live on or in the bottom itself. The shelf benthos shares some of its species with the lower tidal belt, and others with the deep-water benthos. The depletion of light, land influences, and

shallow-water life toward greater depths is gradual, but the edge of the continental shelf is a convenient boundary.

36. The deep ocean benthos communities extend from the edge of the continental shelf to the deepest ocean trenches. These communities are wholly heterotrophic, composed of animals on and in the mud surface, and bacteria. The "muds" are the oceanic oozes, consisting mainly of the skeletons of organisms or a red or brownish clay of volcanic origin. Calcareous oozes (skeletons of foraminifera, or of planktonic snails—pteropods) predominate in tropical waters, siliceous oozes (radiolarian, or diatom skeletons) in cooler waters, and the red clay in the deepest waters. Some of the animals rise from "roots" in the mud, among them sea whips and plumes, sea fans, sea anemones, sea lilies, sponges, and brachiopods; others (clams, snails, various kinds of worms) are embedded in the mud; still others are motile on the mud surface—crabs and other crustaceans, starfish, brittle stars, sea cucumbers, and sea urchins—or swim above it. The softness of the ooze requires special adaptations in some of these animals—anchorage or "rooting" in some, long appendages that can support them at the surface of the ooze in others. The communities of the deep ocean benthos are not only strange but rich, and their richness seems a product of evolution in defiance of the conditions of permanent cold and dark, intense pressure, and poverty of food reaching the benthos from the surface waters far above.

Major Ecoclines

These biome-types represent in a different way one of the conclusions of Chapter 3: the great diversity of the living world. We should like, though, to find simpler ways of expressing the relationship of community structure to environment. For this we may use concepts from gradient analysis—ecocline and pattern—to relate communities to climate on a world-wide scale. Four ecoclines on land will be described; but the principles apply to aquatic and shore communities, and a gradient of coastal communities (marine rocky shore biome) will be described first.

1. Rocky ocean shores, intertidal levels. A complex rhythm of rising and falling tides alternately submerges shore communities in sea water and exposes them to the air, and from low to high tide levels a gradient of increasing exposure to drying (and other environmental fluctuation) affects organisms. At low tide levels there occurs a dense and rich community of many species of algae and marine animals that are only briefly exposed. Many of these species are shared with the permanently submerged communities of the upper continental shelf. Some of these

species extend upward into a lower mid-tidal belt of varied algae and certain animals (sea anemones, sea urchins, starfish, oysters, and so on) tolerant of longer, but limited exposure to air. Although some of these organisms occur also in upper mid-tidal levels, communities of the latter are dominated by animals more distinctly adapted to longer exposure—notably mussels, barnacles, and limpets. The mussels and barnacles can survive by closing their shells when the tide is out, while a limpet protects itself against drying by the close fit of its shell to a particular place on a rock surface. Of these the barnacles extend farthest upward, and the upper edge of the barnacle community merges into an upper tidal belt in which snails adapted to prolonged exposure are the most abundant animals, with crusts of blue-green algae and lichens on rock surfaces. Organisms of this meager community become increasingly sparse upward, through supratidal levels never submerged but affected by wave splash and salt spray, to rock surfaces that receive salt spray but are occupied by plants and animals of terrestial derivation.

2. Climatic moisture gradient, southern United States—Appalachian forest to desert (Figure 4.9A). Westward from the Southern Applachians rich mixed, broad-leaved deciduous forests of humid climates in the mountains change in character until more open oak-hickory forests are encountered. The stature and density of these decrease until oak woodlands (or in many areas a sharp forest edge produced by fire) give way to prairie. Heights of the prairie grasses decrease westward from tall-grass prairie to the short-grass plains below the Rocky Mountains and the desert grasslands west of the Rockies in New Mexico. The dry grasslands give way to sagebrush semidesert (or to creosote bush desert in southern Arizona). Height and coverage of the desert shrubs decrease toward still more arid climates.
3. Climatic moisture gradient in tropical South America (Figure 4.9B). Tropical rain forests of large, evergreen, broad-leaved trees change into semievergreen seasonal forests in drier climates where the rainfall is deficient during part of the year. Toward drier climates the proportion of evergreen trees decreases while that of deciduous trees increases (in some areas). Along the same gradient the trees become smaller and the canopy more open, until the seasonal forests become woodlands. These in turn become thornwoods, usually of more open structure with grassy undergrowth, and from these the sizes of trees decrease into denser, lower thorn scrub with spinose shrubs and cacti. The thorn scrub opens to become semidesert scrub, and toward still drier climates coverage decreases until the scrub becomes true desert.
4. A temperature gradient toward higher elevations in tropical mountains (Figure 4.9C). The trees of the tropical rain forests decrease in stature from the base of the mountains through the lower montane and upper

montane rain forest as one climbs. The montane rain forest is essentially a temperate rain forest of cooler, but very humid, climates within tropical latitudes. Sizes of trees decrease further into mossy forest or a dense montane thicket with rosette trees and many epiphytes, and beyond this into an elfinwood of small trees and shrubs. Elfinwood of decreasing height gives way to the alpine shrubland of the paramo in South America, with its tussock grasses and distinctive rosette shrubs of the sunflower family. Toward still higher elevations the rosette shrubs may decrease and disappear as the paramo gradates through alpine grassland into the alpine desert of highest elevations.

5. Temperature gradient from the tropics northward in forest climates (Figure 4.9D). Tropical semievergreen forests dominated by broad-leaved trees, some of them deciduous, smaller than the rain forest trees and with fewer epiphytes, change northward through less rich subtropical forests and transitional semideciduous forests to temperate broad-leaved forests. (This sequence can be followed from Venezuela along the Caribbean Islands to Florida and the southern Appalachian Mountains.) Warm-temperate forests of a continental climate, dominated by broad-leaved deciduous trees, occur in the southern Appalachians and were used for the beginning of ecocline 2. Northward these give way to other temperate deciduous forests less rich in species, and these (in the area of the Great Lakes) to forests of deciduous trees mixed with needle-leaved evergreen trees. Still farther northward evergreen trees (spruces and firs) are dominant in boreal forests (taiga) similar to the subalpine forest zone described at the beginning of the chapter. As in the mountains, the forest opens out into treeless vegetation; in the Far North this is the arctic plain or tundra which extends northward to the lands bordering the Arctic Ocean.

There are some conclusions to be drawn from such ecoclines:

1. Along a gradient from a "favorable" environment to an "extreme" environment there is normally a decrease in the productivity and massiveness of communities. The decrease in amount of organic matter per unit area is expressed in decrease of height of the dominant organisms and percentage of the ground surface covered. Thus on land a climatic ecocline may lead from a high forest with a dry-weight biomass exceeding 40 kg/m^2, canopy tree height of 40 m, and coverage (counting overlap of different strata of trees, shrubs, and herbs) well above 100 per cent, to a desert with a biomass less than 1 kg/m^2, plant height below 1 m, and plant coverage less than 10 per cent.
2. Related to these are gradients in physiognomic complexity. Toward increasingly unfavorable environments there is a stepping down of community structure and a reduction of stratal differentation, with generally

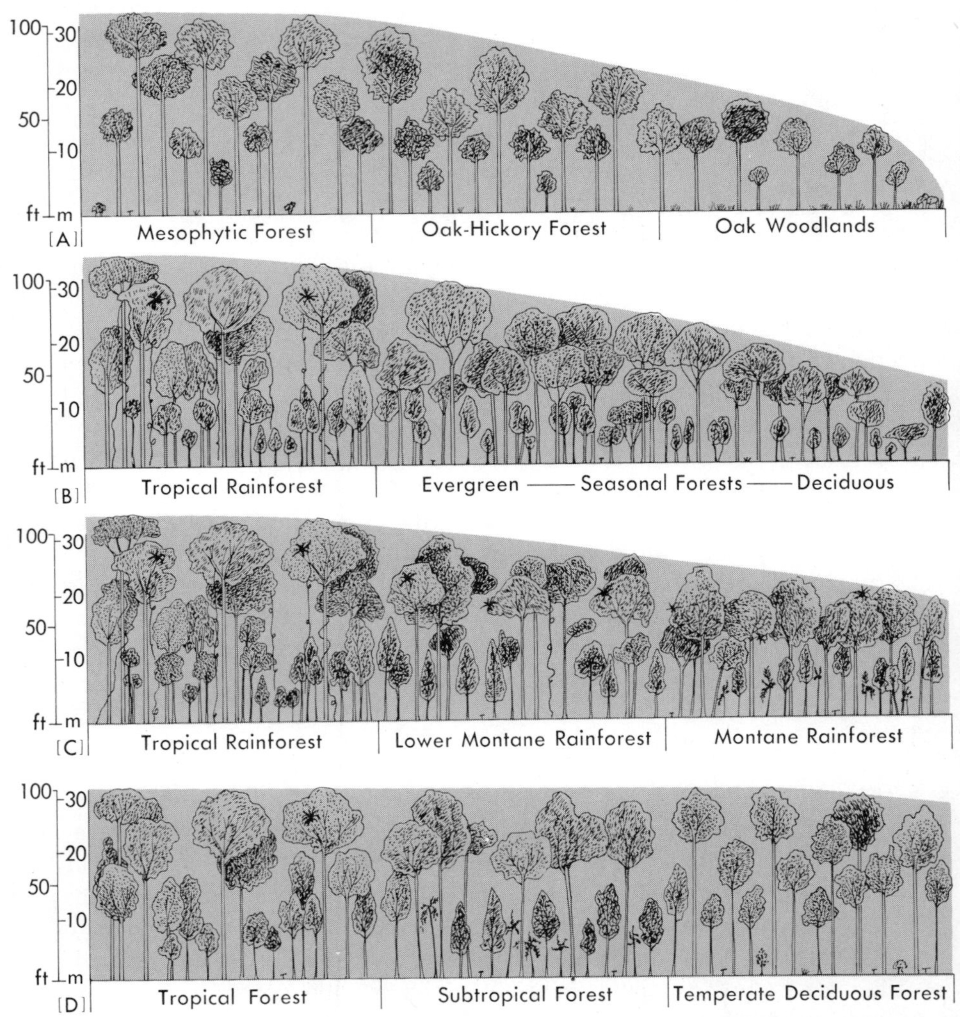

Figure 4.9. Profile diagrams for four ecoclines. **A:** Along a gradient of increasing aridity from mesophytic (moist) forest in the Appalachian Mountains westward to desert in the United States. **B:** Along a gradient of increasing aridity from rain forest to desert in South America. **C:** Along an elevation gradient up tropical mountains in South America from tropical rain forest to the alpine zone. **D:** Along a temperature gradient from tropical seasonal forest northward in forest climates to the arctic tundra. [**B** and **C** are modified from Beard, 1955.]

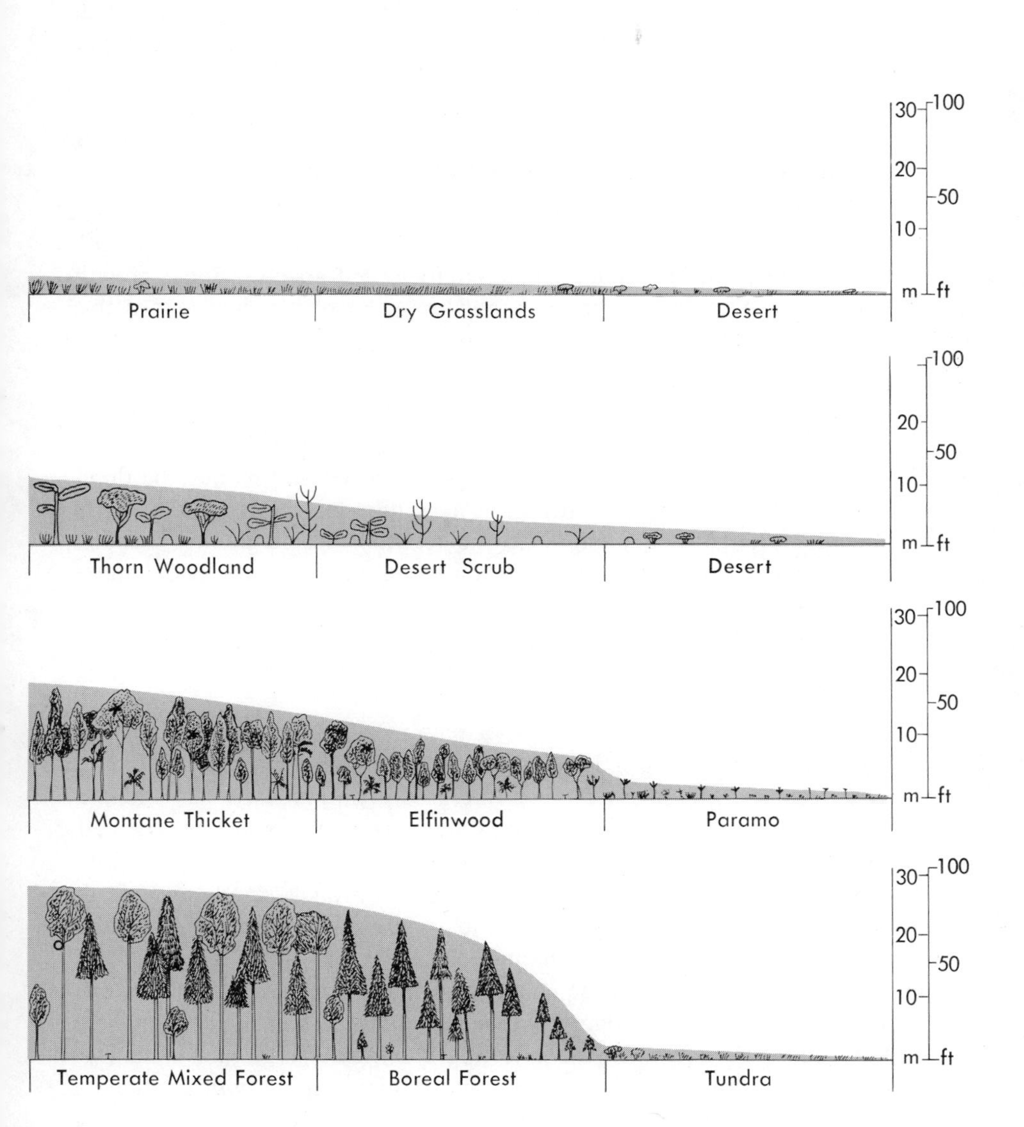

Prairie
Dry Grasslands
Desert
Thorn Woodland
Desert Scrub
Desert
Montane Thicket
Elfinwood
Paramo
Temperate Mixed Forest
Boreal Forest
Tundra
30
20
10
m
100
50
ft

smaller numbers of growth-forms arranged in fewer and lower strata. (But the richness in plant forms of some of the warm semidesert scrubs is an exception to this.)

3. Trends in diversity of structure are broadly paralleled by those in diversity of species. In general (but with exceptions as regards both particular ecoclines and particular groups of organisms), alpha and beta species diversities decrease from favorable to extreme environments, whether the latter are extremes of drought, or of cold, or of adverse soil chemistry, or (for the sea coast) of tidal exposure.
4. Each growth-form has its characteristic place of maximum importance along the ecoclines—rosette trees in some tropical forests, semishrubs in desert and adjacent semiarid communities, and so on. (Some growth-forms, however, such as grasses and grasslike plants, have more than one area of importance along the major ecoclines.) A growth-form, like a species, has dual aspects of adaptation to niche and habitat. Growth-forms are both broad niche categories among plants and broadly significant expressions of plant adaptations to physical environment.
5. The last observation implies that the same growth-forms may be dominant in similar environments in widely different parts of the world. Because of this fact, along with the relationships in points 1 and 2, similar environments on different continents tend to have communities of similar physiognomy, communities that we place in the same biome-type. This adaptive convergence at the level of the community is one of the major generalizations about the geography of life.

We can express this observation as a pattern of biome-types in relation to climate (Figure 4.10). Boundaries between types cannot be located exactly because: (1) Many biomes intergrade continuously. (2) Adaptations of the different growth-forms on different continents are not perfectly convergent. Some of the Australian eucalypt trees, for example, can form forests or woodlands in climates that support only grasslands and shrublands on other continents. (3) Climate is not solely responsible for determining what formation or biome occurs in an area. Different soil conditions or exposure to frequent fires can determine which of the two or more formations, both adapted to the climate of an area, can occur there. (4) A major aspect of climate affecting community structure cannot be included in Figure 4.10; this is the contrast of maritime with continental climates. The same total annual precipitation can support an evergreen forest in the maritime climate of California, a broad-leaf deciduous forest in the continental climate of Missouri. A smaller amount can support shrubland in coastal southern California, pygmy conifer woodland in the continental climate of Nevada, or dry grassland on the Great Plains.

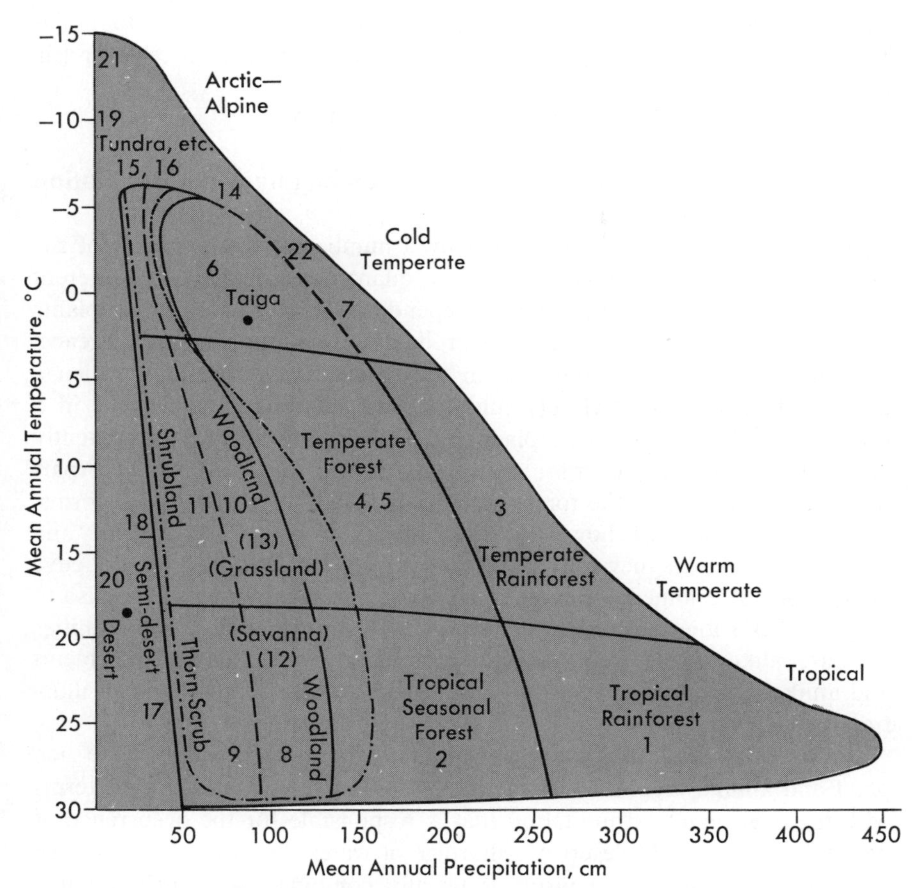

Figure 4.10. A pattern of world biome-types in relation to climatic humidity and temperature. The numbers refer to biome-types described in the text. Boundaries between types are, for a number of reasons, approximate. In climates between forest and desert, maritime versus continental climates, soil effects, and fire effects can shift the balance between woodland, shrubland, and grassland types. The dot-and-dash line encloses a wide range of environments in which either grassland, or one of the types dominated by woody plants, may form the prevailing vegetation in different areas. [Cf. Lieth 1956, Dansereau 1957: 100, and Holdridge 1947, 1967.]

The pattern of Figure 4.10 is a considerable simplification, and yet it expresses some of the broad relations of natural communities, and hence of possibilities for man's use of the land, to world climates (see also the map inside the back cover). Ecoclines 2 and 3 (Figures 4.9A and B) cross the pattern horizontally from forest to desert, ecocline 2 at temperatures near 16°C and ecocline 3 near 25°C. Ecoclines 4 and 5 (Figures 4.9C and D) cross it from bottom to top in rain forest climates and forest climates, respectively. The vegetation pattern for the Santa Catalina Mountains (Figure 4.4) crosses obliquely between the dots indicated in taiga and in desert.

Community-wide Adaptation

Most biomes on land are dominated by a number of plant species of the same growth-form. In many cases these plants are members of different genera and families, and they have independently evolved from other plants of different growth-forms—they are products of evolutionary convergence. Convergence is most conspicuous in the tropical rain forest, in which a hundred tree species of closely similar form and foliage may occur in a forest tract. Competition for a place in the canopy in this most consistently favorable environment exerts a constraint on form in these plants. Apart from the canopy trees, the forest includes a variety of other growth-forms. The trees that succeed, however, are of only a narrow range of forms and leaf types, much as they may differ in soil requirements and reproductive adaptations. Phenomena of evolutionary convergence are impressive also in communities subject to severe environmental problems. Two communities that are alike in the importance of community-wide adaptive problems and unlike in most other respects—the desert and the plankton—can illustrate such phenomena.

Few deserts are really barren of life, and some desert scrubs have rich plant and animal life. All the organisms must in some way come to terms with the environmental limitation that is responsible for the occurrence of desert (excluding cold deserts)—shortage of water. In deserts the shortage of water available to organisms is usually combined with high daytime temperatures and low air humidities, which tend to cause rapid loss of water by evaporation. Desert organisms must maintain a balanced water budget: intake must equal loss if the amount of water in the organisms is to remain sufficient to support life. The balance must be maintained in an environment that makes the balance difficult on both sides—chronic limitation of income combined with continued tendency to overexpenditure by evaporation and other loss.

Adaptations among desert animals include:

1. Increasing water intake by eating plant tissues with high water content, such as cacti, by drinking dew, or traveling to waterholes. Desert toads are able to take in rainwater or dew through the permeable skin of the abdomen while sitting on a wet surface.
2. In some arthropods, direct uptake of water from the air, when relative humidity is over 80 per cent.
3. Efficient use of the metabolic water from respiration of food.

4. Reduction of water loss by excretion and egestion of concentrated urine and nearly dry feces.
5. Impermeable body coverings.
6. Behavior that reduces water loss, such as inactivity and use of shade, and often sheltering underground, during the times of highest temperatures.

Desert arthropods reduce loss by impermeable coverings and accept rise of body temperatures to levels that would be lethal to vertebrates; small mammals must be nocturnal to avoid combined heat stress and water loss, which for them is intolerable above ground in the daytime; large mammals that cannot hide below ground must control body temperature by surface evaporation of water and must replace this water.

Adaptations that appear in different desert plants include:

1. Deep or wide-ranging roots for effective water uptake.
2. Water storage tissues in cacti and other succulent plants.
3. Reduction of water loss from leaves by waxy protective coatings and surface hairs, by gray leaf color in some (reducing the effect of sunlight in heating the leaf), and by effective stomatal function.
4. Reduction of water loss by reducing amount of leaf surface, or by shedding leaves (or branches) in the dry season.
5. Use of the stem surface for auxiliary photosynthesis or all photosynthesis.
6. Modifications of photosynthesis. In some (C_4) plants CO_2 is fixed to 4-carbon compounds in separate chloroplasts from those in which the Calvin cycle of photosynthesis occurs. In these plants photosynthesis can be more efficient to higher temperatures with less water loss than in other (C_3) plants. In some succulents "crassulacean acid metabolism" occurs, with reversed diurnal function of stomata. At night the stomata are open and CO_2 is taken in and fixed (as malate). The CO_2 is then available for photosynthesis during the daytime, when stomatal openings are closed and water loss is less than in other plants.
7. Tolerance of the tissues to reduced water content even, in some club mosses and ferns, to an air-dry condition.
8. Ability, because of the tolerance of water loss and consequent higher osmotic concentration in tissues, to take up water from relatively dry soils.
9. Timing of growth to use water when available, with avoidance of exposure to loss during the dry season.

The last adaptation appears, for example, in annual herbs, which grow rapidly to flower and fruit during the rainy season and spend the dry season as seeds, and in perennial herbs, which survive the dry season as roots that grow above-ground foliage during the rainy season. Survival for the hundreds of species of annual plants of deserts requires systems of inherited metabolic triggers for germination and flowering, neatly adjusted to control the timing of these in relation to amount and duration of rainfall and seasonal temperatures and day lengths.

Protoplasm is a little heavier than water, per unit volume. Organisms of the plankton, which live suspended in the water of lakes and the seas, consequently tend to sink until they fall below the lighted zone that supports most active plankton life. There is, in fact, a steady loss of organisms from the upper plankton by sinking, and the living and dead plankton organisms that sink to the depths of lakes and the oceans are the nutrient rain that supports much of the life of the depths. Plankton organisms can variously solve or reduce their sinking problem:

1. Small size, which slows rates of sinking in bacteria and unicellular algae —the dust-particle approach.
2. In larger forms, development of bristles and projections that reduce sinking rates—the feather approach.
3. Flattened or umbrella-shaped forms that reduce sinking rates as in jellyfish—the parachute approach.
4. Reduction of the specific gravity of the organism by various devices— oil droplets in copepods and diatoms, gas bubbles in blue-green algae and protozoans, large amounts of water in gelatinous sheaths and the tissues of jellyfish.
5. Active swimming upward, which is necessary, despite other adaptations, in most plankton animals.

Two principles can be observed in these community-wide adaptations. First, the problem is normally met by no single device but by some combination of the approaches mentioned. Thus a plankton jellyfish combines parachute form with low specific gravity and upward swimming. A cactus combines stem photosynthesis and low surface area with stomatal and metabolic adaptation to use the CO_2 taken in at night, and with rapid water uptake during rain and storage in succulent tissue. The latter adaptation has required in addition the evolution of spines to provide some protection against animals that eat succulent tissue in adaptation to their own problems of water balance. Adaptation is usually based not on a single device but on a pattern or design of interrelated adaptive devices. Second, the species in a community show some convergences but also the most varied patterns of adaptation, using different combinations of devices to solve the same environmental problem. Such diversity of adaptive pattern is related to the diversity of the niches occupied by the species. Freshwater rapids, sandy beaches, rocky marine shores, and the tundra are other ecosystems with diverse adaptive patterns in response to severe environmental problems.

Succession

As a lake fills with silt it changes gradually from a deep to a shallow lake or pond, then to a marsh, and beyond this, in some cases, to a dry-land forest. When in an area of forests a farm field is abandoned a series of plant communities grow up and replace one another—first annual weeds, then perennial weeds and grasses, then shrubs, and trees—until a forest ends the development. If a landslide exposes a surface of rock in the mountains, the surface may be successively occupied by a sparse cover of lichens; a spreading moss mat; grasses that enter and become a meadow; a shrub thicket that overtops and suppresses the grasses; a first forest stage of smaller trees that seed into the shrub thicket, grow through it, and replace it; and a final stage of larger trees that take dominance from the first trees and may form a larger and potentially permanent forest community.

Such processes of community development are called successions. In the first example the principal cause of the change in the community was a physical process—the filling in of the lake with silt. In the second example, a principal cause was the growth of plants on an existing soil. In the third the succession proceeded by a back-and-forth interplay between organisms and environment: one dominant species modified the soil and the microclimate in ways that made possible the entry of a second species, which became dominant and modified environment in ways that suppressed the first and made possible the entry of a third dominant, which in turn altered its environment. Causes of successional changes are, to varying degrees, external to the community or internal to the community; many successions involve both kinds of causes and reciprocal influences. In any case a gradient of changing environment and a gradient of changing species populations and community characteristics parallel one another. A succession is an ecocline in time.

We can illustrate successions with two very different cases. Bog lakes occur in large numbers in areas of the north-central and northeastern United States and in northern Eurasia. Bog lakes in general are small water bodies that develop in depressions in glaciated areas in a cool, humid, northern forest climate. (Bogs, biome-type 22, are characterized by wet mats of vegetation, in many cases dominated by mosses of the genus *Sphagnum,* and the formation of layers of undecomposed organic soil or peat. Bog communities are of much wider occurrence than the bog lakes we are describing here.) The most distinctive feature of a bog lake is the mat of vegetation, or quaking bog, floating on the water (Figure 4.11). The mat grows in toward the center of the lake, and the open water is thus

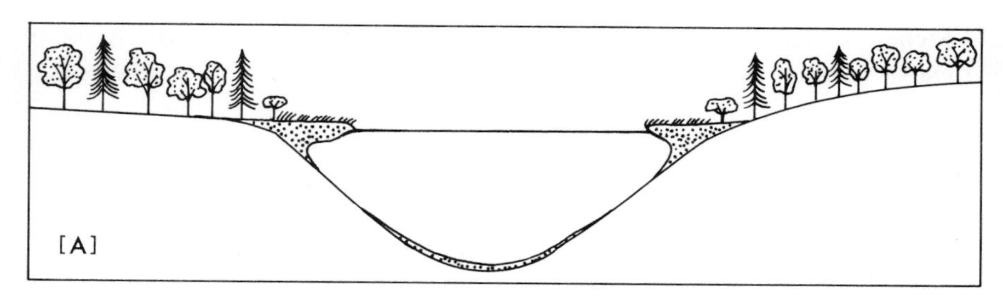

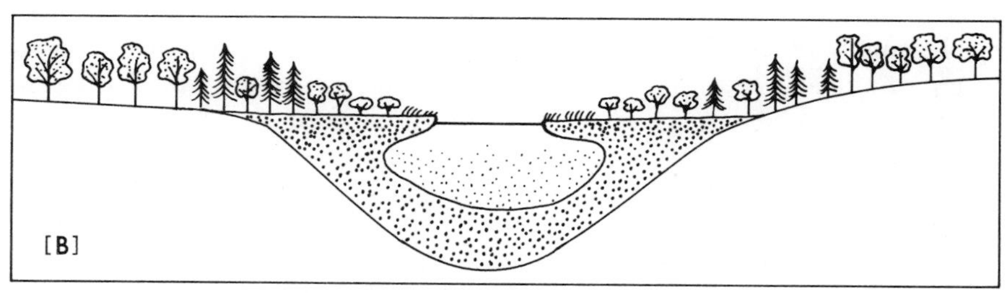

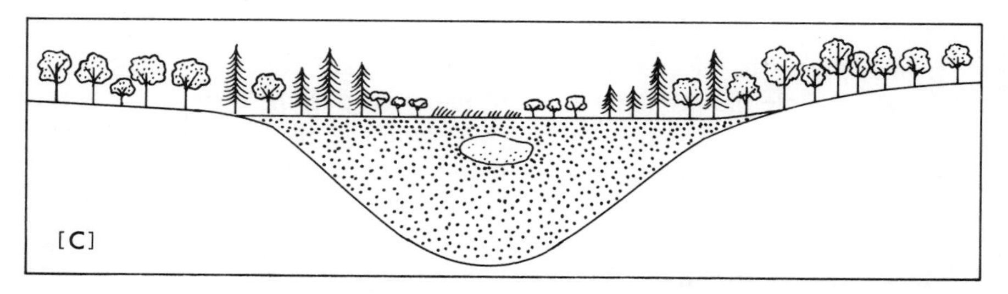

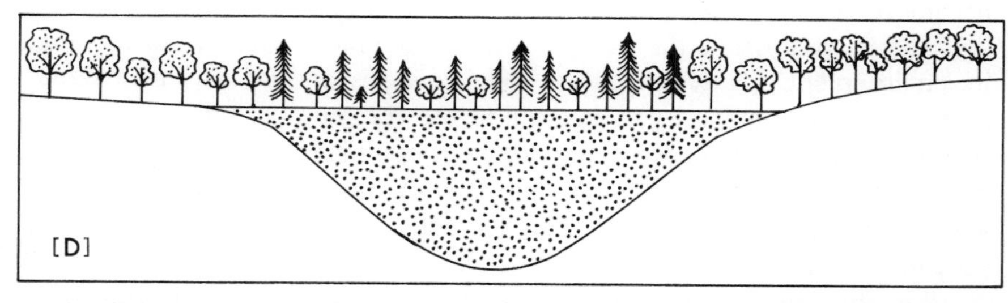

Figure 4.11. A bog lake succession. A floating mat of vegetation advances out over the water surface in a small lake in a cool, humid climate **[A]**. As the mat advances farther and the lake ages **[B]** and **[C]**, scarcely decomposed organic matter (peat) accumulates in the lake basin, until after some thousands of years the lake will be converted to forest **[D]**.

gradually constricted through the centuries, or the few thousands of years since glaciation. Given sufficient time, the lake basin is filled with organic matter and converted to "dry" land.

Although bog lakes differ from one another, we describe a typical bog succession for the north-central United States. The stages of the succession are arranged as rings of vegetation around the lake:

1. The open water of a well-developed bog lake is coffee colored, stained with organic matter or humus derived from the vegetation mat. Light penetration into the water is short, and the water is low in content of inorganic nutrients and in most cases acid (pH 4 to 6). Plant and animal life in the bog water are sparse.
2. Around the open water a floating mat of vegetation is formed by grass-like sedges and the peat mosses (*Sphagnum* spp.) With these occur other distinctive plants—a taller cottongrass, a number of orchids, and carnivorous plants (pitcher plant, sundew) that supplement their limited supply of nutrients by catching and digesting insects. The mat grows, both by increasing its thickness at a given point, and by the growth of its leading edge out onto the water.
3. The thickening of the mat permits its occupation by shrubs, first small shrubs of the heath family less than a meter tall, then a variety of taller shrubs 2 to 4 m tall. Many of the light-requiring plants of the open bog disappear in the shade of the shrubs, and the mat of soil and roots becomes more massive until the bog no longer quakes.
4. The shrub stage is invaded by trees of several species of which the tamarack (*Larix laricina*) is one of the most characteristic. Under the shade of the trees most of the shrubs die and are replaced by a wet-forest undergrowth.
5. These first trees are generally replaced by other species in time. Thus the tamaracks may be replaced by spruces, or by the beeches and maples that dominate the prevailing forests of the area in Figure 4.11.

The bog mat advances by its own developmental processes—the growth of plants, the thickening of the mass of roots and dead remains, the replacement of plants of lower stature by taller and heavier plants. Peat from the mat accumulates on the bottom of the lake, and the lake is thus filled by the growth of the peat upward, as well as by growth of the mat inward and downward. The mat is in time grounded, with a continuous deposit of peat supporting a mature and stable forest.

Dry-land successions can be illustrated for the oak-pine forests of Long Island, New York, which have been subject to extensive clearing for farms and to frequent fires. After either destruction of the forest by fire or abandonment of a farm field, a succession leads back to forest again. Successions that occur on land where a community had previously occurred, and where soil is present to be occupied by new communities, are termed *secondary* successions. Successions that occur on a bare area without soil (such as an area of rock in the mountains) are termed *primary* successions, and they are in general slower than secondary successions because a new soil must be formed. The oak-pine forest at Brookhaven National Laboratory, Long Island, has been intensively studied by G. M. Woodwell and others; and we shall characterize the secondary succession of this forest. There are differences between successions following fire and those following farm abandonment; but the general pattern may be described by stages:

1. In the first and second years annual weeds dominate the newly formed communities; crabgrass (*Digitaria*) is the principal species in the first year, and sorrel (*Rumex*) in the second.
2. After the second year perennial herbs form a meadow in which goldenrods (*Solidago*) are prominent at first, while broomsedge (*Andropogon*) and other grasses are dominant later (until 15 to 20 years).
3. The meadow stage may be followed by a shrub stage, from 15–20 until 30–35 years, dominated first by low shrubs (blueberries, *Vaccinium,* and huckleberries, *Gaylussacia*), then by a taller shrub (bear oak, *Quercus ilicifolia*). In the succession on farmed land the shrub stage may be skipped, and pines may seed into and grow in the meadow; in the succession following fire the meadow stage may be skipped.
4. By 30 to 35 years trees become dominant, first the pitch pine (*Pinus rigida*), that forms a young pine woodland with shrubs, then scarlet and white oak (*Quercus coccinea* and *Q. alba*). By 50 years after the succession began, these form a young oak-pine forest with a well-developed shrub undergrowth.
5. Given time and freedom from fire or other disturbance, this young forest matures into an oak forest with few or no pines and a lower coverage of shrubs. Such a forest might be essentially mature by 200 years after the fire or farm abandonment.

The development of some community characteristics can be followed through this succession. Some of these characteristics—productivity, biomass, and mineral ratio—will be discussed further in following chapters. A most fundamental characteristic of communities is their productivity, or amount of organic matter formed per unit area per unit time. Net productivity is low (175 $g/m^2/yr$) in the first year of the oak-pine succession, increases to a fairly stable level of about 500 $g/m^2/yr$ in the meadow stage, and then increases more steeply through the shrub and young tree stages to 1200 $g/m^2/yr$ in the young oak-pine forest, 44 to 55 years. This level is apparently stabilized and persists into the mature forest (Figure 4.12). Biomass, the total dry weight of organic matter present, increases more slowly toward its stable level, for biomass represents productivity that has accumulated in the community (as plant tissues, particularly woody tissues). Biomass is around 1 kg/m^2 in the annual and meadow stages and increases through the shrub stage to 12 kg/m^2 in the young forest (55 years). It would be expected to increase to 30 to 40 kg/m^2 in the fully mature forest. The growth of the forest community can be expressed also by the ratio of biomass to annual net productivity (biomass accumulation

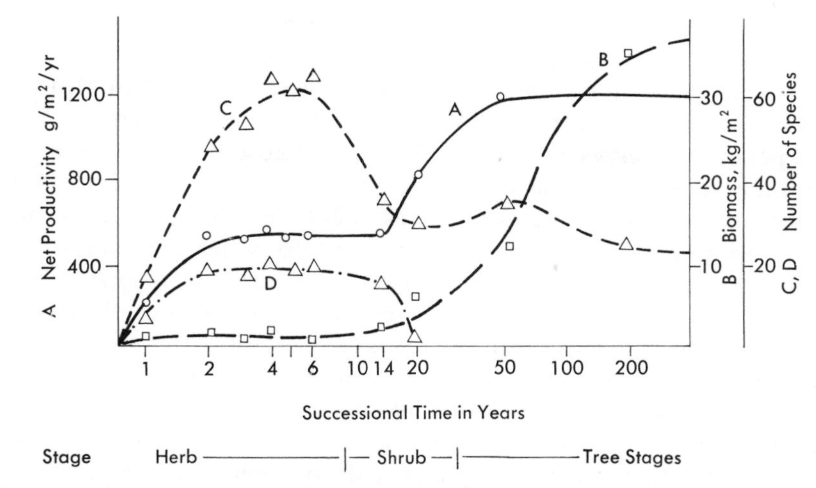

Figure 4.12. Productivity, biomass, and diversity during succession in the Brookhaven oak-pine forest, New York. These three community characteristics tend to increase during many successions, but such trends can be complex in detail. **A:** Net primary productivity increases to a stable level in the herb stages, 2–6 yrs, then increases as woody plants enter the community, 14–50 yrs, to a stable level that may persist into the climax. **B:** Biomass is low through the herb stages and then increases steeply with the accumulation of woody tissues of shrubs and trees; a stable climax biomass is probably not reached until after 200 yrs. **C:** Species diversities—numbers of species in 0.3 hectare samples—increase into the late herb stages, decrease into shrub stages, 14–20 yrs, increase again into a young forest, 50 yrs, and from this decrease into the climax. **D:** Numbers of exotic species; these are present only in the herb and early shrub stages. (Time is on a logarithmic scale to expand the earlier and contract the later part of the succession.) [B. Holt and G. M. Woodwell, unpublished.]

ratio). These ratios increase from about 1.0 in the annual stage, to 2–4 in the meadow stage, to 4–7 in the shrub stage, to 10 in the 55-year forest and probably 25–35 in the mature forest. (No mature forest on the same soil as the successional stands was available for study; the statements about maturity are based on oak forests in other areas.)

As the succession proceeds, an increasing proportion of the community's biomass is above ground. The mass of roots is about three times the above-ground mass in the meadow stage; but in the shrub and young forest stages the above-ground mass is 2 to 3 times that of the roots and in a mature forest it may be 5 to 6 times that of the roots. Surface litter, the dead plant remains on the soil, increases from less than 0.1 to 0.4 kg/m^2 in the annual and meadow stages, to 1.6 kg/m^2 in the young forest, to probably 2.0 kg/m^2 in the mature forest.

Growth in the stock of inorganic nutrients in the plants roughly parallels the growth in plant biomass. There are, however, differences in the nutrient contents of plants in different stages. The earlier herbs, for which nutrients in the soil are abundant, are higher in inorganic nutrient content per unit plant dry weight (mineral ratio about 3.0 per cent) than the later woody plants. The pines, in contrast, have relatively low content of nutrients per unit mass, compared with the oaks of the mature forest (mineral ratios of 0.38 per cent versus 0.64 per cent). The pines, like some other successional trees, are adapted to rapid growth, forming light wood with a fairly low content of inorganic nutrients. The oaks of the mature forest grow more slowly, with heavier use of nutrients per unit dry weight, accumulating as the forest matures a considerable pool of nutrients held in plant tissues. The accumulation and stabilization of this nutrient pool is one of the major implications of forest succession.

Plant populations change and species replace one another as the succession progresses. The appearance of species population densities along the time axis of succession should resemble that along spatial gradients (Figure 4.2), but with a slowing of population change as the community matures. Figure 4.13 illustrates the changes in populations of birds in a forest succession on an abandoned farm field in the Piedmont of Georgia, a succession similar in many ways to that on Long Island. Species diversities of plants and animals tend to increase as the height and complexity of the plant community increases, but the diversity change is not a simple one. At Brookhaven (and probably in many temperate successions ending with a closed forest canopy) plant diversity decreases in the later stages of the succession. The young forest may still have a mixture of species rep-

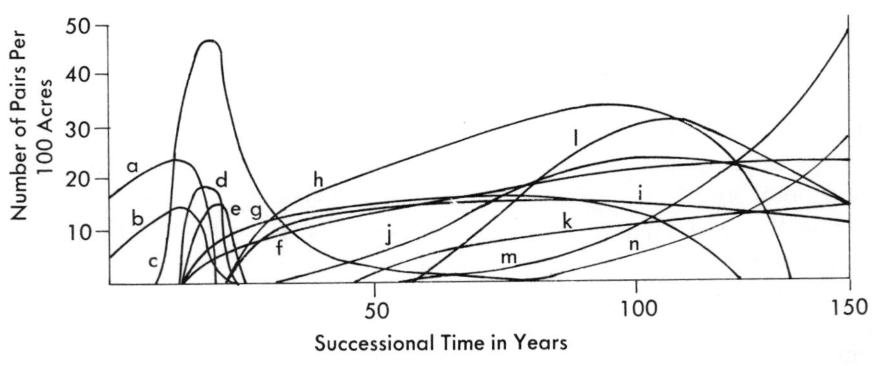

Figure 4.13. Bird populations in successional time. Species with at least 15 pairs per 100 acres (40 ha) are shown for an old field-to-forest succession in the Piedmont region of Georgia: (a) grasshopper sparrow, (b) meadowlark, (c) field sparrow, (d) yellowthroat, (e) chat, (f) cardinal, (g) towhee, (h) pine warbler, (i) summer tanager, (j) Carolina wren, (k) tufted titmouse, (l) hooded warbler, (m) red-eyed vireo, and (n) wood thrush. [Johnston and Odum, 1956, Table 1; and Ricklefs, 1973, Fig. 39.7.]

resenting different successional stages, and it may offer undergrowth plants a wider range of conditions of light and shade, and less intense chemical effects of leaf litter on the soil, than the mature forest.

Diversity and dominance may also fluctuate in earlier stages of succession, as shown in Figures 4.12 and 4.14 for the succession in the Brook-

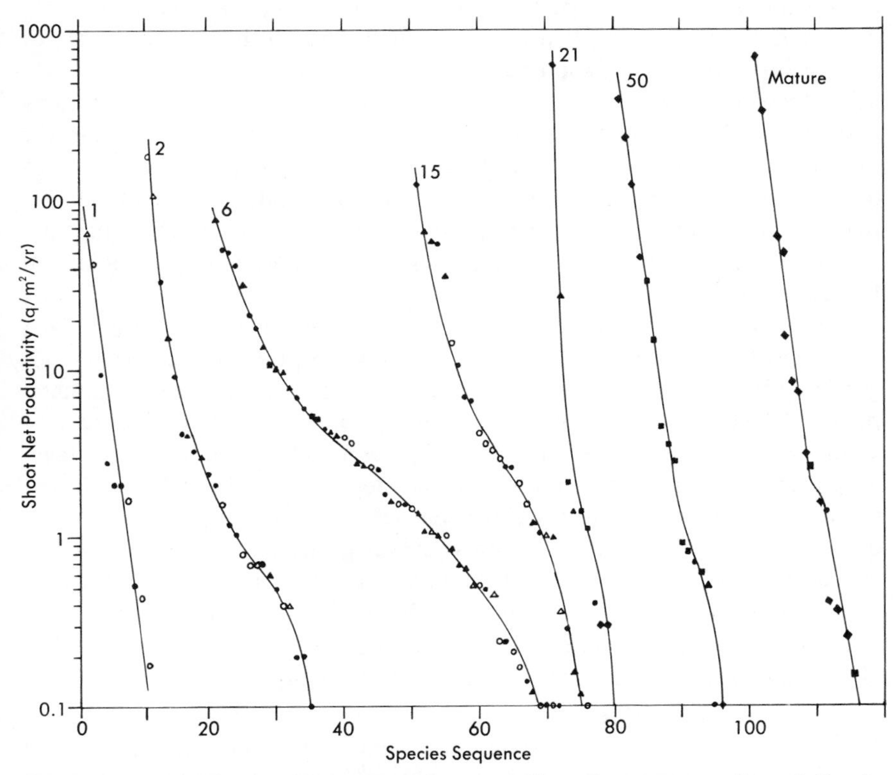

Figure 4.14. Importance-value curves for plants in a forest succession at Brookhaven, New York. Above-ground net productivities of plant species are plotted for the sequences from the most to the least important species, as in Figure 3.14; see also Figure 4.12. As species diversity increases during the herb stages of succession, 1–6 yrs, a steep importance-value sequence approaching a geometric slope (1 yr) becomes a sigmoid, lognormal distribution (6 yrs). Addition to the community of shrubs that overtop and suppress the herbs (15 yrs) leads to a steep importance-value sequence (21 yrs). The young forest (50 yrs) is somewhat richer in species and is less steep. The data for a mature forest are from an oak-hickory forest at Oak Ridge, Tennessee. Circles are herbs, triangles shrubs, and squares trees; open circles and triangles are for exotic and closed ones for native species. The curves start at various points on the species sequence in order to fit them into the same figure. Thus the curve for the mature oak forest starts at point 101 and ends at point 116 on the species sequence. [B. Holt and G. M. Woodwell, unpublished.]

haven forest. The successional meadow is actually richer in species than the forest. Part of this richness in species in the meadow stage, though not the major part, is made up of exotic weed species, introduced to North America from Eurasia. In many cases introduced species become important in unstable successional communities, but are not able to compete with the native plants of mature and stable ones. The first-year old field community has a geometric slope of importance values, species number $S = 10$, and equitability $E_c = 4.3$. Species are added through the first five years of succession, and diversity and equitability rise to $S = 52$ and $E_c = 17.1$. By 15 years shrubs have been added, steepening the curve and by shading and other effects suppressing many of the herb species. By 21 years the diversity has fallen to $S = 11$ and $E_c = 2.7$. In the young forest of 50 years diversity is increasing again with the addition of woody species and shade-tolerant herbs. From the young forest diversity and equitability may decrease into a mature forest with a denser canopy more strongly dominated by scarlet oak.

We have seen that many forest successions involve stages characterized by growth-forms of plants. The pioneer stages (in primary successions, not generally in secondary successions) may be lichens and mosses, or aquatic plants. Whether or not these occur, herbs, shrubs, and temporary tree species replace one another until the succession ends in the final tree stage. Successions that are not in forest climates depart from this scheme. In the more severe environments of arctic and desert communities, extended sequences of species and growth-forms replacing one another do not occur; instead the species that will form the final vegetation seed back into a disturbed area and mature. Despite the differences among successions, however, and the exceptions to any generalizations about successions, a number of trends or progressive developments apply to many successions.

1. There is usually progressive development of the soil, with increasing depth, increasing organic content, and increasing differentiation of layers or horizons toward the mature soil of the final community.
2. The height, massiveness, and differentiation into strata of the plant community increase.
3. The pool or stock of inorganic nutrients held in the vegetation and soil increases, and an increasing fraction of this stock is held in the tissues of the plants.
4. Productivity, the rate of formation of organic matter per unit area in the community, increases with increasing development of the soil and of community structure and increasing use of environmental resources by the community.
5. As the height and density of above-ground plant cover increase, the microclimate within the community is increasingly determined by characteristics of the community itself.

6. Species diversity increases from the simple communities of early succession to the richer communities of later succession or the mature community.
7. Populations rise and fall and replace one another along the time gradient in a manner much like that in stable communities along environmental gradients. The rate of this replacement in many cases slows through the course of succession as smaller and shorter-lived species are replaced by larger and longer-lived ones.
8. The relative stability of the communities consequently increases. Early stages are in some cases of evident instability, with populations rapidly replacing one another; the final community is usually stable, dominated by longer-lived plants that maintain their populations with community composition no longer changing directionally.

Climax

The mature community that ends a succession is called a *climax*. Central to the concept of climax is the community's relative stability. Stability is often an uncertain word in ecology; given the fluctuation of environment, we cannot expect the community to be simply constant in time. What we can expect of the climax is the condition we have termed steady state, as applied to different aspects of the community. First, the species populations of the community should be in steady state, with a balance between birth rate and death rate, between income of new individuals by reproduction and outgo of individuals by death. Given such balance in its populations, the species composition of the climax remains relatively stable. The plant species must be "tolerant" in the sense that they are able to reproduce and grow in the shade (and other effects) of the plants already present. The stability is, of course, relative, allowing for some fluctuation as discussed in Chapter 2. The first three species described in that chapter illustrate different ways climax populations maintain themselves—relatively continuously in the white oak, in a somewhat cyclic pattern in the spruce, and with irregular fluctuation around a mean in the bluestem.

Second, steady states should apply to aspects of over-all community function to be discussed in following chapters. The productivity of the community should be at a steady level, with total or gross primary productivity equaled by total respiration or breakdown of the organic matter formed. With the income and outgo of organic matter in balance, the total living mass of the community (biomass) is in steady state. Paralleling the balance of organic matter is an energy balance—between sunlight energy captured by photosynthesis, and total respiratory energy release. A further aspect of steady state applies to uptake of inorganic nutrients from the soil by roots, and return of nutrients to the soil by litter fall and other processes. When these are in balance, the community's nutrient stock is in steady state. At the same time the soil organic matter itself is in steady state, with the rate of decomposition of soil organic matter balanced by the rate of addition of organic matter to the soil. The climax community is thus to be conceived as an open system in steady state, through which individual organisms, energy, nutrients, and organic matter flow while the community remains relatively constant in time. The climax community is self-maintaining: adapted to essentially permanent steady-state function in relation to its environment, potentially immortal if not disturbed.

Such is the concept, but the application to communities in the field is not so simple. We need to consider how climax communities can be defined, recognized, and interpreted. First, there is the problem of definition, in the sense of the criteria by which the climax community is to be distinguished from the successional communities that preceded it. The climax should be more stable, but this is not always the case. Some successional communities (lichens during an early stage of primary succession on rock, for example) seem to us more stable than some climax communities (such as the prairie grasses). We can say only that the climax is relatively stable compared with the successional stages that led up to it, and that changes in the climax community are fluctuations around a mean, whereas those during succession are trends of directional change. We might define the climax also by the achievement of maximum levels of the successional trends—productivity, biomass, nutrient stock, species diversity, soil development, and dominance by the climax species. The problem is that the maximum levels of these are reached at different times. Maximum productivity and diversity may occur during succession, and the climax species may be dominant before the community has reached its full stature, biomass, and nutrient content. To define climax by soil maturity simply shifts the problem to another part of the ecosystem, and soil maturity is no more easily defined than climax. There are probably some cases in which biomass decreases from late successional communities into the climax. Maximum biomass seems, however, the most generally useful definition of full climax status. In terrestrial communities maximum biomass will normally be based on near-equality of total organic synthesis and breakdown.

Given a landscape of various communities, how is one to recognize the climax? Evidences include:

1. Detailed observations on communities, looking for signs of fire or other disturbance and indications of population stability or instability. Certain species are known to be successional and unable to form self-maintaining communities; other species are known to be capable of climax self-maintenance. (There are also species that are successional in one environment, but climax in another.) In forests, distinctive curves may be formed when numbers of tree stems are plotted against diameter classes (or, if possible, ages). In Fig. 4.15, a self-maintaining popula-

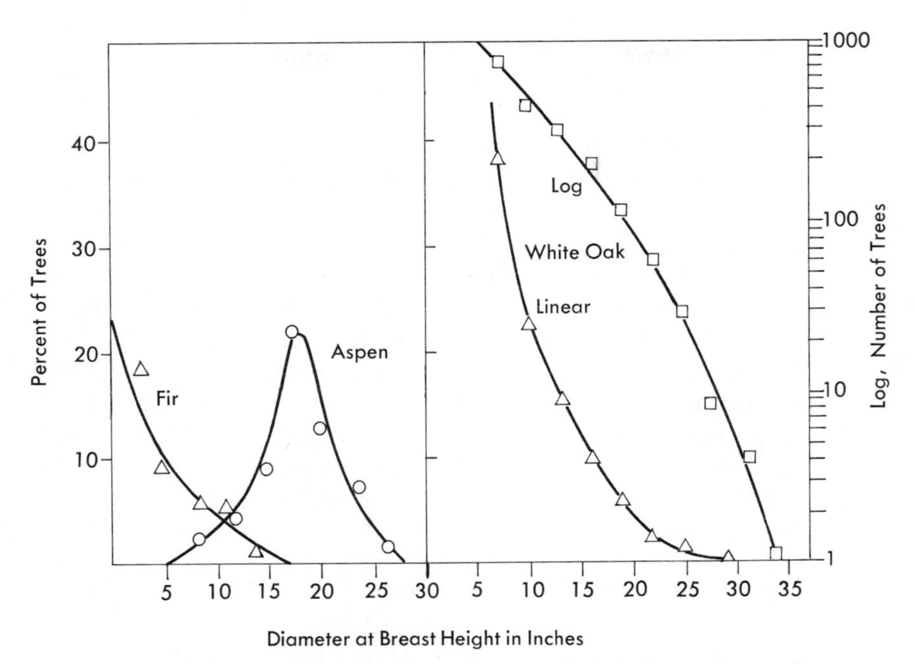

Figure 4.15. Stand curves for a successional and a climax forest. Left: In a successional forest in Arizona mountains aspen (*Populus tremuloides*) is failing to reproduce and is being replaced by fir (*Abies concolor*), which predominates in the smaller and younger trees in the stand. Right: White oak (*Quercus alba*) is maintaining itself in a climax oak-hickory forest. On a linear plot (left) per cents of trees by diameter classes form a descending *J*-curve; on a logarithmic scale (right) numbers of trees in diameter classes form a convex curve. The latter becomes a straight line when tree diameters are replaced by tree ages on the horizontal axis (Figure 2.1). [Left, unpublished data of W. A. Niering and R. H. Whittaker; right, data of Miller, *Illinois Nat. Hist. Surv. Bull.*, **14**:8 (1923).]

tion (the white oak of Chapter 2) is compared with two species in a successional forest—aspen, a temporary forest dominant, and young individuals of white fir, a climax species, which are invading the stand to replace the aspen. The J-curve illustrated by white oak and white fir can occur in either young or mature and self-maintaining forests, but the form of the aspen curve indicates this population is not maintaining itself. In a climax forest the seedlings and young trees should represent the present forest dominants, not other species replacing them.

2. Piecing together successional series. Detailed observations may permit one to fit the different kinds of communities together into successional series. Our account of the Brookhaven succession was based on such fitting together of stands of different ages. In doing this one should recognize that successional communities and climax communities will differ in different kinds of habitats in a given area. When a successional

series is fitted together, however, the climax community can be identified by its maximum stature and biomass for the series and its evidences of population self-maintenance. Climaxes can thus be recognized—if they still exist in the area.

3. Climax convergence in corresponding habitats. The climax community is adapted to its particular environment, its habitat or biotope. Corresponding habitats—such as, southwest-facing slopes on granite between 500 and 1000 m elevation—should support similar climax communities. Successional communities, in contrast, if they have been affected by different kinds and times of disturbance, will differ even in similar habitats. For recognition of climaxes in the field we may use all these evidences. We look for communities that appear undisturbed and are dominated by species capable of, and showing evidence of, self-maintenance, communities that represent maximum biomass and stature in the kind of habitat in question, and that are similar to other such communities in similar habitats in the area.

The broader question of how to interpret a landscape in terms of climax communities involves the manner in which two kinds of observations are accommodated to one another. The first of these is the fact that the successions in an area, even those in quite different kinds of habitats, are to some extent convergent. In a forest area, for example, successions in ponds and bogs, burns and old fields, rock surfaces and river valley silt, will all end in forests. Many successions in a given area converge from disparate beginnings toward similar, though not identical, end stages. From this observation one might infer that all successions in a given area will ultimately converge toward a single climax community, which is determined by the climate of the area and may be termed *climatic climax*. This interpretation (the monoclimax theory of Clements) was of great importance in the development of ecology and is still, with modifications appropriate to the second observation, applied by some ecologists. The modifications must provide for other communities in an area, communities that are not climatic climax but appear stable. These may be termed "subclimaxes," or "proclimaxes" of various sorts, in Clements' treatment.

The second observation is that the vegetation of most areas is complex, including, even when undisturbed by man, a number of types of stable communities. Different stable communities normally occur on south-facing and north-facing slopes, and on rocks that differ in composition and form soils with different characteristics (for example limestone versus granite, or sandstone versus shale). The convergence of successions is partial; substantial differences remain among the climaxes that develop in different habitats. An area will contain a number of kinds of climax communities forming a mosaic corresponding to the mosaic of habitats, as well as various successional communities. In this interpretation (polyclimax theory) the stable and undisturbed community-type that is most extensive in different habitats, including those of intermediate environmental conditions for the area, may be termed "climatic climax." Community stability and regional prevalence are thus recognized separately, but the climatic climax combines these. Other local climaxes occur in habitats of more extreme topographic situations and distinctive soil effects.

There is a third observation to be considered. Undisturbed communities generally intergrade with one another along habitat gradients. Climax vegetation thus does not consist of a definite number of distinct kinds of climaxes. We can state as a third climax interpretation the climax pattern concept. A climax is a steady-state community the characteristics of which are determined by the characteristics of its own habitat or biotope. Habitats and the communities that develop in them intergrade as ecoclines. Despite the occurrence of disturbances and discontinuities, climax communities should be interpreted as not so much a mosaic, as a pattern of intergrading communities corresponding to a pattern of environmental gradients. The pattern will include a range of different kinds of communities, in adaptation to the range of topographic positions and soils in an area. In the pattern, however, one can usually recognize a central or most extensive (steady-state, undisturbed) community-type that comprises the largest share of climax stands in the area and occupies the largest share of habitats that are not special or extreme for the area. This community-type is the prevailing or climatic climax and may be considered to express the climate of the area.

One does not escape some effects of judgment in interpreting climax and successional relationships, yet the concepts have real usefulness. They permit us to relate to one another, along time gradients of succession and along habitat gradients, the varied communities of an area. They help us to understand the developmental relationships of communities in flux under human disturbance, and to relate the disturbed communities to undisturbed ones. They permit better control of variables in research on communities. Thus we can control habitat variables while studying the time variables of succession, by comparing with one another successional communities that occur in a given kind of habitat and are developing toward a particular kind of climax. Also, we can control the time variables of succession while studying effects of habitat, by comparing climax communities (or corresponding successional stages) in different habitats. The concepts permit us also to recognize broader relationships of communities to environment. It is by using prevailing climaxes as means of abstraction from the full complexity of actual community patterns that we can relate communities to climate in major ecoclines and geographic patterns.

SUMMARY

In almost any area of the earth's surface not too much modified by man a complex pattern of natural communities and environments, or habitats of these communities, can be observed. Habitats and communities intergrade along environmental gradients; a gradient of environments together with the corresponding community gradient is an ecosystem gradient, or ecocline. The research technique called gradient analysis deals with relations of gradients of environment, species populations, and community characteristics to one another in ecoclines and in community patterns. When species are studied along ecoclines, their populations have scattered positions along the gradient and broadly overlapping, bell-shaped distributions. Species appear to evolve toward difference of habitat, as well as of niche, and consequently toward scattering of their population centers in relation to environmental gradients. The richness in species of the living world is a product of evolution toward both niche differentiation and habitat differentiation, in the wide range of environments and geographic areas of the world.

The variation in communities of a given area is generally affected by two or more environmental gradients and ecoclines—for example, elevation and the topographic moisture gradient in mountains. The ecoclines can be used as axes in relation to which the communities of the area form a pattern, and in this pattern of communities and ecosystems we can relate to one another the patterns of (1) environmental gradients and habitats, (2) species distributions, which form together a complex population continuum, (3) characteristics of communities, and (4) the types of communities we choose to recognize. Community patterns of this sort are often used to analyze stable, mature, or climax communities only. In any particular habitat in the landscape, however, the climax community may have been destroyed or may not yet have developed. In this habitat the communities go through a progressive development of parallel and interacting changes in environments and communities, a succession. Through the course of succession community characteristics—production, height, mass, nutrient stock, species diversity, and relative stability—and soil depth and differentiation all tend to increase (although there are exceptions). The end point of succession is a climax community of relatively stable species composition and steady-state function, adapted to its habitat and essentially permanent in its habitat if undisturbed.

The complexity of communities in a landscape thus involves a pattern of intergrading climax communities, the continuities of which may be interrupted by local discontinuities of topography and soil, as well as by disturbance and successional communities. It is generally possible to recognize, however, a climax community that is central to the range of variation and is most widespread among the stable communities of the area; this principal community type for the area is the prevailing or climatic climax. The prevailing climax is adapted to the climate of the area, and by means of prevailing climaxes we can relate communities to environments on another level, in terms of broad geographic and altitudinal ecoclines in relation to climates. The gradient from forests to oak woodlands, to grasslands, and finally to deserts along the rainfall gradient westward in temperate climates of the United States is such a major ecocline. Each of the four major community-types along this ecocline, when these are defined by vegetation structure and not by species composition, is a formation or biome. Convergent adaptations to environment appear among the organisms of a given community, and also in the kinds of communities that appear in similar environments in different parts of the world. A grouping of communities of different continents that are of similar structure in adaptation to similar environments is a formation-type or biome-type. The climatic adaptation of natural communities throughout the world can be interpreted in terms of a pattern of biome-types (and of the major ecoclines that connect these) corresponding to a pattern of climatic gradients.

References

Species and Community Gradients

Beals, E. W. 1969. Vegetational change along altitudinal gradients. *Science* **165**:981–985.

Bond, R. R. 1957. Ecological distribution of breeding birds in the upland forests of southern Wisconsin. *Ecological Monographs* **27**:351–384.

*Curtis, J. T. 1955. A prairie continuum in Wisconsin. *Ecology* **36**:558–566.

Curtis, J. T. and R. P. McIntosh. 1951. An upland forest continuum in the prairie-forest border region of Wisconsin. *Ecology* **32**:476–496.

Gleason, H. A. 1926. The individualistic concept of the plant association. *Bulletin of the Torrey Botanical Club* **53**:7–26.

Goff, F. G. and G. Cottam. 1967. Gradient analysis: the use of species and synthetic indices. *Ecology* **48**:793–806.

Greig-Smith, P. 1964. *Quantitative Plant Ecology*. 2nd ed. London: Butterworths. xii + 256 pp.

Knight, D. H. and O. L. Loucks. 1969. A quantitative analysis of Wisconsin forest vegetation on the basis of plant function and gross morphology. *Ecology* **50**:219–232.

McIntosh, R. P. 1967. The continuum concept of vegetation. *Botanical Review* **33**:130–187.

Merriam, C. H. 1898. Life zones and crop zones of the United States. *U. S. Biological Survey Bulletin* **10**:1–79.

RAMENSKY, L. G. 1924. [The basic lawfulness in the structure of the vegetation cover.] In Russian, *Věstnik opÿtnogo dêla Sredne-Chernoz. Obl.*, Voronezh, pp. 37–73; German abstract in *Botanische Centralblatt*, N. F. **7**:453–455, 1962; excerpt in Kormondy, E. J. 1965. *Readings in Ecology*, p. 151–2, Englewood Cliffs, N. J.: Prentice-Hall.

TERBORGH, J. 1971. Distribution on environmental gradients: theory and a preliminary interpretation of distributional patterns in the avifauna of the Cordillera Vilcabamba, Peru. *Ecology* **52**:23–40.

WHITTAKER, R. H. 1951. A criticism of the plant association and climatic climax concepts. *Northwest Science* **25**:17–31.

†WHITTAKER, R. H. 1967. Gradient analysis of vegetation. *Biological Reviews* **42**:207–264.

WHITTAKER, R. H. and C. W. FAIRBANKS. 1958. A study of plankton copepod communities in the Columbia Basin, southeastern Washington. *Ecology* **39**:46–65.

*WHITTAKER, R. H. and W. A. NIERING. 1965. Vegetation of the Santa Catalina Mountains, Arizona. (II) A gradient analysis of the south slope. *Ecology* **46**:429–452.

Ordination and Community Patterns

*AYYAD, M. A. G. and R. L. DIX. 1964. An analysis of a vegetation-microenvironmental complex on prairie slopes in Saskatchewan. *Ecological Monographs* **34**:421–442.

BEALS, E. 1960. Forest bird communities in the Apostle Islands of Wisconsin. *Wilson Bulletin* **72**:156–181.

BRAY, J. R. and J. T. CURTIS. 1957. An ordination of the upland forest communities of southern Wisconsin. *Ecological Monographs* **27**:325–349.

CURTIS, JOHN T. 1959. *The Vegetation of Wisconsin: An Ordination of Plant Communities*. Univ. Wisconsin, Madison: xi + 657 pp.

FAGER, E. W. and J. A. MCGOWAN. 1963. Zooplankton species groups in the North Pacific. *Science* **140**:453–460.

GAUCH, H. G., JR. and R. H. WHITTAKER. 1972. Comparison of ordination techniques. *Ecology* **53**:868–875.

GAUCH, H. G., JR., G. B. CHASE and R. H. WHITTAKER. 1974. Ordination of vegetation samples by Gaussian species distributions. *Ecology* (in press).

*GEMBORYS, S. R. 1974. The structure of hardwood forest ecosystems of Prince Edward County, Virginia. *Ecology* **55**:614–621.

HILL, M. O. 1973. Reciprocal averaging: an eigenvector method of ordination. *Journal of Ecology* **61**:237–249.

LOUCKS, O. L. 1962. Ordinating forest communities by means of environmental scalars and phytosociological indices. *Ecological Monographs* **32**:137–166.

*JAMES, F. C. 1971. Ordinations of habitat relationships among breeding birds. *Wilson Bulletin* **83**:215–236.

NOY-MEIR, I. 1974. Catenation: quantitative methods for the definition of coenoclines. *Vegetatio* (in press.)

WARING, R. H. and J. MAJOR. 1964. Some vegetation of the California coastal redwood region in relation to gradients of moisture, nutrients, light, and temperature. *Ecological Monographs* **34**:167–215.

WHITTAKER, R. H. 1956. Vegetation of the Great Smoky Mountains. *Ecological Monographs.* **26**:1–80.

WHITTAKER, R. H. 1960. Vegetation of the Siskiyou Mountains, Oregon and California. *Ecological Monographs.* **30**:279–338.

†WHITTAKER, R. H., editor. 1973. Ordination and Classification of Communities. *Handbook of Vegetation Science* **5**:1–737. The Hague: W. Junk.

Community Classification

BECKING, R. W. 1957. The Zürich-Montpellier School of phytosociology. *Botanical Review* **23**:411–488.

BRAUN-BLANQUET, JOSIAS. 1932. *Plant Sociology: the Study of Plant Communities.* New York: McGraw-Hill. xviii + 439 pp.

FRYDMAN, I. and R. H. WHITTAKER. 1968. Forest associations of southeast Lublin Province, Poland. *Ecology* **49**:896–908.

HANSON, H. C. 1953. Vegetation types in northwestern Alaska and comparisons with communities in other Arctic regions. *Ecology* **34**:111–140.

JANSSEN, C. R. 1967. A floristic study of forests and bog vegetation, northwestern Minnesota. *Ecology* **48**:751–756.

KENDEIGH, S. CHARLES. 1961. *Animal Ecology.* Englewood Cliffs, N. J.: Prentice-Hall. x + 468 pp.

MILLER, A. H. 1951. An analysis of the distribution of the birds of California. *University of California Publications in Zoology* **50**:531–644.

SHIMWELL, DAVID W. 1971. *The Description and Classification of Vegetation.* London: Sidgwick & Jackson. xiv + 322 pp.

SUKACHEV, V. N. 1928. Principles of classification of the spruce communities of European Russia. *Journal of Ecology* **16**:1–18.

THORSON, G. 1957. Bottom communities (sublittoral or shallow shelf). *Geological Society of America Memoir* **67**, Vol. **1**:461–534.

†WHITTAKER, R. H. 1962. Classification of Natural Communities. *Botanical Review* 28(1):1–239.

†Whittaker 1973.

Formations and Biomes

*BEARD, J. S. 1973. The physiognomic approach. *Handbook of Vegetation Science* **5**:355–386.

DANSEREAU, PIERRE. 1957. *Biogeography: An Ecological Perspective.* New York: Ronald. xiii + 394 pp.

DAUBENMIRE, REXFORD. 1968. *Plant Communities: A Textbook of Plant Synecology*. New York: Harper & Row. xi + 300 pp.

†EYRE, S. R. 1968. *Vegetation and Soils: A World Picture*. 2nd ed. London: Arnold. xvi + 328 pp.

EYRE, S. R., editor. 1971. *World Vegetation Types*. New York: Columbia Univ. 264 pp.

HESSE, RICHARD, W. C. ALLEE and K. P. SCHMIDT. 1951. *Ecological Animal Geography*. 2nd ed. New York: Wiley. xiii + 715 pp.

KENDEIGH 1961.

KÜCHLER, A. W. 1964. *Potential Natural Vegetation of the Conterminous United States*. New York: American Geographical Society, Special Publ. **36.** v + 156 pp., map.

ODUM, EUGENE P. 1971. *Fundamentals of Ecology*. 3rd ed. Philadelphia: Saunders. xiv + 574 pp.

OOSTING, HENRY J. 1956. *The Study of Plant Communities*. 2nd ed. San Francisco: Freeman. viii + 440 pp.

RICHARDS, PAUL W. 1966. *The Tropical Rain Forest: an Ecological Study*. Cambridge Univ. xviii + 450 pp.

RÜBEL, E. 1936. Plant communities of the world, pp. 263–290 in *Essays in Geobotany in Honor of William Albert Setchel,* ed. T. H. Goodspeed. Berkeley: Univ. Calif.

RUMNEY, GEORGE R. 1968. *Climatology and the World's Climates*. New York: Macmillan. x + 656 pp.

SCHIMPER, A. F. W. 1903. *Plant-Geography upon a Physiological Basis*. Oxford: Clarendon. xxx + 839 pp.

TANSLEY, ARTHUR G. 1939. *The British Islands and their Vegetation*. Cambridge Univ. Reprint 1965, 2 vols, xxxviii + 930 pp.

WARMING, E. 1909. *Oecology of Plants: An Introduction to the Study of Plant-Communities*. Oxford Univ. xi + 422 pp.

Major Ecoclines

BEARD, J. S .1944. Climax vegetation in tropical America. *Ecology* **25**:127–158.

*BEARD, J. S. 1955. The classification of tropical American vegetation-types. *Ecology* **36**:89–100.

CARSON, RACHEL. 1955. *The Edge of the Sea*. Boston: Houghton Mifflin. 276 pp.

HOLDRIDGE, L. R. 1947. Determination of world plant formations from simple climatic data. *Science* **105**:367–368.

HOLDRIDGE, L. R. 1967. *Life Zone Ecology*. Rev. ed. San Jose, Costa Rica: Tropical Science Center. 206 pp.

LIETH, H. 1956. Ein Beitrag zur Frage der Korrelation zwischen mittleren Klimawerten und Vegetationsformationen. *Berichte Deutsche Botanische Gesellschaft* **69**:169–176.

Mather, J. R. and G. A. Yoshioka. 1968. The role of climate in the distribution of vegetation. *Annals of the Association of American Geographers* **58**:29–41.

Ricketts, Edward F., J. Calvin and J. W. Hedgpeth. 1962. *Between Pacific Tides*. 4th ed. Stanford Univ. xiv + 614 pp.

Southward, A. J. 1958. The zonation of plants and animals on rocky sea shores. *Biological Reviews* **33**:137–177.

Stephenson, Thomas A. and A. Stephenson. 1972. *Life Between Tidemarks on Rocky Shores*. San Francisco: Freeman. xii + 425 pp.

Troll, C. 1961. Klima und Pflanzenkleid der Erde in dreidimensionaler Sicht. *Naturwissenschaften* **48**:332–348.

Troll, C. 1968. The cordilleras of the tropical Americas. In, Geo-Ecology of the Mountainous Regions of the Tropical Americas, ed. C. Troll, *et al. Colloquium Geographicum,* Geogr. Inst., Univ. Bonn, **9**:15–56.

*Walter, Heinrich. 1973. *Vegetation of the Earth in Relation to Climate and the Eco-Physiological Conditions*. New York: Springer. xiv + 237 pp.

Community-wide Adaptation

Black, C. C. 1971. Ecological implications of dividing plants into groups with distinct photosynthetic production capacities. *Advances in Ecological Research* **7**:87–114.

Brown, George W., editor. 1968. *Desert Biology: Special Topics on the Physical and Biological Aspects of Arid Regions*. New York: Academic. xvii + 635 pp.

Chew, R. M. 1961. Water metabolism of desert-inhabiting vertebrates. *Biological Reviews* **36**:1–31.

†Cloudsley-Thompson, John L. and M. J. Chadwick. 1964. *Life in Deserts*. London: G. T. Foulis. xvi + 218 pp.

*Edney, E. B. 1967. Water balance in desert arthropods. *Science* **156**:1059–1066.

*Hadley, N. F. 1972. Desert species and adaptation. *American Scientist* **60**:338–347.

†Hesse et al. 1951.

Hynes, H. B. N. 1972. *The Ecology of Running Waters*. Univ. Toronto. xxiv + 555 pp.

Idyll, Clarence P. 1964. *Abyss: The Deep Sea and the Creatures that Live in It*. New York: Crowell. xviii + 396 pp.

Jaeger, Edmund C. 1957. *The North American Deserts*. Stanford Univ. vii + 308 pp.

Krutch, Joseph W. 1952. *The Desert Year*. New York: Sloane. 270 pp.

Moore, Hilary B. 1958. *Marine Ecology*. New York: Wiley. xi + 493 pp.

Newell, Richard C. 1970. *Biology of Intertidal Animals*. New York: Elsevier. v + 555 pp.

ODUM 1971, pp. 318–320 on animals of rapids.

RICKETTS et al. 1962.

SCHMIDT-NIELSEN, KNUT. 1964. *Desert Animals: Physiological Problems of Heat and Water*. Oxford: Clarendon. xv + 277 pp.

SHREVE, FORREST. 1951. Vegetation of the Sonoran Desert. *Carnegie Institution of Washington Publication* **591**(1):1–192, and pp. 1–192 in *Vegetation and Flora of the Sonoran Desert,* by I. Wiggins and F. Shreve, Stanford Univ., 1964.

SMAYDA, T. J. 1970. The suspension and sinking of phytoplankton in the sea. *Oceanography and Marine Biology, an Annual Review* **8**:353–414.

SVERDRUP, H. U., M. W. JOHNSON and R. H. FLEMING. 1946. *The Oceans: Their Physics, Chemistry, and General Biology*. New York: Prentice-Hall. x + 1087 pp. (pp. 821–823.)

Succession

CLEMENTS, F. E. 1916. Plant Succession, an Analysis of the Development of Vegetation. *Carnegie Institution of Washington Publication* **242**:1–512.

COOKE, G. D. 1967. The pattern of autotrophic succession in laboratory microcosms. *BioScience* **17**:717–721.

COWLES, H. C. 1901. The physiographic ecology of Chicago and vicinity: a study of the origin, development, and classification of plant societies. *Botanical Gazette* **31**:73–108, 145–182.

CROCKER, R. L. and J. MAJOR. 1955. Soil development in relation to vegetation and surface age at Glacier Bay, Alaska. *Journal of Ecology* **43**:427–448.

†DAUBENMIRE 1968.

DRURY, W. H. and I. C. T. Nisbet. 1973. Succession. *Journal of the Arnold Arboretum* **54**:331–368.

HORN, H. S. 1974. The ecology of secondary succession. *Annual Review of Ecology and Systematics* **5**:25–37.

JOHNSTON, D. W. and E. P. ODUM. 1956. Breeding bird populations in relation to plant succession on the Piedmont of Georgia. *Ecology* **37**:50–62.

*KARR, J. R. 1968. Habitat and avian diversity on strip-mined land in east-central Illinois. *Condor* **70**:348–357.

KEEVER, C. 1950. Causes of succession on old fields of the Piedmont, North Carolina. *Ecological Monographs* **20**:229–250.

KERSHAW, KENNETH A. 1973. *Quantitative and Dynamic Plant Ecology*. 2nd ed. New York: Elsevier. x + 308 pp.

†KNAPP, RUDIGER. 1974, editor. Vegetation Dynamics. *Handbook of Vegetation Science* **8,** 364 pp. The Hague: W. Junk.

MARGALEF, R. 1963. On certain unifying principles in ecology. *American Naturalist* **97**:357–374.

*ODUM, E. P. 1969. The strategy of ecosystem development. *Science* **164**:262–270.

OLSON, J. S. 1958. Rates of succession and soil changes on southern Lake Michigan sand dunes. *Botanical Gazette* **119**:125–170.

WOODWELL, G. M., B. R. HOLT and E. FLACCUS. 1974. Secondary succession: composition of the vegetation and primary production in the field-to-forest sere at Brookhaven, Long Island, N. Y. (unpublished).

Climax

BLUM, J. L. 1956. Application of the climax concept to algal communities of streams. *Ecology* **37**:603–604.

CLEMENTS, F. E. 1936. Nature and structure of the climax. *Journal of Ecology* **24**:252–284.

COOPER, W. S. 1926. The fundamentals of vegetational change. *Ecology* **7**:391–413.

DANSEREAU, P. 1954. Climax vegetation and the regional shift of controls. *Ecology* **35**:575–579.

MULLER, C. H. 1952. Plant succession in arctic heath and tundra in northern Scandinavia. *Bulletin of the Torrey Botanical Club* **79**:296–309.

OOSTING 1956.

SHIMWELL 1971.

TANSLEY, A. G. 1920. The classification of vegetation and the concept of development. *Journal of Ecology* **8**:118–149.

†WHITTAKER, R. H. 1953. A consideration of climax theory: the climax as a population and pattern. *Ecological Monographs* **23**:41–78.

*WHITTAKER, R. H. 1973. Climax concepts and recognition. *Handbook of Vegetation Science* **8**:137–154.

5

Production

The earth is illuminated by sunlight, the energy of which averages 700 calories per square centimeter of the earth's surface per day for all wave lengths on the outside of the atmosphere, about 55 $kCal/cm^2/yr$ in the visible range within the atmosphere at the earth's surface. Man is harvesting organic material as a combine moves across a field of wheat, a trawler lifts its net from the sea, and a logging company works through a forest. Sunlight and harvest are connected by the function of ecosystems in binding energy into organic material—their productivity, on which man, like all animals, is wholly dependent for his life. Productivity seems the most significant single attribute of a natural community. It is appropriate that we consider its measurement, its magnitudes in different communities, its use by plants and other organisms, and its amount for the world as a whole.

Production Measurement

Primary productivity is the rate at which energy is bound or organic material created by photosynthesis, per unit of the earth's surface per unit time; it is most often expressed as dry organic matter in $g/m^2/yr$ ($g/m^2 \times 8.92 = lbs/acre$), or energy in $kCal/m^2/yr$. Productivity, which is a rate, should be clearly distinguished from the amount of organic matter present at a given time, per unit of the earth's surface. The latter is standing crop or *biomass,* and is usually expressed as dry g/m^2 or kg/m^2, or as t/ha. (A metric ton (t) is 10^6 grams, a hectare (ha) is 10^4 m^2; kg/m^2 times ten gives t/ha.) Primary productivity is the consequence of photosynthesis by green plants (including algae in which the green of chlorophyll is masked by other colors). Bacterial photosynthesis and chemosynthesis may also contribute to primary productivity, but these are generally of small significance. The green plants are using for their own respiration part of the organic matter they create. The total energy bound, or organic matter created, by green plants per unit surface and time is their *gross* primary productivity. The amount of organic matter created or energy bound, per unit surface and time, that is left after the respiration of these plants is their *net* primary productivity. Only the net primary productivity is available for harvest by man or other animals. Productivities of heterotrophic organisms —animals and saprobes—in communities are termed *secondary* productivities.

A description of some means of measurement may clarify these concepts. The net primary productivity of a farm field planted in a cereal crop may, for example, be sought. At the end of the growing season the plants, with roots, are removed from a number of sample areas in the field. (Actually, samples should be taken through the growing season to determine loss of old leaves, insect consumption, weed production, and so on.) The plants are dried and weighed by fractions, and the net primary productivity of the field may look like this: Net primary productivity in grams = stems (148) + leaves (72) + flowers and fruits (87) + roots (46) + loss to insects (2) − seed sown (5) = 350 dry $g/m^2/yr$ = 1,500 $kCal/m^2/yr$. (Energy equivalents to dry-weight net production are based on bomb calorimetric measurements for different tissues. Land plant tissues average around 4.25 kCal/dry g, larger aquatic plants 4.5, plant plankton 4.9, and animal tissues around 5.0 kCal/dry g. For gross productivity there are complications affecting the relations of photosynthetic energy to CO_2 assimilation and

ATP bonds, respiration in light and dark, and dry-weight production equivalents that will not be treated here.)

This approach, the harvest method, is simple and reliable when annual plants are in question and these can be fully harvested. The harvest method can also be applied to natural grasslands but with additional difficulties, particularly as regards plant parts (roots, rhizomes, and so on) that live more than one year. Forest plantations, in which the trees have been planted by man and are all of the same age, can be dealt with by the harvest method. Two plantation stands of the same tree species, forty and fifty years old, growing next to one another on the same soil might be studied. In plots of each stand trees are cut and sample roots dug up, and the dry weights of the trees with roots per unit area determined. The difference between these values (which are biomasses) divided by ten is an incomplete expression of mean net primary productivity per year. It must be corrected for annual production and loss of leaves, flowers, and fruits, for animal harvest, for death and loss of branches and roots, and for death of whole trees, if any. Separate measurements may make these corrections possible.

Study of the net productivity of natural forests and shrub communities, which have plants of many different ages, requires a different approach. In the technique of forest dimension analysis:

1. The trees in sample quadrats are measured; diameter and wood growth rate at breast height and tree height are determined for each tree.
2. A set of sample trees is cut and subjected to a detailed analysis, from which are determined the dry weights of fractions—stem wood, stem bark, branch wood and bark, twigs and leaves, root crown and roots, flowers and fruits. By various calculations, the net annual production is obtained for each fraction. Stem volumes, leaf and bark surface areas, chlorophyll content, and other dimensions of interest are determined at the same time.
3. Logarithmic regressions are computed for the sets of trees relating the biomass and production of each fraction to the diameter at breast height, or to other dimensions of the trees.
4. The regressions are used to compute, on the basis of its diameter, the probable biomass and production of each tree in the sample quadrat measurements of step 1. These values summed give biomass and production of trees per unit area in the forest.
5. Biomass and production of shrubs and herbs of the forest undergrowth are separately determined. Larger shrubs can be measured with the same procedures as trees. Measurements for small shrubs and herbs may be based on clippings of their current growth in small sample areas and ratios of current growth weight to plant biomass and production. From these steps result such forest biomass and net productivity values as are illustrated in Table 5.1.

Table 5.1 Net Productivity and Biomass in Temperate-Zone Forests—a Young Oak-Pine Forest at Brookhaven, Long Island, New York, and a Climax Deciduous Cove Forest, Great Smoky Mountains National Park, Tennessee. [Brookhaven data from an intensive study by Whittaker and Woodwell (1968, 1969), cove forest values estimated by Whittaker (1966).]

	Oak-Pine Forest		*Cove Forest*	
	Net Productivity	*Biomass*	*Net Productivity*	*Biomass*
Totals, net productivity ($g/m^2/yr$) and biomass (kg/m^2), dry matter, for trees	1060	9.7	1300	58.5
Totals for undergrowth	134	0.46	90	0.135
Percentages of totals for trees in				
Stem wood	14.0	36.1	33.3	69.3
Stem bark	2.5	8.4	3.7	6.3
Branch wood and bark	23.3	16.9	13.1	10.3
Leaves	33.1	4.2	29.1	.6
Fruits and flowers	2.1	.2	1.8	.03
Roots	25.0	34.2	19.0	13.5
Biomass accumulation ratio	8.5		43.5	
Leaf area ratios, m^2/m^2	3.8		6.2	
Bark area ratio, m^2/m^2	1.5		2.1	
Chlorophyll, g/m^2	1.9		2.2	
Total respiration/gross productivity	0.80		1.0	
Age of canopy trees, yrs	40–45		150–400	
Mean tree height, m	7.6		34.0	

The net productivity of 1200 dry $g/m^2/yr$ for trees and undergrowth together is fairly typical for temperate forests. Mature forests have larger fractions of their productivity in stems, and smaller fractions in branches, twigs and leaves, and roots, as indicated in estimates of these for a cove forest in Table 5.1. A value for gross productivity is also desired, and this may be determined by gaseous exchange measurements. Twigs with leaves are enclosed in transparent plastic chambers, and the CO_2 content of the air entering and leaving the chambers is measured. CO_2 uptake must be corrected for estimated CO_2 release by respiration at the same time. By these measurements, converted from the twigs in chambers to the full mass of foliage on all levels in the forest, and integrated through the year, a total photosynthetic activity or gross primary productivity can be estimated. The gross productivity of the oak-pine forest is calculated as

$$\text{Gross productivity} = \text{net primary productivity} + \text{plant respiration}$$
$$2650 \qquad\qquad 1200 \qquad\qquad 1450\ (\text{g/m}^2/\text{yr})$$

The forest plants thus consume in respiration 55 per cent of their gross productivity; the remaining 45 per cent is available for harvest by animals or decomposition by bacteria and fungi. For the cove forest net production of 1390 $g/m^2/yr$ and an estimated plant respiration of 60 per cent give gross primary productivity of 3500 $g/m^2/yr$.

In some circumstances diurnal curves of change in the CO_2 level in the environment provide an alternative approach. In the oak-pine forest, for example, the air is on some nights trapped beneath an inversion (of temperature increase with height above the ground). CO_2 released by respiration accumulates beneath the inversion, and this accumulation can be measured and its rate related to temperature. By integration of the rates in relation to temperature for different seasons around the year, an approximation of respiration for the whole community is obtained. The value for the oak-pine forest is 2110 $g/m^2/yr$. Estimation of respiration by plants from the gaseous exchange chambers in the forest is 1450 $g/m^2/yr$; the remaining respiration, by animals and saprobes, is consequently 660 $g/m^2/yr$.

Up to 8 per cent of the weight of leaves in this forest is eaten by insects. Since the leaves are 33 percent of net productivity, 8 per cent of leaves consumed is less than 3 per cent of net productivity harvested by leaf-eating insects. In many forests the fraction of leaves eaten by insects is smaller, around 1 to 5 per cent of leaf weight. There is some additional harvest of shrubs by browsing deer, of roots by burrowing animals, of wood by boring insects, and of leaf litter on the ground by soil animals. It is likely that less than 10 per cent of net productivity in the Brookhaven forest is directly consumed by animals, however; whereas the remainder is used by saprobes (bacteria and fungi) or remains as accumulated biomass. We combine these measurements and estimates into a production balance for the forest:

Gross Productivity		Plant Respiration		Saprobe Respiration		Animal Respiration		Biomass Accumulation
2650	=	1450	+	580	+	80	+	540
								($g/m^2/yr$)

Biomass accumulation is the growth in organic matter of the community made possible by the difference between gross primary productivity and total community respiration. This difference is also referred to as *net ecosystem production,* but the latter term is to be clearly distinguished from net primary production. In a climax community such as the cove forest of Table 5.1 net primary production may be high (1390 g/m^2/yr); but net ecosystem production is zero or near zero, for gross primary productivity and total community respiration are essentially equal.

The gas exchange approach can be applied to plankton communities using the light and dark bottle technique. Two bottles, one transparent and one opaque, are filled at a given depth with water containing plankton organisms, closed, maintained at that depth for a time, and brought to the surface to determine the oxygen content of the water. The decrease of oxygen in the dark bottle over the amount in the same volume of free water at the beginning of the experiment represents the respiration by plankton organisms in the bottle. The increase of oxygen in the light bottle represents the release of oxygen in photosynthesis (reduced in measured amount, however, by the fact that the plankton organisms are at the same time using a smaller amount of oxygen for respiration). The sum, oxygen increase in the light bottle plus the amount of decrease in the dark bottle, expresses gross productivity. (The oxygen sum must be multiplied by 0.375 to give an equivalent carbon assimilation, and values must be integrated for different depths, around the daily cycle, and around the year. Growth of and oxygen use by bacteria on the surfaces in the bottles imply increasing error with increasing length of time the bottles are closed.)

Plankton production can be approached also by measurements using CO_2 labeled with radiocarbon, ^{14}C. Plankton samples in water of known CO^2 content are provided with $^{14}CO_2$ and kept at appropriate depths or under controlled light intensities. By separating the plankton from the water and measuring the ^{14}C content of the plankton, carbon uptake during a given period can be determined. The results are complicated by the movement of some of the radiocarbon back into the water as $^{14}CO_2$ from respiration and as organic compounds. The ^{14}C technique consequently gives a measurement between gross and net productivity, closer to gross productivity for short periods (one to three hours), approaching net productivity for longer periods (one to two days). The technique has become the standard means of measuring the productivity of marine plankton, for it is more sensitive than the light-and-dark bottle technique. The diurnal curve approach can also be applied to some aquatic communities. Estimates of production based on chlorophyll content of communities combined with carbon assimilation rates per unit chlorophyll at different light intensities are possible, but are made uncertain by wide differences in assimilation rates in different species and in foliage on different levels of a given community.

Because the energy of gross primary productivity is the basis of all life activity, measurement of that energy is desired. Measurements of gross primary productivity (or of plant respiration, which can be added to net primary productivity to obtain gross primary productivity) are generally difficult. Some results from measuring respiration in land plant communities are shown in Figure 5.1. The fraction of gross primary productivity expended in respiration increases with increasing temperature and biomass in land communities. Some increase with temperature would be expected from effects of temperature on metabolic rates. The increase with biomass results from the increasing ratios of respiring and nonphotosynthetic woody tissue to photosynthetic leaf tissue, in woody plants of increasing size. Respiration in plankton communities appears to be of the order of 30 to 40 per cent of gross primary productivity. The far more numerous measurements of net primary productivity will be summarized below. Despite the interest of gross primary productivity and energy flow, net primary productivity in dry matter g/m²/yr has come to be the focus or hinge-value of production research. From this value other measurements can radiate: gross primary productivity by way of respiration, energy of productivity by way of caloric equivalents, productivity as carbon fixed (used in many aquatic studies) by way of organic dry matter/carbon ratios, community biomass by way of accumulation ratios, nutrient cycling by way of

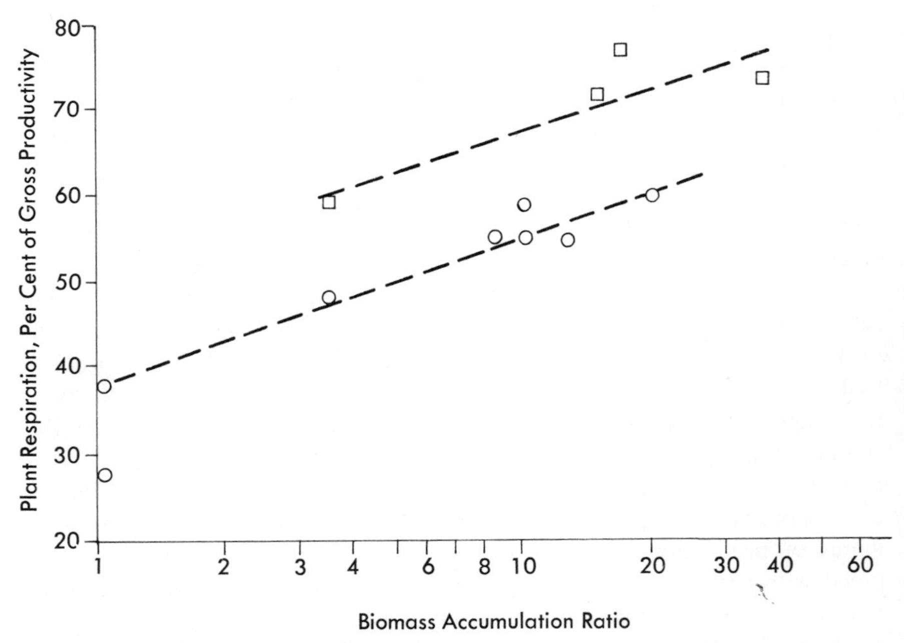

Figure 5.1. Plant-community respiration in relation to biomass accumulation ratio (biomass/net annual primary productivity, above ground, on a logarithmic scale). The circles are for temperate and the squares for tropical communities. The trend lines show increasing fractions of gross primary productivity used for plant respiration from agricultural fields on the left (biomass accumulation ratio treated as 1.0) to forests (biomass accumulation ratios from 10–40) on the right, and higher fractions of productivity spent in respiration in the higher temperatures of the Tropics.

content of elements in organic matter, and secondary productivity by way of animal harvest.

Secondary productivities, of animals and saprobes, are considerably more difficult to obtain than primary productivities. Secondary productivity is defined as the formation of new protoplasm by heterotroph populations, as measured in dry grams of organic matter (or equivalent energy) per unit area and time. Note that the 8 per cent of leaf weight eaten by animals in the Brookhaven forest is not itself a secondary productivity, for only a small fraction of this becomes actual increase in animal organic matter. Of the food ingested by an animal, some is egested unused and has no part in that animal's productivity. In a caterpillar this unused food may be 80 to 90 per cent of the total. Of the food that is digested and assimilated, some fraction is spent on respiration in support of metabolism and activity, and this fraction does not appear as new organic matter. The remaining fraction is incorporated into the organic matter of the animal as secondary productivity, either as growth of an existing individual, or as part of reproduction to form new individuals. The major relationships are:

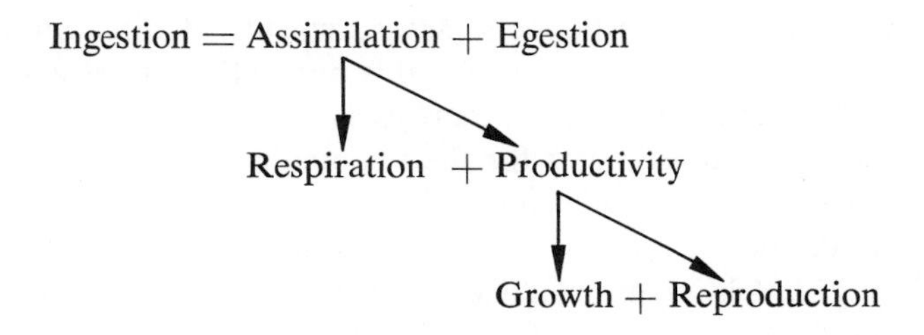

The organic matter in excretion can be difficult to assign in this scheme, because this can be excreted with or without having been part of organic syntheses. Birds and mammals, and some other animals, reach mature sizes at which they cease to grow. When these animals are mature, their total assimilated food energy is spent on respiration in support of life activities, including the maintenance of body temperature in warm-blooded animals; and there is no growth. Secondary production of such an animal as an individual, when not reproducing, is zero. A population, however, includes growing and reproducing individuals; and a population will usually have a measurable secondary productivity. Secondary productivity corresponds to net productivity for plants, but the terms *net* and *gross* are not usually applied to secondary productivity. The equivalent of the gross productivity of plants is total assimilation in animals.

Sometimes an animal population consists of one or more well-defined age classes whose growth can be followed. If, for example, a population of fish of the same age at the time of stocking a pond is observed, then the mean weight increase per fish from one year to the next, times the population density, represents the secondary productivity of the fish. If part of the fish population dies between the two measurements, this loss must of course be corrected for. For many animal populations this approach, which corresponds to the harvest method for plants, is impossible. In most research on secondary productivity, each major animal species must be studied in detail, seeking measurements of population density and age composition, food consumption and utilization, and growth and reproduction. Some of these are very difficult to measure in the field, and secondary productivity estimates are often based on a combination of field population measurements with laboratory studies of metabolism and growth.

For example, a grasshopper population has a density of 3 individuals per m^2 and these individuals (67 mg each) give a biomass of 200 mg/m^2 in the field. Laboratory measurements indicate that a grasshopper eats 0.28 g of grass per g of body weight per day, and that 62 per cent of food eaten is egested and 38 per cent assimilated, with the latter including 34 per cent respired and only 4 per cent becoming growth and secondary productivity. The grasshopper population's secondary productivity is $200 \text{ mg} \times 0.28 \times 0.04 = 2.2 \text{ mg/m}^2\text{/day}$. All these measurements are variables with time and age of the grasshoppers. Measurements at different times were combined to obtain annual values for this population: total assimilation of 1.1 $g/m^2/yr$ and secondary productivity of 0.12 $g/m^2/yr$. With similar measurements and estimates for the remaining species in the community, an old-field grassland in Tennessee, it was found that the arthropods consumed 10.6 per cent of the above-ground net primary productivity of 270 $g/m^2/yr$ (1274 $kCal/m^2yr$). The use of the 29 $g/m^2/yr$ consumed for secondary productivity of these herbivores (and carnivores feeding on them) was:

Herbivore				Carnivore
Ingestion	Egestion	Respiration	Secondary Productivity	Secondary Productivity
29	15	8	6	0.4 $g/m^2/yr$
135	65	38	32	2.3 $kCal/m^2/yr$

Secondary productivities of bacterial and fungal populations are also desired. It is possible to measure or infer their respiration, but determining the growth and reproduction of microscopic cells and filaments dispersed in the soil or water of natural communities is exceedingly difficult. It is convenient to end this section rather than discussing further such measurements, the problems of which have not really been solved.

Water, light, carbon dioxide, and soil nutrients are necessary for production on land, and temperature and successional processes affect rates of productivity. CO_2 is normally available in amounts around 0.03 per cent of the gases in the atmosphere; small variations in CO_2 content of the air are not known to have significant effects on terrestrial productivity. Sunlight varies in intensity, quality, and duration with latitude, altitude, and climate, but apart from length of growing season, differences in sunlight are thought less significant than other factors affecting productivity. Of concern are primarily the effects of moisture and temperature, and secondarily those of nutrients and succession.

Plants on land need large amounts of water for transpiration because the stomata must be open if CO_2 is to be taken in, and water will be lost if the stomata are open. Water is needed also as part of protoplasm and for metabolic reactions including photoysnthesis, but much the largest part of the water taken up by the plant is lost by transpiration. Land plants seem extravagant, in fact, in their use of water for transpiration. Many plants transpire 700 to 1000 g or more of water for every gram of net production. Some plants of dry environments have special photosynthetic adaptations (C_4 and crassulacean acid metabolism) and make do with less water loss—50 to 300 g per g of net production—but even these seem "efficient" in water use only by comparison with other land plants. The availability of water for such loss is a major determinant of productivity on land. In arid climates there is a nearly linear increase in net primary productivity with increase in annual precipitation (Figure 5.2, below 500 mm precipitation). In more humid forest climates the linear relation no longer applies, for there is a plateau beyond which productivity shows little increase, if any, with increased precipitation. The productivity supported by a given amount of precipitation may be affected by seasonal distribution of the precipitation, by mean temperature and the annual temperature cycle, and by nutrient levels and successional status. These and other factors produce the wide scattering of points in Figure 5.2. Because the rate of loss of water from the soil by evaporation and transpiration is less in a cool climate, a given amount of rainfall can support vegetation of less arid character in a cool climate than in a warm one. Primary productivity can be effectively related also to the actual evapotranspiration, the total amount of water that leaves the soil by plant transpiration plus evaporation from the soil surface (Figure 5.3). In a dry climate evapotranspiration may essentially equal precipitation; in wet climates evapotranspiration is less than precipitation but increases with increasing temperatures.

Many mature, stable temperate forests of favorable environments have net primary productivities of 1200 to 1500 $g/m^2/yr$. Some forests of cold or dry environments produce less than that, some young forests, and prob-

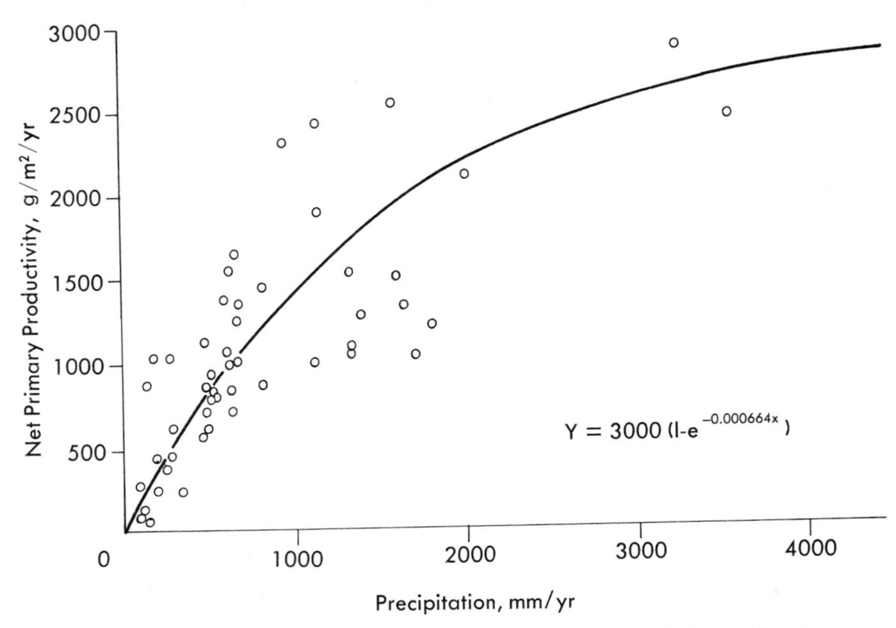

Figure 5.2. Net primary productivity, above and below ground, in relation to mean annual precipitation. [Lieth, *Human Ecol.,* **1**:303 (1973).]

ably some swamp and flood-plain forests, produce more. Many tropical forests probably produce more, up to and in some cases beyond 3000 $g/m^2/yr$. We can take 1000 to 2000 $g/m^2/yr$ as a "normal" range of net primary production in a favorable climate on land. (Forest is the normal vegetation of a favorable climate on land, unless that vegetation is affected by special habitat effects or disturbance.) We can then generalize about terrestrial net primary productivity in terms of four ranges. The normal range of 1000 to 2000 $g/m^2/yr$ includes many forests and some grasslands and highly productive temperate crops. Most non-forest communities, however, are limited to lower productivities by drought, or cold and related climatic factors, or in some cases by nutrient conditions. The middle range, 250 to 1000 $g/m^2/yr$ includes many kinds of communities thus limited—woodlands, shrublands, and grasslands, together with most cereal crops. The low range of more severe limitation of productivity, 0 to 250 $g/m^2/yr$, includes desert, semidesert, and part of the arctic tundra growing in less favorable situations. At the opposite extreme is a very high range of 2000 to 3000 $g/m^2/yr$, or somewhat more. This range includes some rain forests, marshes, and successional communities of favorable environments, as well as some of man's communities of intensive cultivation—notably sugar cane and rice. These communities all have their reasons for very high productivity, for they occur in environments that in different ways

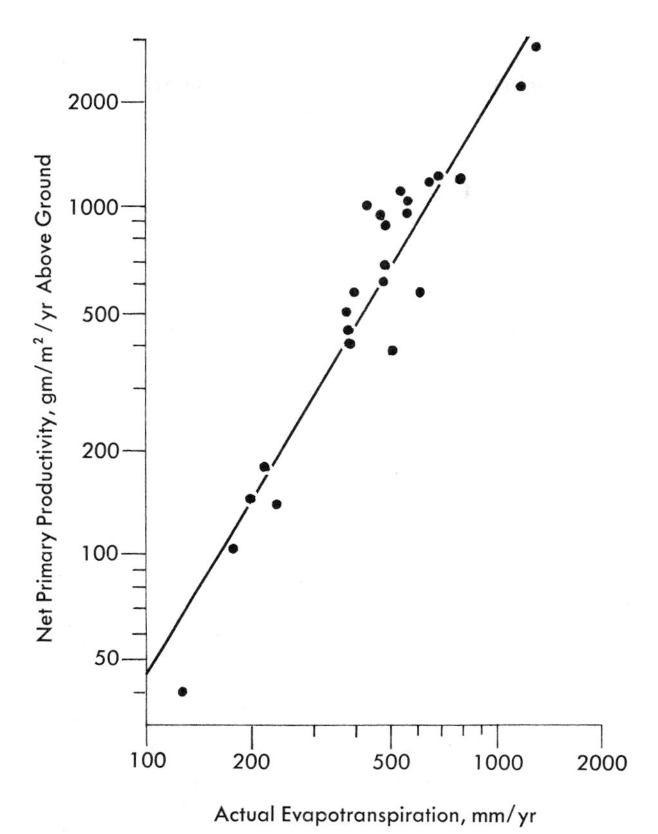

Figure 5.3. Net primary productivity in relation to actual evapotranspiration. [Rosenzweig, 1968.]

combine high moisture availability, relatively warm temperature, and continuing nutrient replenishment.

Productivities decrease along the temperature gradient from the Tropics to the Arctic. Figure 5.4 is an effort to produce a generalized fit of terrestrial net primary productivity to mean annual temperature. As in Figure 5.2, a normal ceiling of about 3000 $g/m^2/yr$ for net primary productivity in natural communities is used as one basis of the fit. As in Figure 5.2, a curve of uncertain form is fitted to points that are widely scattered in response to factors other than temperature.

In the middle of the curve of Figure 5.4, productivity somewhat more than doubles with an increase of 10°C in mean annual temperature. Decreasing length of the growing season toward colder climates may have as much to do with this as temperature itself. The curve of Figure 5.4 may not express the rates of van't Hoff or Arrhenius relations of metabolic reactions to temperature. In each temperature zone the organisms have

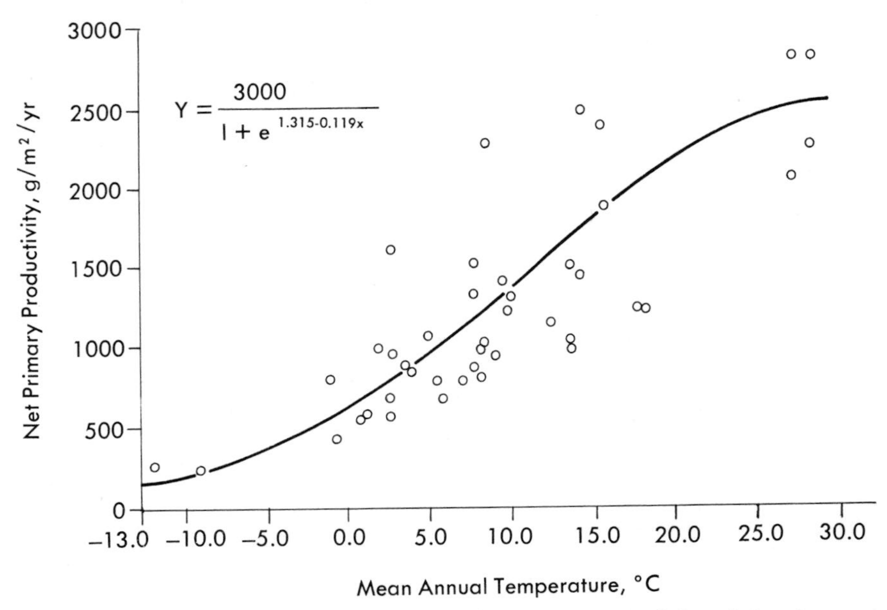

Figure 5.4. Net primary productivity, above and below ground, in relation to mean annual temperature. [Lieth, *Human Ecol.*, **1**:303 (1973).]

evolved adaptations of metabolic rates and life-cycle timings to the temperatures and seasonal cycles they encounter. When communities that are adapted to different temperatures but are otherwise similar are compared, the increase in the combined metabolism of the plants of the community—primary productivity—with a 10°C increase in mean annual temperature may be much less than the two- to threefold increase stated by the van't Hoff relation. Figure 5.5 illustrates more detailed relations of forest net production to altitude and temperature in one area, the Great Smoky Mountains. From these data it appears that:

1. Net productivities of stable forests of relatively favorable moisture conditions, curves *A* and *B*, are convergent over a wide range of warm-temperate to cool-temperate climates.
2. There is a point beyond the latter at which productivity decreases more rapidly toward the low values of alpine and arctic climates. Because of this, mean productivity approximately doubles from the highest-elevation to the lowest-elevation forests studied.
3. Although net productivity of climax deciduous and evergreen forests is similar in warm-temperate climates, in the transition to the arctic and alpine climates evergreen forests (taiga, biome-type number 6) have an adaptive advantage expressed in higher productivity (curve *B*, 1500 to 2000 m).

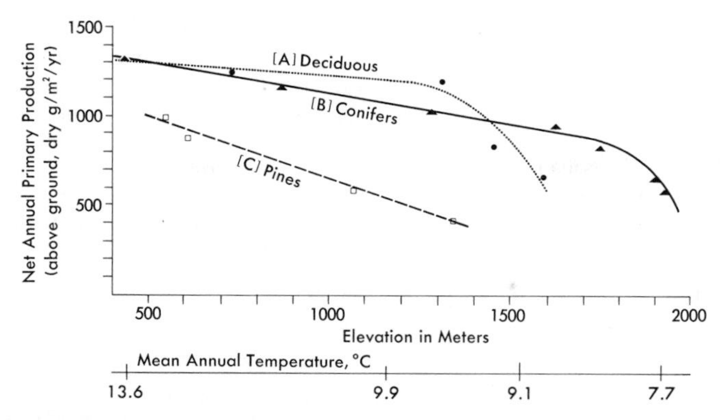

Figure 5.5. Net primary productivity along an elevation gradient in the Great Smoky Mountains, Tennessee, for **A:** Broad-leaved deciduous forests of moist environments. **B:** Evergreen coniferous forests (hemlock, spruce, and fir) of moist environments. **C:** Pine forests of dry environments. [Production data from Whittaker, 1966; temperatures from Shanks, *Ecology,* **35**:354 (1954).]

4. In the pine forests of dry environments (curve *C*) the interaction of moisture and temperature factors (and possibly consequent differences in soil fertility) produces a steep decrease of net productivity with elevation.

The manner in which these relations are to be extended to the Tropics is not yet established. However, as shown in Figure 5.1 the fraction of gross productivity that is expended in plant respiration increases from 50 to 60 per cent in most temperate forests to 70 or 80 per cent in tropical forests. It seems likely that net productivities of many climax tropical forests do not exceed those of temperate forests of favorable environments, but that average and maximum net productivities are higher in the Tropics. The gross productivities of climax tropical forests should be considerably higher, and both gross and net productivities may be very high in successional forests in the Tropics.

There is a general tendency for biomass, as well as productivity, to decrease along moisture and temperature gradients. Biomasses of mature forests are mostly in the range of 20 to 60 kg/m², woodlands 4 to 20, shrublands 2 to 10, grasslands 0.5 to 3, and deserts and tundras 0 to 2. Biomasses of mature forests are not generally higher in the Tropics; biomass ranges are similar for tropical and temperate forests except for the largest of all forests, which are certain temperate rain forests (biome-type 3). The low biomasses of grasslands result from the short lives of the above-ground parts of the fire-adapted plants of grasslands. Apart from grasslands, the ages of dominant plants in general increase from deserts to climax forests (though some woodland, shrubland, and desert plants are quite old). The relation of biomass to productivity is conveniently expressed as the biomass accumulation ratio—the ratio of dry-weight biomass to annual net primary productivity. Such ratios (for above-ground parts of plants) increase through the sequence of terrestrial biome-types; normal ranges are from 2 to 10 in desert, 1 to 3 in grasslands, 3 to 12 in shrublands, 10 to 30 in woodlands, and 20 to 50 in mature forests. Relations of productivity and biomass accumulation to the moisture and temperature gradients underlie some of the physiognomic trends described in Chapter 4.

The high productivities of some land communities have equivocal meaning in relation to human problems. They suggest on the one hand that appropriate techniques—supplying adequate water and nutrients—could produce very high yields in most land environments without the need to modify natural conditions of light, temperature, and CO_2 level. It is possible in theory to increase very much the productivity of the land surface for human needs. The limited occurrence of very high productivities implies on the other hand that these require special circumstances that escape the normal restrictions on terrestrial production. Only by substantial technological effort, economic investment, and energy expense, usually involving irrigation or continuing nutrient addition, or both, applied to particular crop strains of favorable characteristics with control of loss to pests, are the high yields realized in agriculture. The best lands are now under intensive cultivation; for many areas, especially of the Tropics, Far North, mountain slopes, and arid lands, there is either no known technique for productive cultivation, or no present prospect of irrigation. Whole complexes of reasons involving the controls of productivity, limitations of environments and agricultural technology, and political, cultural, and economic circumstances are responsible for the actual limitations on world food production. Hopes based on high yields in favorable circumstances have had a seductive effect in diverting attention from the difficult, laborious, research-demanding, expense- and organization-requiring, slow effort to increase food harvest for human beings, and the real limitations on that effort. It is true both that man could feed well a human population stabilized at a realistic level, and that he is failing now to feed adequately a growing world population even at its present level.

Marine Productivity

Because abundant water generally implies high productivity on land, one might expect aquatic communities to be highly productive. They are, in many cases, not. Because the plankton of the open oceans occupies a larger area of the earth's surface than all other kinds of photosynthetic communities together, its productivity is of special interest. The photosynthetic plankton or phytoplankton is a thin suspension of cells in the surface waters of the sea. This ecosystem is especially suited to quantitative description of the way steady-state nutrient flow controls its productivity. J. H. Steele and G. A. Riley have developed equations describing nutrient movement in relation to marine plankton productivity. We use their approach as a simple example of a mathematical model applied to an ecological problem, but the discussion can be followed without reference to the mathematical statements in brackets. We shall consider relationships affecting productivity in the column of water below a square meter of the surface of the open sea.

Only that uppermost part of the column that is illuminated by sunlight counts as an environment of primary productivity. Ocean waters are often highly transparent; visible amounts of sunlight penetrate well below 100 m depth in some seas, especially in the Tropics. The depth of the compensation point—the light intensity at which plant respiration equals photosynthesis—may be used as the lower limit of the productive, lighted, "euphotic" layer. The light intensity at the compensation point varies with species and physiological state of plants, environmental factors other than light, and span of time considered, but 1 per cent of full sunlight is a conventional approximation. On this basis the depth of the productive layer is between 30 and 120 m in most open ocean waters. A 1 per cent limit of light intensity corresponds to that on the floor of many forests, but terrestrial communities, even forests, are shallower photosynthetic systems than the marine plankton. Depth of light penetration in the open ocean is in large part determined by absorption by the plankton itself; more correctly it is determined by absorption by the seston—plankton plus dead organic particles in the water—as well as by the water itself. On the whole, the more productive the plankton is, the greater the mass of the corresponding seston, and the shorter the penetration of sunlight into the water. Depth of light penetration decreases from tropical open oceans to temperate open oceans and from these to inshore waters in which both high plankton production and particles derived from other sources limit light penetration. [The depth can be estimated from the relation of the extinction coefficient, k, to depth in meters, L, of a limiting light intensity, I_d, below a surface sunlight intensity, I_0, through water with a chlorophyll content, C_h, of the seston in μg/liter of water. The extinction equation and some observed relationships of k, C_h, and L are $I_d = I_0 e^{-kL}$, $k = 0.04 + 0.0088C_h + 0.054C_h^{2/3}$; hence, if $I_d = 0.01 \times I_0$, $L = 4.605/k$.]

Neither on land nor in the sea is photosynthesis simply proportional to light intensity. At lower light intensities, however, photosynthesis approaches a direct linear relation to light intensity. Above these low light intensities a point of light saturation occurs beyond which photosynthesis does not increase with increased light intensity. At still higher light intensities, such as those close to the ocean surface at midday, photosynthesis is inhibited and occurs at lower rates than at intermediate light intensities. For this discussion of the broad relationships of marine plankton production, the complex interrelations of light, chlorophyll, and photosynthesis with depth will be shortcut. We shall consider that summer sunlight intensities are of similar magnitudes in temperate and tropical waters, that over-all efficiencies of light use by plankton (considering all depths together) are similar in temperate and tropical waters of similar productivities, and that productivity is controlled by factors other than light intensity. The effective factors are, primarily, nutrients.

As described in Chapter 4, the plankton has a problem with sinking. Despite all adaptations, a certain fraction of the plankton organisms and their dead remains must sink below the lighted zone, carrying with them nutrients incorporated in protoplasm and skeletons. The loss of nutrients from the lighted zone is intensified by another phenomenon. The warm waters of the open ocean in the Tropics and in the temperate-zone summer are (like those of many lakes in summer) stratified: a layer of warmer and less dense water floats on top of the colder and denser water of the depths, separated by a zone of relatively rapid temperature change per unit depth, the thermocline. Because water density decreases upward in the thermocline, its waters are stable. Vertical movements of water by waves and other forces affect primarily the warm waters above the thermocline. The thermocline is consequently a relative barrier to the return of nutrients from the lower levels to the upper warm and lighted waters. The depth of the warm water above the thermocline, and of the lighted water above the 1 per cent level may correspond roughly, although they will not necessarily do so. But the sinking of plankton implies the depletion to low concentrations of the nutrients in the warm and lighted, productive surface waters, and consequently the low productivity of the open oceans. [The rate of sinking of seston is highly variable with kind and size of organisms and particles, and it increases with warmer water temperature and consequent lower viscosity. Mean rates of sinking, v, ranging from 3 m/day in cooler to 6 m/day in warmer waters, are thought reasonable. The rate of loss of plankton or nutrients by sinking, as a decimal fraction per day, can be expressed as v/L.]

Plankton productivity in the stratified open ocean waters does not decline asymptotically to zero in consequence of the loss of nutrients with sinking, however. Despite the relative stability of the thermocline, there is some return of nutrients resulting from turbulent movement of water at and below the lower limit of the lighted zone. Plankton productivity declines not to zero but to a steady-state level at which the loss of nutrients by sinking and the return of nutrients by water mixing are in balance. A low mixing rate in stratified water necessarily implies low steady-state productivity. Productivity can also be limited, however, by a high mixing rate in unstratified water, in which mixing depletes the photosynthetic plankton by diluting it from the lighted surface water into the underlying dark water. [The mixing rate, m, is the decimal fraction of the productive water displaced downward below the lighted zone (and consequently replaced by nutrient-bearing water from below) per unit time. Mixing rates between 0.02/day and 0.05/day support maximum plankton production; mixing rates one or two orders of magnitude less, 0.005 to 0.0005/day, are normal in the stratified open ocean. For production calculations the effect can be expressed as $m(p_0 - p)$, the mixing rate times the difference between nutrient concentration of the dark water, p_0, and of the lighted water, p, both in μg/liter. At steady state $pv/L = m(p_0 - p)$.]

Steady-state nutrient levels thus determine the rate of productivity in the stratified open ocean. Because the critical nutrient elements are taken up by plankton until only very low concentrations remain in the water, these concentrations themselves do not simply determine productivity. The effective relationship is that between the *rate* at which nutrients become available and the rate of productivity. The rate at which nutrients become available depends strongly on mixing, but also partly on the rate of nutrient circulation in the plankton ecosystem itself—the rate at which the nutrient leaves phytoplankton cells and becomes available for new uptake by phytoplankton cells. There are a number of routes for this circulation: direct loss from phytoplankton cells into water, death and decomposition of the cells, and grazing by animals followed by excretion or by death and decomposition. Grazing and excretion by animals may be more significant in the plankton, and the role of bacteria and the fungi of decay less prominent, than in terrestrial communities. [Let c be a conversion factor between phytoplankton biomass in grams of carbon per m^3 of water and a given nutrient in μg/liter of water, e an excretion rate as a decimal fraction of the nutrients taken up by grazing animals, g a grazing rate in liters of water from which phytoplankton are harvested per day by a unit biomass (one gram of carbon per cubic meter of water) of plankton herbivores, h the biomass of the herbivore population in gC/m^3, and P the phytoplankton biomass, also in gC/m^3, then the nutrient regeneration within the lighted zone by this route will be, $r = ceghP$. If, for example, with phosphorus as a nutrient present at low concentrations in the water $c = 0.774$, and if $e = 0.85$ and $g = 3.4$, then $r = 2.2hP$. The rate of change of h can be expressed as $dh/dt = h(gP - r_h - fC)$, in which r_h is the respiratory rate of the herbivores, and f is a feeding coefficient as a decimal fraction of herbivore biomass harvested per unit biomass of carnivores, C. Estimates of the steady-state relations of herbivore to phytoplankton biomass, h/P, range from near equality in some conditions with low feeding rates of carnivores, $f = 0.0025$, to one third or one half with higher feeding rates of the order of $f = 0.01$.]

There is evidence that phosphorus is the most critical nutrient element in some open ocean waters. Nitrogen may be more critical than phosphorus elsewhere in the open ocean and in coastal waters, and other elements may be critical in some times and places. To some extent, however, other elements, including nitrogen that is available in soluble compounds, tend either to be present in less critically short supply or to parallel phosphorus in their depletion from the lighted zone. A first estimate of open ocean production may consequently be based on phosphate relations in the plankton ecosystem. At low levels of phosphate in the water, the rate of photosynthesis is directly related to phosphate concentration in the water. [The phytoplankton growth rate may be designated P_r, the decimal fraction of its biomass that the plankton produces per day as net primary production. For phosphate levels between 0.05 and 0.40, P_r may be estimated as 0.66 times the value of p in μg/liter. Equations describing phytoplankton production for the open sea in a simplified, approximate, but reasonable way are:

$$dP/dt = P(P_r - gh - v/L - m)$$

change in phytoplankton biomass, P, equals net production minus grazing minus sinking minus downward mixing of phytoplankton cells, and

$$dp/dt = cP(egh - P_r) + m(p_0 - p)$$

change in phosphorus level, p, in the water equals release by recycling minus uptake in production plus net return in mixing. For the latter, in steady state,

$$cPP_r = ceghP + m(p_0 - p).$$

The rate of incorporation of phosphorus in phytoplankton production equals the sum of the rates of recycling of phosphorus through herbivores, $ceghP$, and of phosphorus return by mixing from the water below the lighted zone, $m(p_0 - p)$. PP_r is the desired phytoplankton net production in grams of carbon per unit m^3 of water per day.]

[The calculations cannot be described further (see Riley 1965). As an example, for the area of the North Sea studied by Steele in summer, with a mean mixing coefficient of 0.0042, $p_0 = 0.7$ μg/liter, and estimated chlorophyll content of 0.84 μg/liter (compared with a mean observed value of 0.77), calculations give a mean phosphate concentration in the lighted zone, $p = 0.265$, and the corresponding $P_r = 0.175$, and phytoplankton biomass $P = 0.030$ gC/m^3. Estimated phytoplankton production summed for the 37 m column of lighted water is 0.26 grams of carbon below a square meter of the ocean surface per day.] Compared with the estimated productivity of 0.26 grams of carbon per meter square of the ocean surface per day in an area of the North Sea, an independent measurement by the ^{14}C technique averaged 0.29 gC/m^2/day. An approximate conversion factor of 2.2 from grams of carbon to dry grams of organic matter in net production gives 0.64 dry g/m^2/day. Production estimates and measurements in most tropical open oceans in summer are lower, of the order of 0.05 to 0.20 $gC/m^2/day$.

Yearly cycles of productivity in temperate open seas include winter periods of reduced light intensity and production, and spring periods of increased mixing and higher production. Many tropical areas also have periods of higher mixing rates. An estimated mean annual net productivity for open waters of the North Sea area is 68 $gC/m^2/yr$, or 150 dry $g/m^2/yr$. A comparable value for the Sargasso Sea is 72 $gC/m^2/yr$ or 160 dry $g/m^2/yr$, values for other tropic open oceans outside upwelling areas are 40 to 120 dry $g/m^2/yr$. Temperature has complex and conflicting effects on open ocean production. The increased stratification and sinking rates that go with higher temperatures act to reduce productivity in the Tropics, whereas the low winter light intensities correlated with low temperatures act to reduce productivity in the Temperate Zone. Mean photosynthetic rates and feeding rates for plankton herbivores may not show a strong relation to latitude, because organisms of different climatic belts and seasons are adapted to effective function in the temperatures of their environments. Similar year-round productivities may result from lower production rates by a smaller plankton biomass in a deeper lighted column through a longer warm season in the Tropics, and higher summer production in a shallower lighted layer combined with low winter production in the Temperate Zone. In the open oceans at all latitudes, productivities are predominantly low (Figure 5.6).

Because of the difficulties of the radiocarbon method, estimates of marine production may be subject to revision. It is easy, however, to characterize communities that would be supported by net productivities of 40

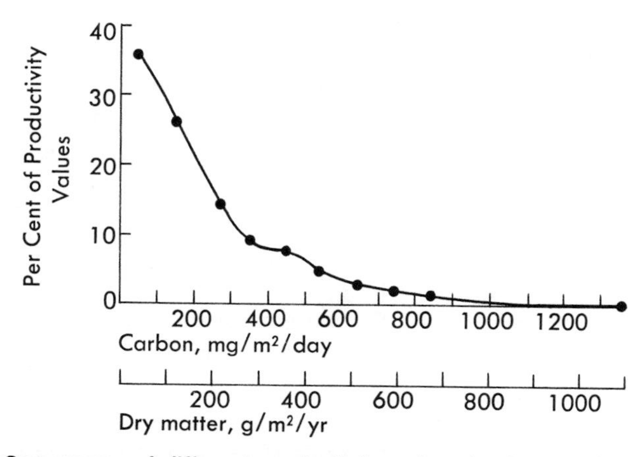

Figure 5.6. Occurrence of different productivity values in the world's oceans. The measurements are strongly skewed toward the lower end of the scale with the largest per cents, representing the greatest areas of the oceans, giving plankton primary productivities below 250 $g/m^2/yr$. [Koblentz-Mishke et al., 1970.]

to 200 dry $g/m^2/yr$ on land: semidesert. As the northern continents and Australia have arid interiors of low productivity, so the oceans have nutrient-poor "interiors" of low productivity. Where upwelling—movement of cold, nutrient-rich water from the depths to the surface of the ocean—occurs, productivity is higher. Upwelling occurs (1) in an area surrounding the Antarctic continent, (2) in certain arctic areas of the North Atlantic and western North Pacific, (3) near the equator in the Pacific Ocean, (4) in areas of the Indian Ocean where seasonal upwelling is related to monsoon winds, and (5) along the western sides of continents where a current flowing toward the equator tends (because of the Coriolis force) to curve away from the shore, drawing water upward from the depths near the shore. With or without upwelling, productivity is in general higher in the shallower waters on the continental shelves. The reasons that the inshore waters are more productive (although their lighted layer is less deep) include more effective turnover of nutrients from deeper waters and bottom sediments to the lighted layer, drainage of nutrients into the sea by rivers and estuaries, and the contribution of attached seaweeds to production.

Plankton productivities of upwelling areas that are not on continental shelves are around 400 to 600 $g/m^2/yr$, and similar values occur in many areas on the continental shelves. Where, however, upwelling occurs along coasts on continental shelves, productivities ranging upward to 1000 $g/m^2/yr$ and more are observed. Productivities in estuaries are high—near 1000 $g/m^2/yr$ in measurements in Long Island Sound. In some inshore waters high productions by attached algae and eelgrass are added to plankton productions. The giant kelp beds of the California coast have some of the highest productivities known for any ecosystem; tropical coral reefs also are among the most productive of ecosystems. In both these cases the productivity is supported by nutrients carried past the organisms by moving water, and in both cases some fraction of the measured productivity is probably secondary or heterotrophic (based on use of organic matter in the water) rather than primary or autotrophic. Marine productivities cover the same wide range of values as terrestrial productivities—from values corresponding to most arid deserts (less than 2 dry $g/m^2/yr$ in the weak light beneath the arctic icecap) to values above 3000 dry $g/m^2/yr$ in some communities of especially favorable circumstances. Fresh-water ecosystems show a correspondingly wide range of production values. The highest productivities occur on the margins of the land and water—certain wet or hydric communities on land, and certain inshore and shallow communities in water.

The highly productive marine waters are essentially the fringes of the oceans. The major commercial fisheries of the world depend on the harvest of these fringes. The fish harvested are carnivores of high positions in food chains. For reasons to be explained in a following section man can harvest through these carnivores only small fractions—much below 1 per cent—

of net primary productivity. Direct harvest of marine plankton by ships and nets is exorbitantly costly, and extensive underwater "farming" on the continental shelves does not yet appear economically or technologically feasible. No technique for more successful harvest of marine production of food than that through marine animals is yet in prospect. A large share of the most productive marine fish populations are already being harvested; only limited increase in harvest seems likely. Reductions in the harvest of some marine populations must be expected because of the effects of pollution, excessive harvest, and destruction of breeding and feeding areas in estuaries (Chapter 7). The oceans are great and taken as wholes greatly productive, but the idea that they offer abundances of food awaiting harvest to feed hungry nations is an illusion.

Pyramids and Efficiencies

Three major ways (or groups of ways) in which aquatic animals obtain food have been mentioned: (1) The filtering of the living particles of plankton from the lighted surface waters by plankton animals. A large part of this filtering in the sea is done by copepods (Figure 1.1), which are microcrustaceans (mostly 0.5 to 8 mm long) with appendages adapted as miniature rakes or combs that strain particles from the water. Such a straining process is largely unselective except for the size of particles. What the copepod collects with its rake is consequently seston, a mixed harvest of living cells and dead cells and fragments. Despite the apparent indiscriminateness of this process, there is room for niche differentiation in behavior and in the size and kind of particles eaten among these animals. (2) Use of the nutrient rain of seston particles that settles through the water toward the depths of the water body. Some of these particles feed plankton organisms of the unlighted deeper water; others feed bottom organisms with ciliated tentacles, water tubes, and other devices for the concentration of organic material from water. The animals of the bottom community, the benthos, show a wide range of techniques of nutrition as one aspect of the impressive diversity of species, forms of organisms, and animal phyla in this community. (3) Predation by aquatic animals that capture and feed on other animals.

The food energy of one kind of organism is consumed by another kind of organism. Diatoms, and other plant plankton cells, are raked from the water and eaten by copepods. Small fish, say sardines, eat the copepods. These sardines are eaten by larger fish and these in turn are eaten by still larger predatory fish—such as tuna or shark. An atom of organic carbon may by this time be incorporated in the protoplasm of a fifth organism in the sequence from diatom to shark. A shorter sequence may lead from phytoplankton to a larger crustacean plankter (krill or euphausiid shrimp) to one of the baleen whales that feeds by straining these crustaceans out of the water.

Such sequences represent the familiar idea of food chains. Numbers of links in such chains are variable, but three to five links are common. Positions of links in the chains are named:

1. Producer, the photosynthetic plant or first organism of the sequence
2. Herbivore or primary consumer, the first animal, which feeds on plant food
3. First carnivore or secondary consumer, an animal feeding (as predator, parasite, or scavenger) on a plant-eating animal
4. Secondary carnivore or tertiary consumer, feeding on the preceding
5. Tertiary carnivore.

These positions along food chains are *trophic levels*. Trophic levels were used in the early study of productivity by R. L. Lindemann; and work by E. P. Odum, H. T. Odum, and others has developed understanding of community function on the basis of the trophic levels and related concepts. The boundaries of the levels are not sharp. Many animals take any food that is suitable in size range and other characteristics, and consequently they take food from more than one trophic level.

The productivity of the animal plankton is less than that of the plant plankton. Such must be the case for more than one reason. Only the fraction of primary productivity that is net, remaining after plant respiration, can be harvested. Of this net primary productivity only a fraction can be harvested live by animals, if they are not to destroy by overharvest their own food sources. Only part of the plant material eaten by animals will be digested and assimilated. For most animals, both the siliceous walls of diatoms and the cellulose walls of land plants are useless as food even though the latter are organic material. For these reasons productivity of the second trophic level, the herbivorous animals, is generally one tenth or less that of the plants on the first level. For similar reasons production of the first carnivores must be less than that of the herbivores, and production of the secondary carnivores must be less than that of the first carnivores.

We can formulate some of these relationships as efficiencies of food use for individual populations. The grasshopper we described under measurement of production assimilates 38 per cent of the organic matter and energy of the grass it eats; the grasshopper has an assimilation efficiency of 38 per cent. Only 4 per cent of the organic matter eaten becomes new organic matter of growth and reproduction by the grasshopper. The population's gross growth efficiency (new protoplasm in growth and reproduction/food consumed) is 4 per cent. The ratio of growth (and reproduction) to assimilation is 4 per cent/38 per cent, and this ratio (10.5 per cent) is the grasshopper population's net growth efficiency. The gross growth efficiency is of course the product of the assimilation efficiency times the net growth efficiency. Assimilation efficiencies for terrestrial animals are mostly between 20 and 60 per cent for herbivores, 50 and 90 per cent for carnivores. For marine animal plankton assimilation efficiencies are widely variable, with most values probably between 40 and 80 per cent. For *Daphnia,* a small crustacean common in fresh-water plankton, gross growth efficiencies of 4 to 13 per cent and net growth efficiencies of 55 to 59 per cent were reported. Gross efficiencies for a snail and a mite were 6 per cent and 4 per cent; a mean gross growth efficiency of 16 per cent has been suggested for marine animal plankton. The key relationship for comparing productivities is the gross growth efficiency, since this relates secondary production by animals to the food harvest from a lower trophic level. A value of 10 per cent is often used as a convenient middle figure, with the expectation that efficiencies for herbivores on land may be generally lower than this and those for animal plankton and some carnivores may be higher.

These relationships imply that there must be a steep, stepwise decrease in productivity up the sequence of trophic levels, forming the pyramid of productivity (Figure 5.7A). Two other pyramids appear as corollaries. The numbers of individual organisms will generally decrease up the sequence of levels to form the pyramid of numbers that Charles Elton first recognized (Figure 5.7C). Pyramids of numbers, however, are subject to reversal when many small organisms feed on one large organism of a lower level, as thousands of insects may feed on a tree, or hundreds of parasitic worms on one host. Biomasses for trophic levels also decrease up the sequence (Figure 5.7B). Biomass pyramids are less frequently reversed, but animals at times exceed plants in mass in plankton communities. It is

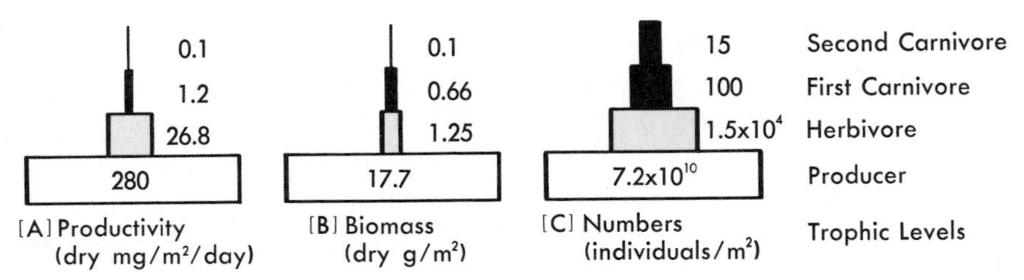

Figure 5.7. Community pyramids for an experimental pond. Productivity was estimated from rate of phosphorus uptake in a shallow pond of low nutrient content. The fourth trophic level was estimated as a fraction of carnivores feeding on both the second and third levels. Widths of steps for numbers of organisms are on a logarithmic scale. [Data of Whittaker, *Ecol. Monogr.,* **31**:157 (1961).]

the pyramid of productivity that has fundamental significance; the pyramids of numbers and biomass are less fundamental and less reliable consequences of it.

A ratio of a level of the pyramid of productivity to the preceding level is also an efficiency. The first efficiency of interest is the relation of the productivity of the first trophic level to the energy of sunlight, which supports it. In the North Sea plankton community, the energy of incident sunlight in the visible range at the ocean surface in summer is about 0.329 cal/cm^2/min or 473 cal/cm^2/day. Part of this light is reflected by and absorbed in the water. Only the light that reaches plant cells can be used by the phytoplankton for photosynthesis. Only that fraction of the visible spectrum that plant pigments can absorb, and only that fraction of this that is actually absorbed and made available to the photosynthetic process, can result in productivity. The efficiency of photosynthesis is necessarily low. In the North Sea plankton community, a net productivity of 0.64 g/m^2/day converts to energy of 0.314 cal/cm^2/day, and this relative to incident sunlight in the visible spectrum is an efficiency of only 0.066 per cent or 1/1500.

In a forest, the foliage of different plants is staged in depth as has been described. Above a square meter of a deciduous forest floor there are 4 to 6 square meters of surfaces of leaves (counting only one surface of a leaf) containing about 2 g of chlorophyll (Table 5.1). There is an additional 1.5 to 2.0 m^2 of bark surface (computing this as if the bark were smooth). Leaf and bark surface together often extinguish the light to and beyond the compensation point. In many forests, light intensity on the forest floor is 1 to 2 per cent of incident sunlight at midday; in some forests it is around 0.2 per cent. Efficiencies in relation to light and the forest's light-receiving structure can be expressed in several ways. A net primary productivity of 1200 g/m^2/yr in the oak-pine forest is equivalent to 510 cal/cm^2/yr; in relation to an annual sunlight energy of the visible spectrum of 56,000 cal/cm^2/yr, the net production efficiency is 0.91 per cent. The corresponding efficiency of gross primary productivity in relation to year-round incident sunlight in the visible range is 2.0 per cent. A gross productivity of 2650 g/m^2/yr implies also an annual photosynthetic capture of 3000 kCal energy per square meter of leaf surface, for productive output of 700 dry g/m^2 of leaf surface, and 1400 dry g/g of chlorophyll. The net production efficiencies of 0.91 and 0.066 per cent are representative values for ecosystems of moderately high and relatively low productivities.

Efficiencies of the second trophic level in relation to the first can be variously expressed, using either the gross or net productivity of the producers, and for the second level combinations of respiration, yield (as loss from the trophic level by predator harvest, death, and removal or emigration), excretion, and dry weight growth of herbivorous animals, or of these plus the bacteria and fungi feeding on plant material. One appropriate expression of efficiency of the second trophic level compares herbivore assimilation, or respiration plus yield (including excretion with this), with gross primary productivity. This "ecological efficiency of energy flow" gives values ranging from about 10 per cent down. Efficiencies of the third level, comparing respiration and yield for primary carnivores and herbivores, can (but will not necessarily) be somewhat higher: values may range from 15 per cent down. The efficiency of food assimilation by a carnivore can be higher than that of an herbivore because of the closer match of the carnivore's food to its own chemical composition and needs.

There is thus a tendency for ecological efficiencies relating a trophic level to the preceding level to increase up the pyramid: from green plants to herbivores and (in some cases) from herbivores to carnivores. Net or "production efficiencies" for the second trophic level (herbivore secondary productivity/net primary productivity) and higher levels (secondary productivity of a given animal trophic level/that of the preceding level) are lower than the corresponding gross or ecological efficiencies. This is so because the percentage of food energy respired tends to rise from plants, to herbivores, to predators. Because respiration percentage may increase (and net growth efficiencies consequently decrease) along food chains, production efficiencies need not increase up the pyramid. They may in fact decrease.

Efficiencies related to human harvest should be differently stated. What man should take is not productivity as such, but a yield—a fraction of productivity that it is feasible to remove and use without destroying the basis of the productivity. The efficiency of human harvest is most appropriately stated as dry mass of yield, relative to dry mass of net primary productivity, per unit area and time. Man can harvest around 30 per cent of net primary productivity when plant material is taken as either the grain of cereal crops or the wood from a forest. Higher efficiencies are possible in some favorable circumstances, but lower ones apply to many environments and crop species. If harvest is in the form of the meat of herbivorous animals, the harvest efficiency must be lower than the trophic-level efficiency, hence less than 10 per cent. Pyramid relations imply that if the meat of aquatic carnivores is used, the harvest efficiencies in relation to primary production will range downward from 1.0, 0.1, and 0.01 per cent for primary, secondary, and tertiary carnivores respectively. For harvest of animal populations also, yield is a figure different from and lower than productivity. Only a fraction of a population can be harvested on a sustained basis—that fraction that represents a surplus above the individuals necessary to maintain the reproduction and growth of the population itself. This limitation on harvest is one of the factors determining trophic level efficiencies, and when man's harvest exceeds this limitation, decline in productivity and yield may result—as observed in overfishing and overgrazing (Chapter 7).

Detritus and Reducers

Productivities and efficiencies have so far been discussed as if only two major groups of organisms were involved—green plants (producers), and animals (consumers). Also, our discussion of pyramids has an unstated implication that needs to be considered. Only about 10 per cent or less of the net primary productivity on land is directly harvested by animals; something must happen to the other 90 per cent. The roles of producers and consumers in communities are clearly different; the producers use the energy of sunlight to create food, and the consumers harvest and use part of this food. The fact that the latter harvest only a part implies the significance of the third major group of organisms that use the rest. Organisms of the third group are the reducers, or transformers, or decomposers—the saprobes or organisms of decomposition and decay, the bacteria and fungi. These organisms live on or within organic food supplies, especially dead tissues, and feed by absorption of soluble organic food. In many cases digestive enzymes are excreted from the bacterial cells or fungal filaments, and these enzymes digest organic matter into soluble forms that are then absorbed. Through such digestion and the effects of their own respiration, the reducers break down organic matter first into soluble forms and finally to inorganic remnants. In the course of this they use organic energy not harvested by the consumers for their own growth and metabolism, and release inorganic nutrients back into the environment.

The dead organic material in ecosystems (excluding that in solution) is *detritus*. Detritus on land includes dead leaves that fall on the ground to form the leaf-litter; but it includes also the dead stems and branches on a forest floor, dead roots and humus particles below ground, and the dead remains of animals. In the marine plankton, the detritus is made up of the dead remains of plankton and other organisms, together with bacteria on and in these remains, and small particles that form by processes such as adsorption of organic matter on the surfaces of bubbles. In lakes and streams the greater part of the detritus may be derived from vascular plants that grow along the shore or in shallow water and only a smaller part may be plankton remains. In coastal waters of the ocean the largest part of the detritus may be the remains of algae in the shallow waters and of vascular plants in coastal bays.

The masses of detritus are large, compared with living communities. In water bodies the mass of detritus particles may exceed that of living organisms severalfold, or tenfold, or more; and the mass of dissolved organic matter in the water may exceed the particulate detritus by one hundredfold. We can, if we wish, extend the biomass pyramid (Figure 5.6B) downward two additional steps for this particulate and dissolved material. In a grassland the mass of dead grass remains may be comparable to that of the living grass; in some grasslands the dead mass is larger. In cold-temperate forests the mass of litter on the soil surface may be equivalent in amount to several years' productivity, and it may be a significant fraction (one tenth to one fifth) of the more obvious living mass (and heartwood) of the trees. In many land communities the humus—finely divided organic matter, colloidal and soluble—in the soil exceeds the surface litter in mass.

These masses of dead organic matter are resources awaiting harvest. The harvest in many communities is based on a collaboration of animals and reducers in detritus food chains. Earthworms feed in part by emerging from their burrows to eat dead leaves, in part by passing soil through their guts to digest some of its organic matter. In a tropical forest termites are detritus feeders, consuming dead wood with the aid of their symbiotic protozoans, while the leaf litter in the same forest may be attacked primarily by fungi. In a temperate forest the leaf litter is occupied by the cryptozoa as a distinctive community of animals—springtails, mites, millepedes, and other groups—, and some of these animals, together with bacteria and fungi, consume the litter. Many of these animals are not really feeding on the plant tissues themselves; they are grazing on bacteria, or browsing on or sucking from fungal filaments. The animals that do eat dead plant tissue in many cases obtain much of their food by digesting the bacteria and fungi in that tissue. At the same time these animals are physically breaking up the plant tissue into smaller particles that can be more effectively attacked by fungi and bacteria. Experiments have shown that the decomposition of dead leaves on the forest floor by bacteria and fungi may be delayed for months if the leaves are enclosed in mesh bags that exclude the animals while permitting bacteria and fungi to reach the leaves. Fecal pellets are important units of detritus. Detritus, already partly decomposed, is eaten by an animal that uses part of the dead organic matter and reducer cells as food, and packages the remainder as fecal pellets. Bacteria and fungi grow again in these and decompose the organic matter further. A fecal pellet may later be eaten by another animal of the same or a different species, for a second harvest of some fraction of the pellet's food value.

Such is evolution's response to detritus as a resource: complex processing systems of many bacterial, fungal, and animal species differently using different kinds and fractions of detritus in different timings and interactions with other species. Bacteria, fungi, and animals are intimately braided into detritus food nets. Diverse food chains are possible, such as: Plant tissue (death)—earthworm—(death)—bacteria of decay. Plant tissue (death)—first fungal invasion—millepede (feces)—fungus in feces—springtail (feeds on fungus)—predatory mite—centipede as predator (death)—bacteria of decay. Food chains may or may not include a top carnivore, but they may include, and will normally end with, reducers. Because animals collaborate with bacteria and fungi in detritus food webs, the terms "consumers" and "reducers" become blurred. We may better speak here of three interrelated and intergrading groups: live-tissue consumers, detrital consumers (including scavengers and animals feeding on bacteria and fungi), and reducers or decomposers (bacteria and fungi of decay). The crucial role of the reducers should not be overlooked. Detritus processing ends with mineralization—the break-down of organic matter to inorganic substances, including the inorganic nutrients, that are released into environment. Much, or most, of this final decomposition to inorganic matter is by bacteria and fungi.

There are not two major groups of organisms in natural communities—plants and animals—but three—plants, animals, and saprobes. These three (though parasites occur in all three) correspond in general to the producers, consumers, and reducers that are the functional kingdoms of natural communities. The three represent major directions of evolution, and in each the kinds of organization evolved are closely related to the mode of nutrition. Plants feed primarily by photosynthesis, and the higher forms have evolved toward organizations that include leaves or blades as organs of photosynthesis, supported by stems or stipes, rising from roots or holdfasts that provide anchorage and (in vascular plants) water and nutrient uptake, while the vascular tissues serve for transport between these organs. Animals feed primarily by ingesting food that is digested in and absorbed from an internal cavity. Their organization has consequently evolved toward differentiation of both a digestive tract together with supporting circulatory and excretory systems, and the sensory, nervous, and muscular-and-skeletal systems that permit motility and response to food. Saprobes feed by absorption and have need for an extensive surface of absorption but (apart from reproductive systems) little need for other structural differentiation. The principal kinds of organization evolved among saprobes are the unicellular of bacteria and yeasts, the chytrid of the lower fungi, and in higher fungi the mycelial, with a network of syncytial protoplasmic filaments or hyphae.

The functional kingdoms of communities are not to be identified with the kingdoms of systematics, but they are a large part of the evolutionary basis of the latter. In the traditional two-kingdom system the animal kingdom includes organisms, both unicellular and multicellular, characterized by ingestion and motility, and the extended plant kingdom includes the primarily photosynthetic groups and (somewhat arbitrarily grouped with them) the predominantly absorptive bacteria and fungi. In current alternatives the bacteria (with the blue-green algae) become the kingdom Monera, characterized by procaryotic organization, whereas the higher plants and animals become kingdoms characterized by the directions of evolution stated above. It is logical to recognize the higher fungi as a kingdom coordinate with these, whereas the eucaryotic unicellular organisms become the kingdom Protista.

The reducers are the least visible of the three fractions of the community. Bacteria are part of the plankton, and some lower fungi (chytrids) occur, but these are inconspicuous. The few mushrooms that may be seen on a forest floor scarcely suggest the rich saprobic life of the soil, which includes many kinds of bacteria and fungi of diverse nutritive relations to living and dead plant roots, leaf litter and soil organic matter, dead stem and branch wood, and other food sources. In most terrestrial communities secondary productivity by reducers should much exceed that by consumers, though there are no good measurements of reducer secondary productivity for natural communities. Even the biomass of the reducers is difficult to measure. The biomass of the reducers is small, however, in relation to their productivity and significance for the community. These organisms function as enzymes of the community: small masses of reducers (with high production consequent on rapid growth and reproduction balanced with mortality) transform far larger masses of organic matter through sequences of reactions to inorganic remnants.

In so doing the reducers disperse back to the environment as heat the energy of photosynthesis in the organic compounds they decompose. They have thus a major part in one of the most general characteristics of ecosystems—energy flow. The manner of the flow is illustrated in a simplified and generalized way in Figure 5.8. A community, like an organism, is an open energy system. Through the continuing intake of energy by photosynthesis the energy dissipated to environment by respiration and biological activity is replaced, and the system does not run down, through the loss of free energy, to maximum entropy. If energy intake exceeds dissipation, the pool of biologically useful energy of organic bonds in the community

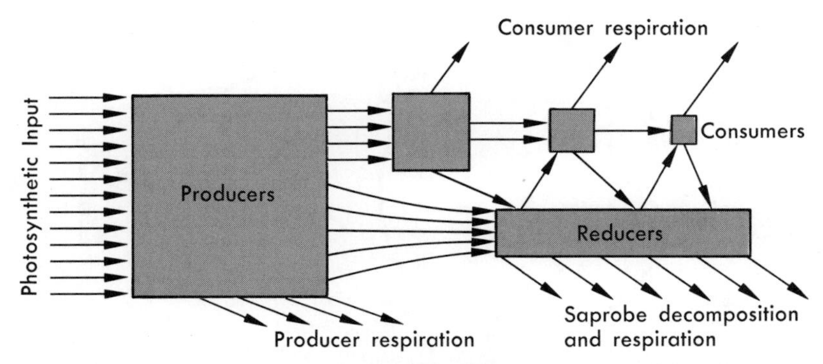

Figure 5.8. Flow of energy in a natural community. In a steady-state community the intake of photosynthetic energy on the left, and the dissipation of energy back to environment toward the right, are in balance. The pool of energy of organic compounds within the community remains constant.

increases as does community biomass, and the community grows; such is the case in succession. If energy loss exceeds intake the community must in some sense retrogress. If energy intake and dissipation are in balance, the pool of organic energy is in a steady state, such as is characteristic of climax communities.

The reducers have a crucial role in the second most general characteristic of ecosystems—the circulation of materials between community and environment. The manner in which the three functional kingdoms relate to this circulation is indicated in a most generalized form in Figure 5.9. Decomposition of organic matter by reducers releases inorganic materials, including nutrients, into the environment—the soil, or the water—from which they may be taken up by plants and recycled through the community. The reducers have thus a major role in the closure of nutrient circuits and the continuing productivity at a relatively high level based on this circulation. The significance of the circulation of materials in ecosystems has been touched on in the discussion of plankton production and will be considered further in Chapter 6.

The Biosphere

Before discussing nutrient circulation, we should draw together observations relating to world productivity. The surface of the earth bears a thin and variable film of organic matter, made up of all the world's living organisms. This, the natural community of the world, is the *biosphere*. We shall consider, as far as the information permits, some characteristics of the biosphere as a living system. Table 5.2 is a summary of primary production and plant biomass for the world.

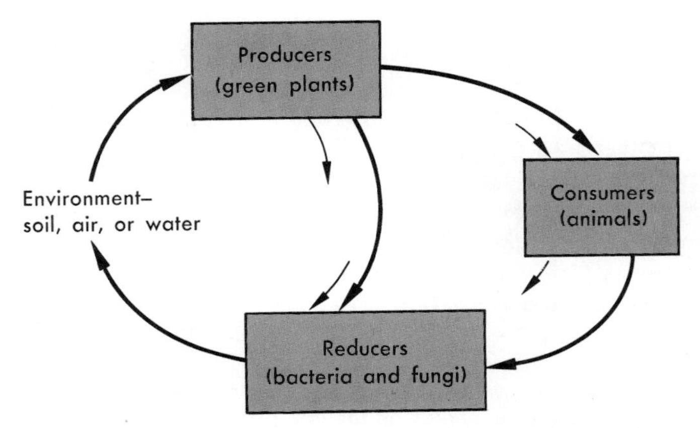

Figure 5.9. Circulation of materials in an ecosystem between environment and organisms, in a much simplified form.

Research on productivity has upset one natural assumption about the biosphere. Drought limits productivity over much of the land surface, but not in the sea. It might then be assumed that the land, much of which is desert, is less productive than the sea, and that the biosphere's production is strongly concentrated in the 70 per cent of the earth's surface that is ocean. It is not so, as we have seen in the discussions of marine and terrestrial productivity. Table 5.2 makes clear the extent of the disparity. "Typical" land vegetation (forest) is about ten times as productive as the plankton of much of the open oceans. Despite the fact that 70 per cent of the globe is ocean, the total net production on land is about twice that in the oceans. Almost half of the biosphere's production is in forests that, before clearing of some of these forests for agriculture, occupied only about 11 per cent of the earth's surface. The contrast in biomass of land and sea is even greater. The total plant biomass on land appears to be about 500 times that in the seas, and more than 1000 times that of all the plant plankton in the oceans. About 90 per cent of the world's biomass is in forests (Table 5.2 is estimated as of 1950).

The key reason for this contrast is stability of surface. The plant plankton, dominated by short-lived cells, has no means for the long-term conservation and accumulation of nutrient elements. In the open oceans these are steadily lost by sinking. A forest occupies a fixed surface on which it can develop a massive structure accumulating nutrients into plant tissue and tending, by partly closed cycles between soil and plant tissue, to hold these nutrients against loss. For the plant plankton, physical processes such as turbulence largely govern nutrient availability and thereby productivity. The forest, in contrast, has evolved a degree of biological control of nutrient availability. Its conservation of nutrients in turn permits the development of a complex community structure including extensive leaf surface for photosynthesis, woody stem and branch tissue supporting the leaves in the light, and roots and vascular tissue that ensure water and nutrient supply to the leaves. In a sense the plankton operates as a "primitive" economy of relative poverty, rapid turnover of materials, and little capital accumulation; the forest operates as an "advanced" economy of high productivity supporting, and supported by, long-term accumulation of organic capital or wealth.

Some other characteristics of the biosphere are estimated in Table 5.3. Most land communities of reasonably favorable environments intercept sunlight with a "leaf area index" or ratio of 3 to 8 m^2 of leaf surface displayed above one m^2 of ground surface. Higher values occur in some communities, especially evergreen needle-leaf forests. The total leaf area estimate for land communities, 644×10^6 km^2, implies a mean leaf area index of 4.4 m^2/m^2. The mean efficiencies for net dry-matter productivity and energy capture per unit leaf surface-area on land are 178 g/leaf m^2/yr and 760 kCal/leaf m^2/yr. For land communities of favorable environments the

Table 5.2 Net Primary Production and Plant Biomass for the Earth. Units are square kilometers, dry grams or kilograms per meter square, and dry metric tons (t) of organic mattter. [Whittaker and Likens in Lieth and Whittaker, 1975.]

Ecosystem Type	*Area * 10^6 km²*	*Net Primary Productivity, per Unit Area † g/m²/yr*		*World Net Primary Production ** 10^9 t/yr*	*Biomass per Unit Area ‡ kg/m²*		*World Biomass ** 10^9 t*
		Normal Range	*Mean*		*Normal Range*	*Mean*	
Tropical rain forest	17.0	1000–3500	2200	37.4	6–80	45	765
Tropical seasonal forest	7.5	1000–2500	1600	12.0	6–60	35	260
Temperate evergreen forest	5.0	600–2500	1300	6.5	6–200	35	175
Temperate deciduous forest	7.0	600–2500	1200	8.4	6–60	30	210
Boreal forest	12.0	400–2000	800	9.6	6–40	20	240
Woodland and shrubland	8.5	250–1200	700	6.0	2–20	6	50
Savanna	15.0	200–2000	900	13.5	0.2–15	4	60
Temperate grassland	9.0	200–1500	600	5.4	0.2–5	1.6	14
Tundra and alpine	8.0	10–400	140	1.1	0.1–3	0.6	5
Desert and semidesert scrub	18.0	10–250	90	1.6	0.1–4	0.7	13
Extreme desert, rock, sand, and ice	24.0	0–10	3	0.07	0–0.2	0.02	0.5
Cultivated land	14.0	100–3500	650	9.1	0.4–12	1	14
Swamp and marsh	2.0	800–3500	2000	4.0	3–50	15	30
Lake and stream	2.0	100–1500	250	0.5	0–0.1	0.02	0.05
Total continental	149.		773	115		12.3	1837
Open ocean	332.0	2–400	125	41.5	0–0.005	0.003	1.0
Upwelling zones	0.4	400–1000	500	0.2	0.005–0.1	0.02	0.008
Continental shelf	26.6	200–600	360	9.6	0.001–0.04	0.01	0.27
Algal beds and reefs	0.6	500–4000	2500	1.6	0.04–4	2	1.2
Estuaries	1.4	200–3500	1500	2.1	0.01–6	1	1.4
Total marine	361		152	55.0		0.01	3.9
Full total	510		333	170		3.6	1841

dry-matter efficiencies are generally 150 to 300 g/leaf m^2/yr, with lower values for evergreen communities; for many communities of arid and cold environments they are 50 to 150 g/leaf m^2/yr. Both leaf area and chlorophyll estimates in Table 5.3 are conservative, excluding green surfaces of stems and branches and the chlorophyll in living tissues other than leaves and in dead organic matter.

Chlorophyll is rather evenly distributed among the more productive land communities, but it is very unevenly distributed between land and water communities. For a wide range of land communities (excluding the least and most productive), chlorophyll contents in leaves of 1 to 4 g/m^2 of ground surface support net primary productivities of 200–2000 g/m^2/yr. The mean productive efficiency for leaf chlorophyll on land, expressed as dry-matter net annual primary productivity in grams, per gram of chlorophyll, is 510 g/g; expressed as energy it is 2200 kCal/g/yr. Efficiencies are higher in forests (mostly 300 to 700 g/g, with lower values in temperate evergreen forests), than in deserts, tundra, and dry grasslands (100 to 300 g/g). For a wide range of plankton communities chlorophyll contents of 0.002 to 0.1 g/m^2 support net productivities of 50 to 1000 g/m^2/yr; for the marine plant plankton Table 5.3 suggests mean net production efficiencies, per gram of chlorophyll, of 3300 g/g and 16,300 kCal/g. The mean chlorophyll amounts on land and sea are 1.5 and 0.05 g/m^2, a thirtyfold contrast. The land (except deserts) has reason to be a deeper green than the sea, and the chlorophyll contrast relates again to stability of surface and extent of photosynthetic apparatus maintained.

Organic matter escapes from the biosphere in various ways and forms a worldwide mass of dead organic substance that can be called the necrosphere. One fraction of this is given in Table 5.3 as the litter on the soil surface of land communities. The amount of this litter per unit area decreases from wet to dry environments (because the productivity contributed to the litter is less in the latter) and from cold to hot climates (because decomposition is faster in the latter). The total mass of surface litter appears to be much less than the "living" land biomass (which includes heartwood and dead branches of living trees), and of the same order as one year's net primary productivity. The mass of humus in the soil is

Unit conversions for Tables 5.2 and 5.3.

* Square kilometers × 0.3861 = square miles.

† Grams per square meter × 0.01 = t/ha, × 0.1 = dz/ha or m centn/ha (metric centners, 100 kg, per hectare, 10^4 square meters), × 10 = kg/ha, × 8.92 = lbs/ acre.

** Metric tons (10^6 g) × 1.1023 = English short tons.

‡ Kilograms per square meter × 100 = dz/ha, × 10 = t/ha, × 8922 = lbs/acre, × 4.461 = English short tons per acre.

Productivities and biomasses expressed as carbon can be multiplied by 2.2 as an approximate conversion to dry matter.

Table 5.3 Other Biosphere Characteristics Related to Productivity. Worldwide totals by ecosystem types and for the earth's surface. Units are square kilometers and metric tons (t) of chlorophyll and dry organic matter. [Whittaker and Likens, 1973, and in Woodwell and Pecan, 1973.]

Ecosystem Type	*Area* * 10^6 km^2	*Chlorophyll* ** 10^6 t	*Leaf Surface Area* * 10^6 km^2	*Litter Mass* ** 10^9 t	*Animal Consumption* ** 10^6 t/yr	*Animal Production* ** 10^6 t/yr	*Animal Biomass* ** 10^6 t
Tropical rain forest	17.0	51.0	136	3.4	2600	260	330
Tropical seasonal forest	7.5	18.8	38	3.8	720	72	90
Temperate evergreen forest	5.0	17.5	60	15.0	260	26	50
Temperate deciduous forest	7.0	14.0	35	14.0	420	42	110
Boreal forest	12.0	36.0	144	48.0	380	38	57
Woodland and shrubland	8.5	13.6	34	5.1	300	30	40
Savanna	15.0	22.5	60	3.0	2000	300	220
Temperate grassland	9.0	11.7	32	3.6	540	80	60
Tundra and alpine	8.0	4.0	16	8.0	33	3	3.5
Desert and semidesert scrub	18.0	9.0	18	.36	48	7	8
Extreme desert, rock, sand, and ice	24.0	0.5	1.2	.03	0.2	0.02	0.02
Cultivated land	14.0	21.0	56	1.4	90	9	6
Swamp and marsh	2.0	6.0	14	5.0	320	32	20
Lake and stream	2.0	0.5			100	10	10
Total continental	149	226	644	111	7810	909	1005
Open ocean	332.0	10.0			16,600	2500	800
Upwelling zones	0.4	0.1			70	11	4
Continental shelf	26.6	5.3			3000	430	160
Algal beds and reefs	0.6	1.2			240	36	12
Estuaries	1.4	1.4			320	48	21
Total marine	361	18.0			20,230	3025	997
Full total	510	244			28,040	3934	2002

*, ** See footnotes to Table 5.2 for unit conversions.

variable and difficult to estimate; but it is believed to be much larger than the mass of surface litter, and probably of the order of 2 to 3×10^{12} t, worldwide. There are other, larger pools of dead organic matter. Dead organic matter in the seas probably exceeds the living by 3 to 4 orders of magnitude. (One estimate of its total is 10×10^{12} t.) Other organic masses are the fossil fuels, petroleum (5×10^{11} t) and coal (5×10^{12} t). These fossil fuels represent accumulations of net ecosystem production in past geological time. Petroleum was probably formed from the fats of diatoms and other marine organisms, gradually accumulated in sediments on the ocean bottom, chemically transformed into hydrocarbons, and concentrated into the particular rock layers man taps with oil wells. Coals were formed in great swamp forests from trees of extinct types, under conditions in which plant tissues failed to decompose as in modern forests. Some net ecosystem production should be accumulating at present as fats on their way to becoming petroleum, and as peat deposits in bogs; but there may be no coal forming in modern forests.

The mass of fossil fuels is far exceeded by still another dead organic fraction: organic matter ("kerogen," etc.) widely dispersed in sedimentary rocks, and probably amounting to 10×10^{15} t, or more than 5000 times the mass of the biosphere's organisms. The oxygen in the atmosphere is a consequence of the photosynthesis that produced this organic matter. Only after the evolution of photosynthetic plants could oxygen accumulate in the atmosphere and support the respiration of animals. Most of the oxygen produced in past geologic periods was used to oxidize inorganic minerals in rocks. The oxygen now in the atmosphere is the chemical complement of some of the reduced carbon of the necrosphere. If that fraction (perhaps 1/15) of the organic matter of the necrosphere could be burned to carbon dioxide and water, the free oxygen of the earth's surface would be exhausted. Thus the biosphere has exerted chemical influences that extend above and below the earth's surface and far exceed in mass the present biosphere itself.

The final columns of Table 5.3 are efforts to estimate secondary productivity and animal biomass. Mean consumptions of net primary production by animals have been estimated as 1 per cent in cultivated land, 2 to 3 per cent for desert and tundra, 4 to 7 per cent for forests, and 10 to 15 per cent for grasslands. Consumption per cents in the oceans are higher, perhaps 40 per cent in the open ocean, 35 per cent in upwelling areas, 30 per cent on continental shelf waters, and 15 per cent in algal beds and estuaries. The over-all mean estimated consumption per cents are 7 per cent on land and 36 per cent in marine communities. Gross growth efficiencies (new protoplasm/food ingested, or secondary production/consumption) are widely variable from zero in some animals no longer growing or reproducing, to 5 to 15 per cent in many herbivores, to higher values in some, especially young carnivores. For the secondary production estimate of Table 5.3, consumption has been multiplied by gross growth efficiencies of 15 per cent for marine communities, grasslands with their grazing mammals, and deserts with their seed eaters, and 10 per cent for all other communities. The results imply that animal secondary production may be less than 1 per cent of net primary production on land, 5 to 6 per cent in the sea. They give the further striking suggestion that about three quarters of world animal secondary production (excluding man and domestic animals) may be in the seas. As we have seen, most of this wealth of marine animal production is in microorganisms inaccessible to human harvest. The production of animals on higher trophic levels is based on harvest by them of part of this secondary production; production of secondary consumers plus all higher animal trophic levels should be around 10 per cent of our estimates for herbivores or primary animal consumers.

These results re-emphasize the importance of detritus feeders. Estimates for bacterial and fungal secondary production become even more tenuous than those for animals. It seems reasonable, though, to treat the biosphere as approaching (apart from the effects of man) a steady state of the whole, in which total respiration of all heterotrophic organisms almost equals total net primary production. (The small difference is the escape of some net ecosystem production from the biosphere into sediments.) Given that near-equality, total reducer assimilation should approximately equal net primary production minus animal assimilation. There are complications to this, and we cannot assess adequately the role of symbionts and the contribution of animals to detritus chains. We judge, though, that as much as 93 per cent of net primary production goes to reducers on land, and perhaps 63 per cent in the oceans. Growth efficiencies (new protoplasmic carbon/substrate carbon) for reducers on land are probably higher over all than gross growth efficiencies of animals. Growth efficiencies may be in the range of 30 to 40 per cent for fungi of decomposition, 5 to 10 per cent for aerobic bacteria, and 2 to 5 per cent for anaerobic bacteria. If we use 20 per cent as a reasonable intermediate growth efficiency for terrestrial reducers, then total reducer production on land is of the order of 21×10^9 t/yr, and 24 times total animal secondary production on land. If we assume a growth efficiency of 5 or 10 per cent for marine reducers, then marine reducer production would be 1.5 to 3.0×10^9 t/yr, or about half to about the same as our estimated marine animal production. Reducer biomasses are largely unknown, but small size and rapid turnover make possible the extensive secondary production by bacteria. Bacterial biomass is probably, both on land and in the sea, much less than that of animals. On land the mass of fungi may not be small compared with that of animals.

The limited data available on animal biomass have been used for the estimates in Table 5.3. In most terrestrial communities, animal biomass is concentrated in small and short-lived arthropods and soil animals—mites, springtails, annelids, and so on—rather than in the more conspicuous vertebrates. (The singing birds in a temperate deciduous forest, for example, have a total biomass of only about 0.1 to 0.2 g/m^2.) Earthworms are apparently the most massive group in forests. Vertebrate biomass may exceed invertebrate biomass in grasslands supporting large populations of grazing mammals. In the sea the greater part of the animal biomass is probably in the smaller and more obscure animal plankton, rather than in the fish. The total animal biomass estimates are quite similar for the continents and the seas. The microscopic cells of plant plankton are relatively easy for animals to harvest. The fairly large fraction harvested (30 to 40 per cent) supports animal production and biomass that are high in comparison with the plant plankton. Tissues of land plants are relatively hard for animals to harvest both because of their structure (wood especially) and the chemical defenses against animals to be discussed in the next chapter. Most plant food on land must be used by reducers. The small fractions harvested by animals (1 to 15 per cent) support animal productivity that is very small compared with that of plants, and animal biomass that is minute (perhaps one two-thousandth) compared with that of the long-lived plants.

The reader may recognize that some of the estimates are developed in the margin between information and ignorance. It seems of interest, however, to characterize the organic film of the world that supports human life, and is now affected by man in ways we shall discuss in Chapter 7. Man's place in the biosphere is no longer small. Biomass estimates (for 1970) are about 52×10^6 t (dry metric tons) for man's world population, and 265×10^6 t for his livestock. If man has not already surpassed the earthworms of the world in biomass, he seems likely to do so. Man harvests (1970) about 1200×10^6 t/yr in cereals and other plant foods—somewhat more than one hundredth of land production, from arable lands producing somewhat more than one tenth of land production. Man is harvesting animals for food to the amount of 72×10^6 t/yr (including milk and eggs) on land, and 16.5×10^6 t/yr from water bodies. About 88 per cent of the latter is taken from the oceans. Human harvest of wood is about 2.2×10^9 m^3/yr, which is equivalent to more than 2×10^9 t/yr of aboveground dry weight cut down in the harvested trees. The cutting and clearing of forests is accelerating, and has probably reduced world forest biomass well below the 1950 values in Table 5.2.

The average energy of sunlight in the visible range available for photosynthesis at the earth's surface is 55 kCal/cm^2/yr. We have given representative efficiencies of primary productivity as 0.91 per cent for a forest and 0.066 per cent for a marine plankton community. Because of the wide expanse of open ocean and the extent of arid and cold desert on land, the mean for the earth's surface is heavily weighted on the lower end. The estimated mean net primary productivity of the biosphere, 336 g/m^2/yr, is equivalent to 1490 kCal/m^2/yr, and implies an over-all efficiency of world net production of 0.27 per cent. The corresponding energy efficiency of gross primary productivity is somewhat more than twice this value—about 0.6. (Gross primary productivity should be less than twice net primary productivity in the sea, but more than twice on land; mean ratios may be 1.5 and 2.7.) Man is releasing energy at an accelerating rate from the accumulated profit of past photosynthesis in coal and petroleum. The total industrial energy release by man, predominantly from fossil fuels, was about 4.7×10^{16} kCal in 1970. This energy release is still much exceeded by the world energy fixation in gross primary productivity, which is about 17×10^{17} kCal/yr. Unlike that world productivity, man's energy release and exploitation of the biosphere is increasing exponentially. World industrial energy release is increasing about 4 per cent a year, and the recent acceleration of American energy use has been near 7 per cent per year. Man's energy release is not yet sufficient to compete with, but is already sufficient to begin to alter the characteristics of, the biosphere.

SUMMARY

Perhaps the most fundamental characteristic of an ecosystem is its productivity: the rate of creation of organic material, by photosynthesis primarily, per unit of area and time. All biological activity of the community, and of human life, is dependent on the energy of gross primary productivity, the energy bound in photosynthesis. The amounts of this productivity are controlled by a number of characteristics of environment, notably nutrient availability, water availability (on land), and temperature; high productivities result from favorable combinations of these characteristics. Large areas of the continents are deserts of low production because of climatic drought or extreme coldness; even larger areas of the oceans are "deserts" of low production because of the deficiency of nutrients in the surface waters of the open oceans.

Three major modes of nutrition and ways of utilizing productivity are represented in the three functional kingdoms of natural communities: the producers, or green plants that create their own food and respire part of it for their needs; the consumers, or animals that feed by ingestion and internal digestion of organic material; and the reducers, or bacteria and fungi, which live by absorptive nutrition, employing external digestion and decomposing organic matter to inorganic products. That part of gross primary productivity that is not respired by green plants is net primary productivity and is available for harvest by consumers and reducers. Organic material and energy are passed along food chains—from plants, through first consumers (or first reducers), to second and third consumers (or reducers).

The steps along food chains, when organisms are grouped by their positions in food chains, are trophic levels.

Because energy is dissipated in respiration on each trophic level, there must be a step-wise decrease in productivity, a pyramid relationship through the sequence of trophic levels. If net primary productivity is rapidly used by consumers, the community may have only a small biomass of organic material accumulated in its plants, as in the plankton. If harvest is mostly delayed, extensive biomass may accumulate in a complex vegetation structure, as in forests. If the rate at which organic matter is produced exceeds that at which it is decomposed, the biomass and structure of the community increase, as in succession. If the rates of photosynthesis and respiration, production and decomposition, are in balance, the community is in a steady state, as is the case in climax communities. Maintenance of a relatively high level of production in this steady state is dependent on the consumers and reducers. These organisms, by using and decomposing plant food, release inorganic nutrients for new uptake by plants, and thus make possible the continuing circulation of nutrients through the ecosystem.

Total net primary production of the biosphere, the community of all organisms of the earth's surface, is about 170×10^9 tons of dry organic matter per year. Land communities average more productive than those of the sea, and about two thirds of global productivity occurs on land. Because of the accumulation of woody biomass on land, the biomass disparity is even greater; biomass on land is about 1800×10^9 tons, and more than one thousand times the plant biomass of the marine plankton. The global efficiency of primary production is about 0.27 per cent for net, and 0.6 per cent for gross primary production relative to the energy of sunlight in the visible range at the earth's surface. Man harvests about 1200×10^6 tons/yr of plant food, and about 90×10^6 tons/yr of animal food from the biosphere. These harvests, and man's release of industrial energy, are still small compared with the biosphere as a whole, but man's pressures on the biosphere are increasing exponentially.

References

General

LIETH, HELMUT and R. H. WHITTAKER, editors. 1975. *The Primary Production of the Biosphere*. New York: Springer (in press).

ODUM, EUGENE P. 1971. *Fundamentals of Ecology*. 3rd ed. Philadelphia: Saunders. xiv + 574 pp.

PHILLIPSON, JOHN. 1966. *Ecological Energetics*. London: Arnold. 57 pp.

*TURNER, FREDERICK B., editor. 1968. Energy Flow and Ecological Systems. *American Zoologist* **8**:10–69.

Production Measurement

EDMONDSON, W. T. and G. G. WINBERG, editors. 1971. A Manual on Methods for the Assessment of Secondary Productivity in Fresh Waters. *International Biological Programme Handbook* **17,** xvi + 358 pp. Oxford: Blackwell.

LIETH and WHITTAKER, 1975.

MILNER, C. and R. ELFYN HUGHES. 1968. Methods for the Measurement of the Primary Production of Grassland. *International Biological Programme Handbook* **6,** xii + 70 pp. Oxford: Blackwell.

NEWBOULD, P. J. 1967. Methods for Estimating the Primary Production of Forests. *International Biological Programme Handbook* **2,** viii + 62 pp. Oxford: Blackwell.

ODUM, H. T. 1956. Primary production in flowing waters. *Limnology and Oceanography* **1**:102–117.

SMALLEY, A. E. 1960. Energy flow of a salt marsh grasshopper population. *Ecology* **41**:672–677.

SMITH, K. L. JR. and J. M. TEAL. 1973. Deep-sea benthic community respiration: an in situ study at 1850 meters. *Science* **179**:282–283.

STEEMANN NIELSEN, E. 1963. Productivity, definition and measurement. In *The Sea* ed. M. N. Hill, vol. **2**:129–164. London: Interscience.

VAN HOOK, R. I. 1971. Energy and nutrient dynamics of spider and orthopteran populations in a grassland ecosystem. *Ecological Monographs* **41**:1–26.

VOLLENWEIDER, R. A., editor. 1969. A Manual on Methods for Measuring Primary Productivity in Aquatic Environments. *International Biological Programme Handbook* **12,** xvi + 213 pp. Oxford: Blackwell.

WHITTAKER, R. H. and G. M. WOODWELL. 1968. Dimension and production relations of trees and shrubs in the Brookhaven Forest, New York. *Journal of Ecology* **56**:1–25.

WHITTAKER, R. H. and G. M. WOODWELL. 1969. Structure, production and diversity of the oak-pine forest at Brookhaven, New York. *Journal of Ecology* **57**:157–174.

Production on Land

BRAY, J. R. and E. GORHAM. 1964. Litter production in forests of the world. *Advances in Ecological Research* **2**:101–157.

CHEW, R. M. and A. E. CHEW. 1965. The primary productivity of a desert-shrub (*Larrea tridentata*) community. *Ecological Monographs* **35**:355–375.

DUVIGNEAUD, PAUL, editor. 1971. *Productivity of Forest Ecosystems:* Proceedings of the Brussels Symposium 1969. Paris: Unesco. 707 pp.

ECKARDT, F. E., editor. 1968. *Functioning of Terrestrial Ecosystems at the Primary Production Level:* Proceedings of the Copenhagen Symposium. Paris: Unesco. 516 pp.

KIRA, T. and T. SHIDEI. 1967. Primary production and turnover of organic matter in different forest ecosystems of the Western Pacific. *Japanese Journal of Ecology* **17**:70–87.

KUCERA, C. L., R. C. DAHLMAN and M. R. KOELLING. 1967. Total net productivity and turnover on an energy basis for tallgrass prairie. *Ecology* **48**:536–541.

*LIETH, H. 1973. Primary production: terrestrial ecosystems. *Human Ecology* **1**:303–332.

REICHLE, DAVID E., editor. 1970. *Analysis of Temperate Forest Ecosystems.* New York: Springer. xii + 304 pp.

†RODIN, L. E. and N. I. BAZILEVICH. 1967. *Production and Mineral Cycling in Terrestrial Vegetation.* Edinburgh: Oliver & Boyd. v + 288 pp.

ROSENZWEIG, M. L. 1968. Net primary production of terrestrial communities: prediction from climatological data. *American Naturalist* **102**:67–74.

WHITTAKER, R. H. 1966. Forest dimensions and production in the Great Smoky Mountains. *Ecology* **47**:103–121.

Marine Production

DUGDALE, R. C. 1967. Nutrient limitation in the sea: dynamics, identification, and significance. *Limnology and Oceanography* **12**:685–695.

KOBLENTZ-MISHKE, O. J., V. V. VOLKOVINSKY and J. G. KABANOVA. 1970. Plankton primary production of the world ocean, pp. 183–193 in *Scientific Exploration of the South Pacific,* ed. by W. S. Wooster. Washington: National Academy of Sciences.

MANN, K. H. 1973. Seaweeds: their productivity and strategy for growth. *Science* **182**:975–981.

RAYMONT, J. E. G. 1966. The production of marine plankton. *Advances in Ecological Research* **3**:117–205.

RILEY, G. A. 1965. A mathematical model of regional variations in plankton. *Limnology and Oceanography* **10**(Suppl.):R202–R215.

*RILEY, G. A. 1972. Patterns of production in marine ecosystems. In *Ecosystem Structure and Function,* ed. John A. Wiens. Oregon State University Annual Biology Colloquia **31**:91–112.

*RYTHER, J. H. 1969. Photosynthesis and fish production in the sea. *Science* **166**:72–76.

†RYTHER, J. H. 1963. Geographic variations in productivity. In *The Sea,* ed. M. N. Hill, vol. **2**:347–380. London: Interscience.

STEELE, J. H. 1958. Plant production in the northern North Sea. *Marine Research, Scientific Home Department* **1958**(7):1–36.

STEELE, JOHN H., editor. 1970. *Marine Food Chains.* Berkeley: Univ. Calif. viii + 552 pp.

STRICKLAND, J. D. H. 1965. Production of organic matter in the primary stages of the marine food chain. In *Chemical Oceanography,* ed. J. P. Riley and G. Skirrow, Vol. **1**:477–610. New York: Academic.

Pyramids and Efficiencies

CHEW, R. M. and A. E. CHEW. 1970. Energy relationships of the mammals of a desert shrub (*Larrea tridentata*) community. *Ecological Monographs* **40**:1–21.

ELTON, CHARLES. 1927. *Animal Ecology*. London: Sidgwick & Jackson. xxi + 207 pp.

ENGELMANN, M. D. 1966. Energetics, terrestrial field studies, and animal productivity. *Advances in Ecological Research* **3**:73–115.

*GOLLEY, F. B. 1972. Energy flux in ecosystems. In *Ecosystem Structure and Function,* ed. John A. Wiens. Oregon State University Annual Biology Colloquia **31**:69–90.

LINDEMANN, R. L. 1942. The trophic-dynamic aspect of ecology. *Ecology* **23**:399–418.

KOZLOVSKY, D. G. 1968. A critical evaluation of the trophic level concept. I. Ecological efficiencies. *Ecology* **49**:48–60.

MANN, K. H. 1965. Energy transformations by a population of fish in the river Thames. *Journal of Animal Ecology* **34**:253–275.

ODUM, E. P. 1971.

ODUM, H. T. 1957. Trophic structure and productivity of Silver Springs, Florida. *Ecological Monographs* **27**:55–112.

SLOBODKIN, L. B. 1962. Energy in animal ecology. *Advances in Ecological Research* **1**:69–101.

WELCH, H. E. 1968. Relationships between assimilation efficiencies and growth efficiencies for aquatic consumers. *Ecology* **49**:755–759.

WIEGERT, R. G. 1965. Energy dynamics of the grasshopper populations in old-field and alfalfa field ecosystems. *Oikos* **16**:161–176.

Detritus and Reducers

ALEXANDER, MARTIN. 1961. *Introduction to Soil Microbiology*. New York: Wiley. x + 472 pp.

BOCOCK, K. L. 1964. Changes in the amounts of dry matter, nitrogen, carbon and energy in decomposing woodland leaf litter in relation to the activities of the soil fauna. *Journal of Ecology* **52**:272–284.

BRAY and GORHAM, 1964.

DARNELL, R. M. 1964. Organic detritus in relation to secondary production in aquatic communities. *Proceedings of the International Association of Theoretical and Applied Limnology* **15**:462–470.

FRANKENBERG, D. and K. L. SMITH, JR. 1967. Coprophagy in marine animals. *Limnology and Oceanography* **12**:443–450.

LAUFF, GEORGE H., editor. 1967. *Estuaries*. Washington: American Association for the Advancement of Science Publication no. **83,** xv + 757.

†MANN, K. H. 1969. The dynamics of aquatic ecosystems. *Advances in Ecological Research* **6**:1–81.

MARGULIS, L. 1975. Five kingdom classification and the origin and evolution of cells. *Evolutionary Biology,* ed. W. Steere, M. Hecht, and T. Dobzhansky Vol. **7** (in press).

OLSON, J. S. 1963. Energy storage and the balance of producers and decomposers in ecological systems. *Ecology* **44**:322–331.

REINERS, W. A. 1973. Terrestrial detritus and the carbon cycle. *Brookhaven Symposia in Biology* **25**:303–327.

RILEY, G. A. 1963. Organic aggregates in seawater and the dynamics of their formation and utilization. *Limnology and Oceanography* **8**:372–381.

*SAVORY, T. H. 1968. Hidden lives (the cryptozoa). *Scientific American* **219**(1): 108–114.

*TEAL, J. M. 1962. Energy flow in the salt marsh ecosystem of Georgia. *Ecology* **43**:614–624.

WHITTAKER, R. H. 1969. New concepts of kingdoms of organisms. *Science* **163**:150–160.

*WITKAMP, M. and J. van der Drift. 1961. Breakdown of forest litter in relation to environmental factors. *Plant and Soil* **15**:295–311.

The Biosphere

BAZILEVICH, N. I., L. YE. RODIN and N. N. ROZOV. 1971. Geographical aspects of biological productivity. *Soviet Geography: Review and Translation* **12**: 293–317.

BROWN, L. R. 1970. Human food production as a process in the biosphere. *Scientific American* **223**(3):161–170.

BREGER, IRVING A., editor. 1963. *Organic Geochemistry*. New York: Macmillan and Pergamon. x + 658 pp.

COLE, L. C. 1958. The ecosphere. *Scientific American* **198**(4):83–92.

EGLINTON, G. and M. CALVIN. 1967. Chemical fossils. *Scientific American* **216**(1):32–43.

GOLLEY, 1972.

*HUTCHINSON, G. E. 1970. The biosphere. *Scientific American* **223**(3):44–53.

WHITTAKER, R. H. and G. E. LIKENS, editors. 1973. The Primary Production of the Biosphere. *Human Ecology* **1**(4):299–369.

*WOODWELL, G. M. 1970. The energy cycle of the biosphere. *Scientific American* **223**(3):64–74.

†WOODWELL, GEORGE M. and ERENE V. PECAN, editors. 1973. Carbon and the Biosphere. *Brookhaven Symposia in Biology* 24, vii + 392 pp. Springfield, Va.: National Technical Information Service (CONF-720510).

6

Nutrient Circulation

In an ecosystem materials circulate from environment through producers, consumers, and reducers back to environment. The processes of transfer and concentration of materials in ecosystems have great, and increasingly urgent, significance to man. We shall illustrate some of the principles of nutrient circulation, especially the ways parts of ecosystems are related to one another by nutrient transfer and moving water. The word "nutrient" can apply to any substance taken into an organism that is metabolized or becomes part of ionic balances (excluding toxins and substances used only as signals for behavior). In this chapter we shall deal with various chemical exchanges between organisms and environment and between different organisms; but we shall emphasize inorganic elements and ions such as calcium and potassium, nitrate and phosphate, that are present in water and soil and may be taken up to become part of community function.

Phosphorus in Aquaria and Lakes

Phosphorus has a major role in determining productivity of plankton communities. It is natural that, for further study of the role of phosphorus in aquatic communities, radiophosphorus (^{32}P) should be introduced and followed as a community-level tracer. Such experiments can be done in the laboratory, using either a plankton-and-water sample from a natural aquatic community or the small-scale community that develops through some weeks in an aquarium provided with water and nutrients and an initial seeding with pond organisms. Into a 200-liter aquarium microcosm of the latter sort 100 μc of ^{32}P-labeled phosphate (as phosphoric acid) were introduced, with results shown in Figure 6.1. After the tracer introduction:

1. There was initial, very rapid movement from water into plankton organisms and turnover from these back to water. In the experiment illustrated one half of the ^{32}P had moved into the plankton (predominantly single-celled green algae) within two hours, and by twelve hours the distribution of ^{32}P was in a steady-state balance between plankton and water. Other experiments, using millipore filters for fuller separation of small plankton cells and particles with bacteria from the water, have

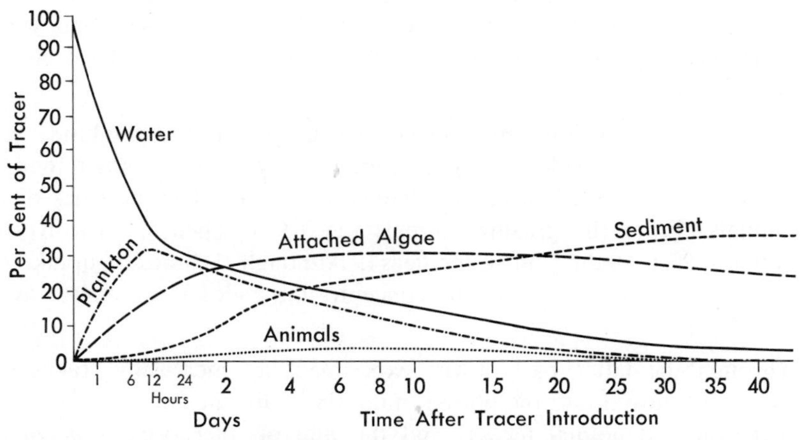

Figure 6.1. Movement of radiophosphorus in an aquarium microcosm. Percentage of the tracer present at a given time (after correction for radioactive decay) is on the vertical axis, time after tracer introduction (on a square-root scale) is on the horizontal. [Whittaker, 1961.]

shown even faster uptake (up to half the amount in the water in three minutes) and balance (93 per cent in the plankton and particles after twenty minutes). Turnover rates (in fractions of the ^{32}P in the plankton returning to the water per unit time) were 0.27/hr and 0.013/min for the two experiments. These rapid movements of ^{32}P are affected by adsorption onto the surfaces of cells as well as absorption through those surfaces. The mass of the plankton is very small in relation to the water in which it is suspended. With the greater part of the phosphate in this small mass, the concentration of the ^{32}P is many thousand times as high in the plankton as in the water. Concentration ratios (expressed as ^{32}P content per unit dry mass of organisms divided by ^{32}P content of an equal mass of water) are in some cases, with low phosphate content in the water, of the order of one to two million-fold.

2. Somewhat more slowly through the first few hours, the ^{32}P moved into the filamentous algae growing on the sides and on the mud-containing bottom trays of the aquarium. Maximum ^{32}P content per unit mass and approximate equilibrium with the water were reached for attached algae within twenty-four hours. (In general, the larger the organism, the larger the mass or pool of phosphate it contains, the lower the turnover rate for this pool, the slower the uptake per unit mass, the later the equilibrium is reached, and the slower the decline from that equilibrium if the experiment is continued long enough for such decline to occur.) As a substantial part of the ^{32}P moved into the attached algae, the ^{32}P in the water-and-plankton together declined. Even though equilibrium of ^{32}P content per unit mass of algae and water was reached during the first day, the total content of ^{32}P in attached algae continued to increase, because of the growth in mass of the algae, until the sixth day. Thereafter, because of the turnover of ^{32}P between algae and water, there was net movement of ^{32}P out of the algae as ^{32}P levels in the water declined.

3. The tracer moved into animals (water fleas grazing the plankton algae, snails grazing the sidewall algae) more slowly than into their food, at rates that varied with size, food habits, and other characteristics of the animals. From the grazing animals the ^{32}P reached the carnivorous animals (fish feeding on water fleas). Although the rate of uptake decreases along a food chain, the concentration ratios in animals may be very high.

4. Through the latter part of the experiment an increasing fraction of the tracer moved into the bottom mud, the sediment (of recently settled plankton and animal feces), and the film of microorganisms on the aquarium walls. Some turnover of ^{32}P between these three parts of the ecosystem and the water continued. There was, however, net movement of ^{32}P "downward"—out of the water and active circulation between

water and organisms, and into the less active or bound forms of the sediment, mud, and surface films. By gradual accumulation three quarters of the ^{32}P had moved into these pools of less active turnover by the end of the experiment, forty-five days after tracer introduction.

The tracer approach can be applied with isotopes other than ^{32}P, but such work is limited as yet. The ^{32}P transfer patterns are representative in broad features of the ways other substances circulate, but details for other isotopes and compounds necessarily vary. It is of interest to see whether the patterns of ^{32}P movement in lakes are similar to those in aquaria; in general they are. ^{32}P introduced into the surface waters of a lake is rapidly taken up by plankton, particles, and bacteria, while only a small amount remains in the water. There is a slower, longer-term movement of ^{32}P into rooted plants along the shores of the lake, into animals, and downward into the sediments. The plants are analogous to the side-wall algae of the aquarium, but phosphorus movement through these higher plants appears to involve two rates: a faster uptake and release by the diatoms and other microorganisms forming the surface film on the plants, and a slower turnover through the tissues of the plants themselves.

The study of phosphorus in lakes provides further observations. In summer many lakes are stratified. An upper layer of warmer water, usually a few meters deep, is separated by a thermocline (a zone of steep temperature change with depth) from the mass of colder water in the depths as shown in Figure 6.2. ^{32}P introduced into the surface water can be traced through the thermocline into the deeper water. The movement indicates that the sinking of plankton and particles carries nutrients downward from the surface waters, as in the open ocean. ^{32}P has also been released from a bottle exploded in the deeper water to observe its spread there. In one experiment the ^{32}P was found to move about 3 meters per day in the deeper water, but to show little movement through the thermocline into the surface water. From the deeper water the ^{32}P can be traced into the mud by settling of particles and uptake by bacteria and other organisms in the mud. Some ^{32}P is released back from the mud into the water. The amount moving back in this way is less than that moving into the mud, however, so that there is progressive net accumulation of ^{32}P in the mud. This ^{32}P accumulation and turnover in the mud is affected by the oxygen content of the deeper water. When free oxygen is present, the phosphorus is more effectively removed from the water and held in the mud than when free oxygen is lacking.

Such observations bear on lake productivity (which is discussed in Chapter 7) in a number of ways. When phosphate concentration in the water is increased by adding phosphate fertilizer, the concentration in the water and plankton thereafter declines (usually within some weeks) back to the level before fertilization. A single fertilization thus produces only

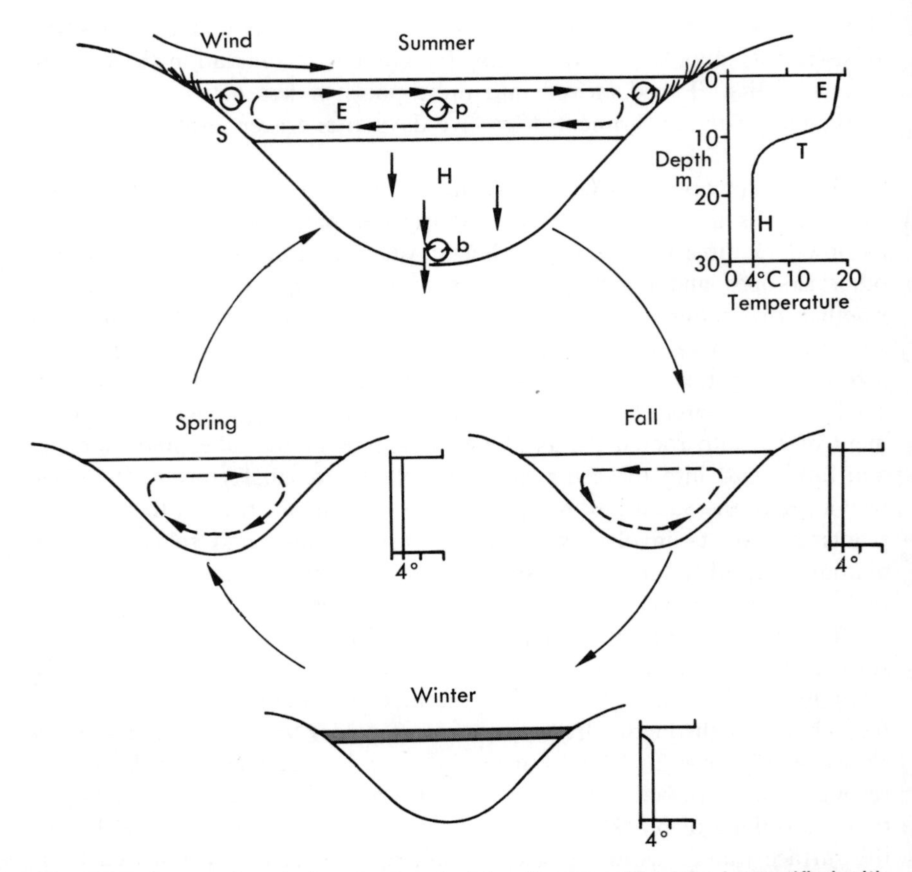

Figure 6.2. Seasonal cycle in a temperate lake. **Summer:** The lake is stratified with a layer of warmer water (*E,* the epilimnion) on top of the colder and denser water of the depths (*H,* the hypolimnion). As shown in the temperature curve on the right, a layer of steep temperature change with depth, the thermocline, *T,* separates the warmer and colder water. Nutrient circulations are indicated by the smaller circles—*p,* the plankton; *s,* the shore community; and *b,* the benthos or bottom community. The friction of the wind causes the upper layer of water to circulate, so that nutrients are transported between the plankton and shore communities. Particles and organisms settle through the deeper water, carrying nutrients to the bottom, where they may be circulated between water and sediment, *b,* but where there is net movement into the sediment. **Fall:** The surface water cools until temperature is the same at all depths, and the whole lake is circulated by the wind. **Winter:** The water is stagnant and near 4°C., except close to the surface ice cover. **Spring:** The surface water warms until temperature is the same at all depths, and the lake is again circulated. Further warming of the surface water leads again to summer stratification.

a temporary increase in productivity. Some of the fertilizer phosphate may leave the lake in the stream that flows from it. The lake retains in its basin, however, most of the phosphorus that enters it, whether it enters as fertilizer, or as the normal, continuing input in streamwater, or from another source. In a typical lake only a fraction (a third, say) of the phosphorus that enters the lake leaves it in the outflowing stream. The remaining phosphorus does not simply accumulate, but after being taken up by lake organisms and used as a basis of their productivity is lost to the lake by the net movement into permanent deposit in the sediments. Because of the steady loss of phosphorus (and other nutrients) into the sediments, as well as stream outflow, the lake is dependent on the flow of nutrients into it from its watershed to maintain its productivity.

The uptake of phosphorus by shore plants indicates the importance of these plants for nutrient movement and productivity. In many lakes much, and in some shallow ponds most, of the primary productivity is by shore plants and not by the open-water plankton. The sinking of plankton and particles removes phosphorus and other nutrients from the surface waters in stratified lakes and limits their productivity. Stratification affects nutrient movement in a number of ways that relate to annual cycles of productivity. Those cycles vary widely in different lakes, but we can describe one that seems typical for some medium-sized, temperate lakes (Figure 6.2).

During the winter the water is both cold (near the temperature of maximum density, 4°C, at most depths) and fairly stable (particularly if the surface water is below 4°C or frozen). Productivity is low, and nutrients tend to accumulate in the deeper water. In the spring, when the waters are at the same temperature (which may be near 4°C) at all depths, there is no density contrast; and the waters of the lake are turned over by the wind. Nutrients are brought to the surface at a time when light intensity is increasing and surface temperatures are 4°C or warmer, and the result is a spring bloom or peak of plankton productivity. In summer the lake may be stratified, with the wind mixing only the warmer, less dense, surface waters while the deeper waters remain cold, generally near 4°C. Through the summer the nutrient content of the surface waters, and the productivity supported by them, decrease. If, however, the lake is a productive one, the large amount of dead organic matter (including some from shore plants and from outside the lake as well as from plankton) that sinks into the deeper waters may exhaust the oxygen there as bacteria use the oxygen to decompose organic matter. The nutrients released by decomposition partly remain in the water and partly accumulate in the mud; but the concentration of nutrients in the water will be higher if the lake is productive and if oxygen is exhausted in the deeper waters. In the fall the surface waters cool; and when water temperature and density are the same at all depths, the lake waters again turn over. Nutrients are then brought to the surface where they may support a secondary, fall peak of productivity. Productivity decreases into the winter but does not end while sunlight reaches the surface water. Even in midwinter and even under ice cover that admits light, some plankton algae carry on photosynthesis.

From the phosphorus experiments in both lakes and microcosms, a number of principles can be stated:

1. The role of organisms in movements. Introduced substances may be rapidly turned over between water and organisms and effectively removed from the water into organisms. The rapidity of these processes and their effect on water content emphasizes the responsiveness of organisms and communities to the chemistry of environment, and the significance of their effects on the chemistry of environment. Uptake and accumulation by organisms (and settling of dead organisms) are responsible also for the slower, longer-term removal of the introduced substance from the water with its plankton, into other parts of the ecosystem.
2. High concentration ratios into organisms. Although concentration ratios vary with many factors, many-thousand-fold concentrations from environment into organisms are common. Such high concentration ratios can occur both for essential nutrients such as phosphate and for toxic materials.
3. Progressive removal from environment, versus steady state. When a substance subject to biological concentration is introduced as a single dose, as in the microcosm experiment described, the substance is first taken up by organisms and then moves progressively out of active circulation and into less active or inactive forms. The concentration of a fertilizer (such as phosphate) introduced as a single dose will decline toward the same steady-state level, determined by the functional characteristics of the ecosystem, as existed before its introduction. When a substance is introduced on a continuing, constant basis, however, this continuing introduction becomes one of the factors determining the steady state for the system. A new pattern of pool contents and transfer rates for the system will develop, based on the new rate of introduction and its relationship to the system's new rates of turnover and removal.
4. Complexity of transfer pattern. A pattern of ^{32}P movement worked out for an aquarium microcosm is illustrated in Figure 6.3. The diagram is by no means a simple water-plant-animal-water circle, yet the diagram is quite incomplete in relation to the complexity of ^{32}P movements in the pond. Rates and routes of transfer vary in most diverse ways for different substances in a given ecosystem and can vary for a given substance in different ecosystems.
5. Transfer unity of the ecosystem. A multiplicity of transfers, a complex exchange of many inorganic and organic substances, interrelate organisms with one another and environment and give the ecosystem its functional unity. The pattern of movement of a substance in a large ecosystem such as a lake is a product of both movements in space (of

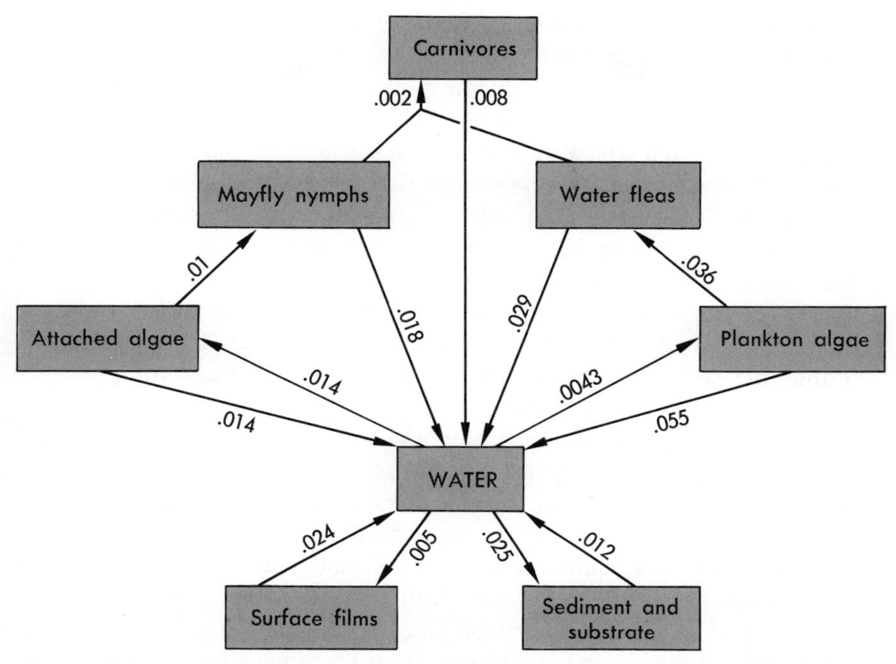

Figure 6.3. Transfer pattern for radiophosphorus in an experimental pond. Numbers are transfer rates, as decimal fractions of the ^{32}P in the box at the tail of the arrow, moving into the box at the head of the arrow per hour. [Whittaker, 1961.]

water or air and organisms) and movements in place (transfer between organisms and environment and along food chains).

Nutrients in Forests

Some aspects of nutrient cycling are more easily studied in a forest than in a water body. The more complex and massive structure of the forest makes it easier to separate different tissues and organisms from one another and measure some of the transfer rates between them. In particular, analyses of nutrient elements in plant tissues may be combined with measurements of biomass and net production of those plant tissues. To these data should be added measurements of nutrient movement into the soil in precipitation, animal harvest of plant tissue and nutrients, and (if possible) information on nutrient release and movement in the soil. Using such information we may observe how nutrient movement in a forest differs from that in a lake even though an underlying principle—the ecosystem as a functional system in which environment and community are linked together by nutrient cycling—remains the same.

For the most part inorganic nutrients enter the organisms of the forest, and the living phase of the ecosystem, only by way of plant roots. These roots extend through the soil, dividing into finer and finer branch roots, in a way that fairly well occupies the volume of soil and provides a large surface area through which nutrients are taken up. The underground, root surfaces of communities are unmeasured but suspected of exceeding the above-ground leaf surfaces. In any case, nutrients mostly enter the community not through the surfaces of the roots themselves (nor, in most plants, through root hairs) but through the still larger surface of fungal hyphae. The surfaces of most roots are mycorrhizal (occupied by symbiotic root fungi), and filaments or hyphae of these fungi extend from the roots out into the soil; these fungi are the agents, or intermediaries, of nutrient uptake into most land plants.

Nutrients are taken up along with soil water; but the nutrient uptake is selective, and not simply osmotic or passive. Energy is expended to take up some nutrients against the concentration gradient (from lower concentrations in soil water to higher concentrations in protoplasm). In contrast to this, only small amounts of other nutrients may be taken up despite their relative abundance in the soil. Nutrients can also be taken up from the surfaces of soil particles, or from decomposition of organic matter, by the mycorrhizal fungus and transferred from the fungus to the plant.

Once in the plant the nutrients are transported inward along branch roots to the stem base and upward through the stem to leaves and other above-ground tissues. We have indicated that about half the net primary productivity of a tree may go into stem and branch wood and bark, and about a third into leaves. The distribution of inorganic nutrients into new tissues is quite different from this. The leaves are complex organs made almost wholly of living cells and their walls, adapted to carry on photosynthesis and other metabolism, with a relatively high content of protoplasm and the nutrients used in protoplasm. Wood, on the other hand, is adapted to support the plant and transport water and other materials. Much of the volume of wood consists of the dead water-conducting tubules of the xylem, and as a whole wood is less dense in protoplasm than are leaves. The contrast in nutrient use in different plant tissues is indicated in Table 6.1. The last column gives the ratio of leaf to wood nutrient concentration. For a number of elements the leaves have 6 to 20 times as high a content per unit dry weight as wood. Considering all nutrients together, about three quarters of the current use of inorganic nutrients above-ground in net primary production goes into the leaves. The nutrient content of cambium and living bark is higher than that wood and dead bark, and the nutrient content of other short-lived parts of plants (flowers, fruits, and bud-scales) is relatively high per unit mass.

There is thus an inverse relation between the durability of tissues of trees and their nutrient contents. Once nutrients are incorporated in the

Table 6.1 Nutrient Contents of Plant Tissues. Contents are given in parts per million of dry weight in sugar maple (*Acer saccharum*) in mixed deciduous forest at Hubbard Brook, New Hampshire. [Likens and Bormann, *Yale Univ. Forest. Sch. Bull.*, 79 (1970); Gosz *et al.*, *Ecology*, 53:769 (1972).]

	Stem Heart-wood	*Stem Sap-wood*	*Stem Bark*	*Branches*	*Current Twigs*	*Summer Leaves*	*Autumn Leaves*	*Leaves / Sapwood*
Nitrogen	970	980	5,500	3,700	13,400	22,000	8,400	22.5
Phosphorus	40	100	300	700	1,800	1,800	400	18.0
Sulfur	90	110	640	450	800	2,100	1,450	19.1
Calcium	3,200	1,000	14,100	4,300	16,200	6,000	5,000	6.0
Potassium	3,000	700	2,900	1,700	9,300	10,100	5,400	14.4
Magnesium	600	200	600	300	1,000	1,200	600	6.0
Manganese	470	140	940	580	1,330	1,740	2,140	12.4
Iron	18	21	55	24	54	120	97	5.7
Zinc	10	8	29	19	62	52	32	6.5
Sodium	10	7	93	7	40	16	18	2.3
Copper	.5	1	6	4	13	9	7	9.0
Ash weight	13,900	4,200	45,400	16,600	58,700	56,500	43,700	13.4

inner dead heartwood they remain there for decades or centuries, until the tree dies and falls and its stem is decomposed. The nutrient cycle from soil into wood and back to soil is slow, but the fraction of nutrients used in this slow cycle is relatively small. Leaves have life spans of a few months in deciduous forests, around a year in tropical rain forests, or a few years in temperate evergreen forests. The greater share of the nutrients are used in these short-lived tissues and are more rapidly cycled between plant and soil. Presumably it is to the plant's adaptive advantage to economize on use of its nutrients in wood, and to direct the greater share of its nutrients through leaves and other more active tissues. Too rapid nutrient cycling through the leaves may not always be to the plant's advantage, however. Pines seem more successful than deciduous trees on some soils of low fertility in the Temperate Zone. The pines may have an advantage from better conserving their nutrient supply, first because leaves are longer-lived on the tree, and second because nutrients are released gradually as leaves fall and decompose slowly. In a deciduous forest, in contrast, many of the nutrients are released in an annual pulse as the leaves fall during a short period in autumn and as these less resistant leaves are more rapidly decomposed.

The return of nutrients to the soil is affected by other processes. As leaves age they change in weight and nutrient content. Shortly before deciduous leaves fall, some of the organic matter that is still in transportable forms moves back from leaves into twigs. At this time concentrations of some inorganic nutrients in the leaves increase, while concentrations of others decrease. The contrast between midsummer and autumn leaves before they fall in one tree species is indicated in Table 6.1. Which nutrients increase and which decrease in leaves before their fall varies with species, but such critical nutrients as nitrogen and phosphorus generally decrease. The late-season transfer from leaves into twigs implies that some nutrients are conserved against loss with the falling leaves.

A process that works against this conservation of nutrients is the rain-washing of leaf (and bark) surfaces. *Leaching* refers to the effect of water in dissolving materials and transporting them away—generally downward either from plant surfaces to the soil, or from upper soil levels to lower ones. Table 6.2 gives data on nutrient leaching for a British oak forest.

Table 6.2. Movement of Nutrients into the Soil in an English Oak Woodland. Values in the first three rows are percentages of the total nutrient weights in the fourth row. [Carlisle *et al.*, *Jour. Ecol.*, 54:87 (1966). See also 55:615 (1967).]

	Nitrogen	*Phosphorus*	*Potassium*	*Calcium*	*Magnesium*	*Sodium*	*Carbon*
Percentage in the precipitation above the canopy	19.1	12.3	7.7	17.8	35.0	61.8	2.4
Added to the precipitation by washing from plant surfaces	−1.4	25.1	65.1	24.1	35.8	35.3	8.0
In litter fall of dead leaves and branches	82.3	62.2	27.2	58.1	29.3	2.9	89.6
Total weight of element added to the soil, $g/m^2/yr$	4.99	0.35	3.86	4.10	1.32	5.72	219

The second row of numbers is the percentage of the total amount of a nutrient that reaches the soil from above, that does so by being leached from plant surfaces and carried downward in rain water. For two of the nutrients —potassium and sodium—a larger fraction of the element reaches the soil by washing from leaf and bark surfaces than in falling dead leaves. Nitrogen shows a reverse effect here, implying that it is being extracted from the rain water flowing across plant surfaces. This extraction may be primarily uptake by organisms (lichens, algae, and bacteria) on bark and leaf surfaces. As in aquatic communities the surfaces of land plants support other organisms that are active in nutrient uptake. The water that reaches the forest floor contains also a significant amount of organic carbon, as indicated in the last column of the table. The greater part of this carbon was in a triple sugar, melezitose, from the secretions of aphids; a smaller part was in the form of phenolic compounds.

A much larger amount of organic carbon reaches the forest floor as litter—dead leaves and fallen branches and other plant parts. Once fallen these tissues are acted on by the detritus-processing system of fungi, bacteria, and animals. Some of the litter may be eaten by animals, but much of it is permeated by the hyphae of fungi. In food relations the fungi form a spectrum from some deriving organic nutrition primarily from the plant with which they are symbiotic, through some that both are symbiotic and act as decomposers (some of which transfer both food and inorganic nutrients to plants), to those that are decomposers without symbiotic relation to plants. Bacteria also are active in the decomposition and are, along with the fungi, part of the sheath of organisms surrounding roots and occupying the soil nearby. (This layer, from the root surface to the soil close to and influenced by the root, is the rhizosphere.) The bacteria and other organisms living around root surfaces use the organic matter that leaks from roots into the soil. Some of the fungi and bacteria are eaten by animals and protists. There is thus a very complex pattern of transfer of organic and inorganic nutrients among the plants, animals, bacteria, fungi, and protists of the soil. Some of these movements have been followed with radioactive tracers (for example, radiocarbon can be tracked from a spruce tree, to a fungus mycorrhizal with it, to a non-green vascular plant, *Monotropa,* deriving its nutrition from the fungus with which it also is mycorrhizal). Most of this wealth of transfer detail is unknown to us and must be left out when we discuss the over-all movement of nutrients from plants to litter and soil and back to plants.

The amount of organic matter returned to the soil by litter fall and by root death is closely related to net primary productivity. In a climax forest the annual addition of detritus to the litter and soil should differ from net primary productivity only by the small amount of direct animal harvest of living plant tissues. The amounts of detritus added to soil and litter each year are similar in many kinds of forests, but the rates of decomposition vary widely. Wood is more resistant to decay than leaves, and evergreen leaves are more resistant than deciduous leaves. Tissues high in structural materials (hemicellulose, cellulose, and lignin) are more resistant than tissues lower in these materials, and a high content of lignins in particular makes tissues resistant to decomposition. Temperature also affects decay rates. If we set aside the different rates affecting different tissues (and different periods in the decomposition of a given tissue), we can characterize litter decay by a rate constant k, and by decomposition half lives—the time after which one half of an original dry mass of litter remains (Table 6.3). Litter half lives are of the order of more than ten years in northern coniferous forests, one to a few years in temperate deciduous and southern pine forests, and fractions of a year in tropical rain forests. These half lives imply quite different rates of turnover of nutrients through the litter in different kinds of forests. They also imply wide differences in the steady-state masses of litter (and pools of nutrients in litter) present on the soil surface in different kinds of communities (Tables 6.3 and 5.3).

Table 6.3 Litter Steady States

L is the annual fall of dead leaves and other litter, and M is the mass of litter present at a given time, decaying with a rate constant k (an instantaneous decay rate analogous to the λ of radioactive decay).
The change of litter mass with time is,

$$\frac{dM}{dt} = L - kM \tag{1}$$

Of an original litter mass, M_o, the amount remaining, M_t (excluding new litter fall) after decay loss through time t is,

$$M_t = M_o e^{-kt} \tag{2}$$

(in which e is the base of natural logarithms).
The amount of litter accumulated after t time units of a succession (if k and L are constant) is,

$$M_t = M_m(1 - e^{-kt}) \tag{3}$$

M_m is the maximum or steady-state litter mass. When $dM/dt = 0$, the litter mass is in steady state and,

$$L = kM_m \tag{4}$$

If $M_t = 0.5M_o$, then t becomes a half life or $t_{.5}$, the time for one half the litter present at a given time to decompose.
It is also possible to define a "residence time" as an average time spent by a molecule in the litter before decomposition as,

$$t_r = M_m/L = 1/k = 1.44t_{.5} \tag{5}$$

Sample Litter Steady-State Relations

	M_m g/m²	L g/m²/yr	k yrs⁻¹	$t_{.5}$ yrs
Tropical rain forest	200	1200	6.0	0.12
Temperate deciduous forest	1200	800	0.67	1.0
Boreal conifer forest	6000	600	0.10	7.0
Temperate grassland	2000	500	0.25	2.8

The different rates of decomposition in turn imply different effects on the soil. In a tropical forest litter decomposes rapidly and relatively completely—to carbon dioxide and water as major end products. Carbon dioxide is released and in solution in soil water becomes carbonic acid, and this weak acid moves down through the soil carrying many of the inorganic nutrients down with it. In a cool-temperate evergreen forest decomposition is slow, because of low temperature, low pH, and the resistant character of the litter. Substantial amounts of the organic compounds produced by incomplete breakdown of organic matter are present in the soil at a given time. Among these a group of phenolic acids are relatively abundant. These acids also are active in the downward transport of nutrients. Leaching in the soil by both inorganic and organic acids can carry nutrients downward in the soil to levels at which they may be no longer available to plant roots. Most forests are affected by this loss from the ecosystem's pool of available nutrients. In order to maintain a steady-state pool of nutrients in a climax forest (or a growing pool in a successional forest) they must be replaced from one or more sources—rainfall, weathering of inorganic soil material, or transport into the ecosystem.

Figure 6.4 draws together information on calcium transfer and pools in a young forest, the Brookhaven oak-pine forest for which productivity measurements have already been given. In this forest the net ecosystem production of 540 g/m^2/yr is accumulating in wood and bark, the 540 g include 1.8 g of calcium being accumulated in wood and bark. In this respect the community is not in steady state; but the addition of calcium in rainfall and weathering should balance that lost in leaching plus that accumulated in new tissue.

Figure 6.4 reasonably represents the pattern of flow of other inorganic nutrients, but the quantitative relations differ from one element to another. As Tables 6.1 and 6.2 show, elements differ in the extent to which they are leached from plant surfaces, and the extent of return from leaves to twigs before leaf fall. It is possible to arrange the various elements in a series according to relative stability versus mobility in the nutrient pool of the plant. Consider first the fraction of the carbon that turns over from the above-ground biomass into the surface litter each year. For the Brookhaven forest this fraction is 0.095/yr; for the Hubbard Brook forest to be discussed below it is 0.043/yr. Parallel fractions can be calculated for each nutrient element, as a fraction returning to the soil by both litter fall and leaching, of the total in the above-ground pool of that nutrient. On the basis of these turnover rates or fractions, a sequence of major elements from most to least mobile in a temperate deciduous forest is: sulfur, potassium, manganese, and sodium, magnesium, nitrogen, and iron, calcium and phosphorus, and carbon (Table 6.5). Nutrients differ also in their mobility in the soil once returned to it or released in weathering. There are differences in leaching effects in different soils, but over-all relative mobilities (based on comparing rock content and river transport) are given in the Polynov series. The sequence from high to low mobility—chlorine, sulfate, calcium, sodium, magnesium, potassum, silica, iron, aluminum—differs from the forest turnover series, for elements given in both.

There are differences also in the way the different plant species in a community turn over the nutrients. Figure 6.5 shows root and shoot pro-

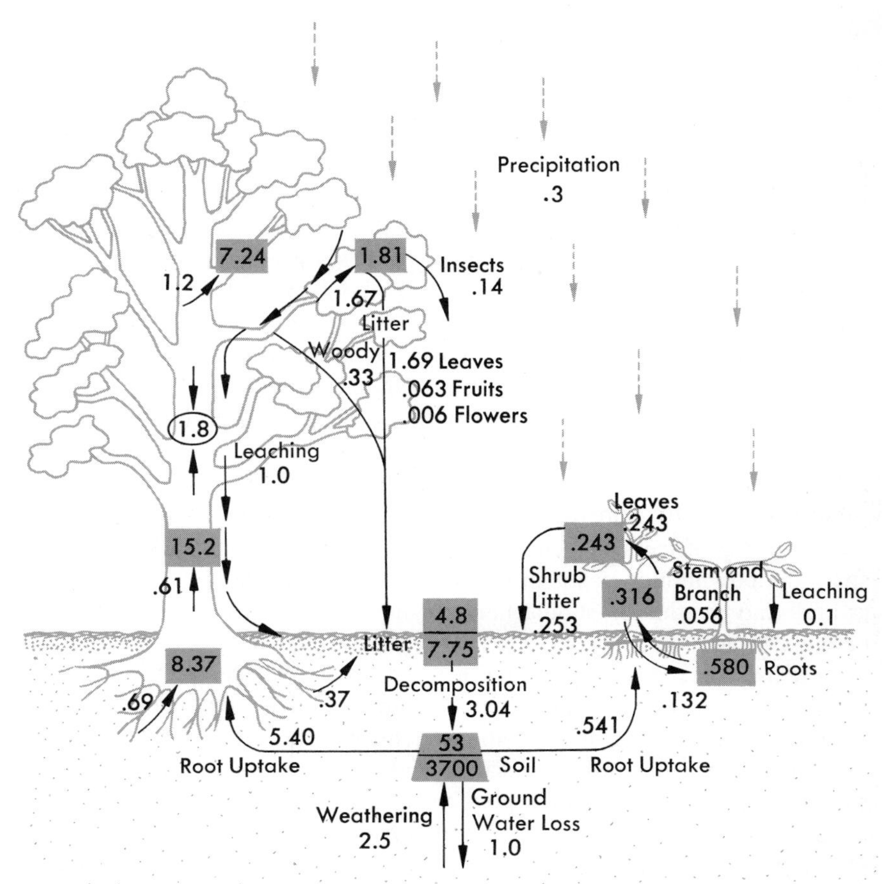

Figure 6.4. Calcium cycle for an oak-pine forest. Calcium storage and flow are shown for the forest at Brookhaven, New York, with a canopy of oaks and scattered pines, an undergrowth dominated by blueberry and huckleberry shrubs (see also Figure 6.5). Boxed numbers are calcium stocks or pools in g/m^2; unboxed numbers are transfers in g/m^2/yr. Arrows into plant pools are calcium used in net primary productivity. The circled 1.8 g/m^2/yr is calcium into storage in net ecosystem production or net community production (accumulation of calcium with increase in the community's woody tissue). Total root uptake includes calcium into net primary productivity, plus that leached from plant surfaces, plus late-season addition of calcium to leaves before they fall (estimated as 0.16 g/m^2/yr for trees and 0.002 for shrubs). The upper soil compartment is for exchangeable and the lower for total soil calcium. The shrub, litter, and soil pools are assumed to be in steady state. Calcium input by precipitation plus weathering equals calcium into net ecosystem production plus groundwater loss. [Unpublished data of G. M. Woodwell and R. H. Whittaker; some values preliminary.]

files for the seven major woody plants of the Brookhaven forest. They differ—in ways that relate to nutrient function. Scarlet oak (*b*) has oblique branches that conduct much of the rainwater inward to the stem, down which the water flows to the base of the tree, where it spreads through a

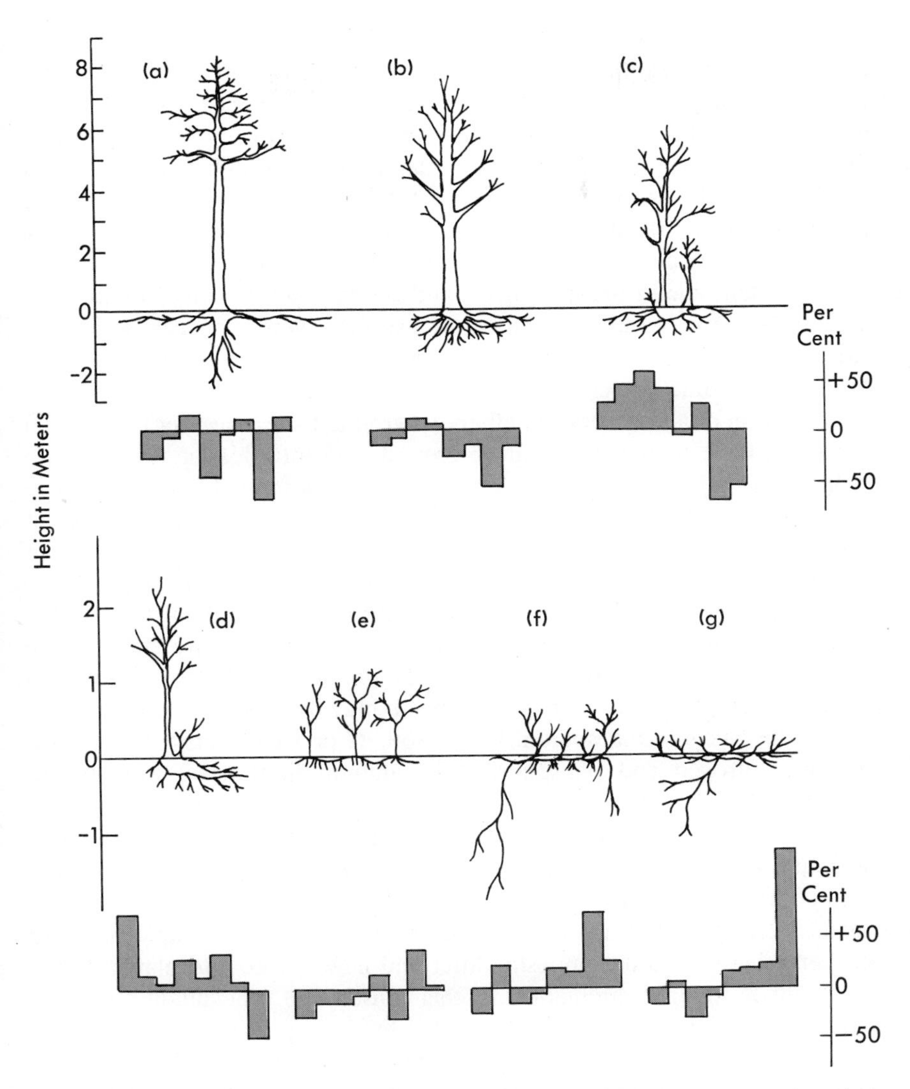

Figure 6.5. Plant profiles and nutrient profiles in the Brookhaven forest, New York. Branch and root forms are shown for three tree species and four shrub species. Below each drawing the bar graph gives relative nutrient concentrations in the same species. Each bar gives the per cent of departure from the community mean for concentration of a given element, in current tissue growth in that species. The community means are average concentrations of an element in eight major woody species in the forest; both averages and departures were calculated for net primary production (woody and leaf tissues together). The eight elements in the bar graphs from left to right are nitrogen, phosphorus, potassium, calcium, magnesium, sulfur, iron, and sodium. The seven species are (a) pitch pine, *Pinus rigida;* (b) scarlet oak, *Quercus coccinea;* (c) white oak, *Q. alba;* (d) bear oak, *Q. ilicifolia;* (e) black huckleberry, *Gaylussacia baccata;* (f) low-bush blueberry, *Vaccinium vacillans;* and (g) dwarf blueberry, *V. angustifolium.* [Whittaker and Woodwell, *J. Ecol.,* **56:**1 (1968) and unpublished data.]

hemisphere of soil densely occupied by roots of the oak. Some of the leached nutrients are thus returned to the mycorrhiza and roots of the tree from which they were leached. Pitch pine (*a*), in contrast, has horizontal branches; rain carries nutrients leached from these directly to the soil. The pine has long roots that spread near the surface through a wide area of soil from which the pine may obtain nutrients, including some of those leached from other plants. The pine has also, in contrast to the oaks, a tap root reaching deeper soil water. White oak (*c*) is intermediate to scarlet oak and pitch pine in branch angles; and the four shrubs (*d-g*) differ from one another in root outlines and uses of surface, versus deeper, soil water and nutrients. Below each profile is a bar diagram showing relative concentrations of nutrient elements in that species. Each bar gives a ratio—concentration of an element (in all above-ground tisues together) in a given species, relative to the mean above-ground concentration of that element in the eight major species of the community. Thus treated, each species in the community has its own nutrient signature, which differs from those of other species. It appears that niche differentiation among the species of this relatively simple forest community is expressed in relationships to nutrient circulation, as well as to vertical height and successional time.

Nutrients differ further in the extent to which they are held in the tissues of plants, or are free in the soil. The greater part of the nitrates and phosphates circulating between tree and soil may be in the trees, the greater part of the calcium and sodium in the soil. (For some nutrient elements, including nitrogen and phosphorus, the amount in the soil available to mycorrhizae and roots may be very much less than the amount present in soil materials but not available at a given time.) Circulation of some of the requisite, but limited, nutrients is relatively "tight" in the sense that most of the ecosystem's available stock of these nutrients is held in the organisms, and the fraction available in the soil is subject to rapid turnover between release from decomposing litter and new uptake into plants. When a nutrient is in tight circulation in this sense: (1) The amount present in the soil at a given time is much affected by the turnover rate in the community, (2) the nutrient may be in sufficiently short supply to affect the productivity of the community, and (3) an addition of the nutrient to the soil as fertilizer may produce a significant, if temporary, increase in productivity.

Communities differ in relative tightness of nutrient circulation, total concentrations of nutrients and proportions of different elements, turnover rates for different elements, and other characteristics. From observing forest nutrient cycling we mention a few themes. One of these—the difference in nutrient function among plant species—refers back to the discussion of niche. Others of these—the implication of organic materials released into environment, the interrelation of vegetation and soil, and the role of the community in holding its nutrients—will be discussed further below.

Allelochemics

The aphid sugar washed from foliage to soil in the oak forest is food for some soil microorganisms. The phenolic compounds released into the soil by leaching and the decomposition of dead plant tissues are potentially food for some saprobes, but inhibitory to others. Among the substances released from roots some are foods for microorganisms surrounding the roots, but some act as chemical signals that either repel or attract other organisms such as soil insects and nematode worms that may feed on the roots. As chemical signals these substances are examples of a large class of interactions termed *allelochemic* effects—chemical effects (other than the supplying of food) of one species on another.

Chemical signals act on different levels in the living world. Substances produced by one tissue that influence another tissue within the same organism are hormones. Those produced by one individual and influencing another individual of the same species are pheromones. Those active between different species are allelochemics. Among the last we may distinguish directions of adaptive advantage. Some allelochemics (allomones) are released by one organism and produce an effect on a second that is advantageous to the first. Others (kairomones) are released by one organism but used by another to its adaptive advantage. Many substances are active in more than one of these ways. A bark beetle (*Ips confusus*) that feeds on ponderosa pine releases a pheromone (actually, three different terpenes) that attracts other beetles of the species to a pine tree suitable for infestation. The pheromone is used also as a kairomone by which beetles of a predatory species (*Enoclerus lecontei*) find and feed on the pine bark beetles. Plants of the mustard family secrete a number of mustard oils that are irritants to animals and prevent most animals from feeding on mustard plants. The same oils are used as chemical cues to the location of mustard plants as food by animals that are adapted to tolerate these chemicals and feed on mustard plants; and one of the mustard oils released from roots stimulates germination of the spores of a fungus, *Plasmodiophora brassicae,* that is parasitic on the roots. The same compounds are thus both allomones that defend the mustards against some organisms, and kairomones that other organisms respond to for their advantage. The flower scents that attract pollinating insects have a dual role of advantage to both plant and insect, assuming the latter receives food from the flower. Insects have hormones that govern moulting and metamorphosis from larva to adult. A number of plant species have evolved the ability to defend themselves by producing these hormones (or chemical analogues) in their tissues, thus causing failure of normal development in most insects that might try to feed on the plant.

All higher plants contain secondary substances: chemicals that are not part of such major organic groupings as carbohydrates, proteins, fats and oils, and nucleic acids; that are of irregular occurrence, present in some plant species but not in others; and that mostly have no known role in the metabolism of the plants in which they do occur. These secondary substances belong to three groups that are widespread among plants and a number of minor groups, some of them of very limited occurrence in particular groups of plants.

1. The phenolic grouping are a wide variety of compounds including the benzene ring of six carbon atoms in their structures. A number of phenolic acids are common in plants, and flavonoids and other compounds based on more than one benzene ring are very widespread in plants. Tannins are polymers, or multiple compounds, of phenolic units; tannins are protein coagulants and are responsible for the bitter flavor of many plant tissues. Lignins are other multiple phenolic compounds that are an important component of wood and a principal reason wood is resistant to decay by bacteria and fungi. Phenolic compounds, including those derived from breakdown of tannins and lignins, are abundant in soils and are a major part of the dead organic matter of the world, particularly coal and the dispersed organic matter in sedimentary rocks. Because of this accumulation in the necrosphere the phenolic compounds as a group are probably the most abundant organic substances in the world.
2. The terpenoid grouping is based on a five-carbon branched chain or isoprene unit; these compounds include simple terpenes and a wide range of double and multiple terpenoid substances, among them steroids and rubbers (which are high polymers of terpenes). Terpenoid substances serve a number of roles in plants, but many of them provide defense against the plant's enemies. Some plant communities, such as the soft chaparral or coastal sage scrub of southern California, produce such an abundance of strong-scented terpenes that the plants are pungent, and the air above them is redolent, with these substances. The effect is attractive enough to a human being, but not to most animals that might consume the foliage of the plants.
3. The alkaloids are a grouping of diverse compounds, related in many cases only by the occurrence of chemical rings including nitrogen as well as carbon atoms. Alkaloids are responsible for the bitter taste of many plant tissues, and some are strongly toxic to animals. Caffeine and nicotine are alkaloids, as are a number of drugs and such effective poisons as strychnine, curare, and lysergic acid.
4. Among the minor groups the mustard oils are sulfur-containing compounds occurring in plants of the mustard family and some others. The organic cyanides are a more striking evolutionary accomplishment, occurring in the rose family and some other plants. These compounds

unite three parts: (1) a phenolic substance that is mildly toxic or inhibitory, (2) the poison cyanogen or cyanide, CN, and (3) a sugar that, attached to this combination becomes a molecular sugar coating, making the compound harmless within the plant. Broken down in the digestive tract of an animal, the compound is not harmless. Almonds, which are seeds of a plant of the rose family, are flavored with a delicate (and trivial) scent of cyanide.

The formation of compounds with sugars (such combinations are termed glycosides) is very common among the secondary substances. Plants must be protected against many of the toxic or inhibitory compounds that they produce. Some of these compounds occur in solution in the vacuoles of plant cells and are thus separated from the metabolic processes in the protoplasm. Other secondary substances are made inactive in the plant as polymers, or are concentrated into dead cells or ducts, or are secreted into glands on the surface of the plant.

Along with these organic compounds, plants use some inorganic substances for defense. Grasses secrete microscopic grains of silica (silicon dioxide, quartz) between their cells, and this plant sand abrades the teeth of animals that eat grass. When communities dominated by grasses evolved, these offered animal consumers both resistance and an opportunity; the latter was met by the evolution of grazing mammals with high-crowned teeth adapted to being gradually worn down by the silica. A number of tropical and southern-hemisphere trees contain silica grains. One of the plants of a tropical woodland in Brazil (*Palicourea rigida*) has leaves so heavily provided with both silica and wax or cutin that they have the texture of plasticized cardboard, and can be clacked together in the living condition. The low content of sodium in many land plants may represent a kind of chemical defense—animals consuming these plants may be relatively starved for this element essential to them. Some plants of salt deserts, in contrast, have such high salt contents in leaf tissues that they are unpalatable to animals. One kangaroo rat (*Dipodomys microps*) has evolved chisel-like teeth with which it scrapes off the surface tissues, before eating the less salty inner tissues of the leaves of saltbush (*Atriplex confertifolia*).

The higher plants have thus shown great chemical ingenuity in evolving defenses against animal consumers and other enemies. Some of these substances have further significance in communities. The concentration of secondary substances in plants is often such that substantial quantities are released into environment, either from the living plant or by decomposition of the litter. In the soil they may have significant effects on other vascular plants, soil animals, bacteria and fungi, and even on the plant population from which they are released. The concentration in the soil of secondary plant substances from a dominant plant may exclude many other plant species from the community. In southern California shrubs of the soft chaparral, a community of low, aromatic shrubs, are able to spread into areas of grassland. During dry periods, on clay soils, plants of the grassland are largely excluded from the shrub patches and from a fringe a meter or more in width around the shrub patches (Figure 6.6B). The exclusion is by some of the same substances that render the community pungent—terpenes, notably cineole and camphor. These volatile chemicals are released into the atmosphere from the shrubs and adsorbed onto soil particles during the dry season, and in the rainy season they inhibit the germination and growth of annual plants. Such effects of chemicals derived from a higher plant in inhibiting the growth of other higher plants are termed *allelopathy*. Substances of established or suspected allelopathic effect are known in a wide range of terrestrial communities—desert and grassland, shrubland and forest—and in all the major groups of secondary plant substances. Water-soluble substances of the phenolic grouping are allelopathic in the California hard chaparral dominated by chamise (*Adenostoma fasciculatum*) and in the effects of walnuts (*Juglans* spp.) and other species.

Some of the shrubs of the chaparral thus dominate the community by chemical influences, as well as by root competition, shade, and microclimatic effects. Allelopathic effects appear to be significant in many communities of strong single-species dominance. In these communities the chemistry of the soil may be much affected by the secondary substance chemistry of the dominant plant species, and only those other species that are tolerant of this chemistry can maintain populations in the community. Forests of strong dominance by a single tree species, particularly a single conifer species, seem often to have a meager undergrowth of low species diversity, in contrast with the richer undergrowth of many forests of mixed dominance. Allelopathic substances also can influence successions. A chemical inhibitor released by one species can help it suppress and replace another species; inhibition of the grasses in this way may facilitate the spread of the soft chaparral in grassland. Allelopathic effects of a successional dominant may make it more difficult for other species to invade and replace it. Certain grasses inhibit nitrogen fixation by microorganisms and appear thus to delay (by poverty in soil nitrogen to which they are themselves adapted) their replacement by other species. In most cases it is not clear that a plant releasing allelopathic substances gains an advantage from inhibiting other plants. It seems more likely that allelopathic substances have evolved as defenses against animals and bacterial and fungal pathogens, and the effects on other plants are only a by-product of this defensive role.

In some unstable communities of strong single-species dominance, allelopathic substances can become autopathic or autotoxic, inhibiting the growth of the dominant species itself. Older patches of the soft chaparral shrubs may show reduced vigor in consequence of the accumulation of their own secondary substances. Certain plants of early successional stages

Figure 6.6. Chemical antagonism between organisms. A: Antibiotic effects between lichens. The lichen in the center (*Rhizocarpon* cf. *rittokense*) has by release of a chemical agent prevented the growth of other lichen species in the bare belt surrounding it, in which the light-colored rock (gneiss) can be seen. Because the chemicals are carried in moving water, the belt is wider on the lower edge of the lichen patch; the patch is 13 cm in diameter, the inhibition zone up to 6 cm wide. [Photo taken at Torssukatak Fiord, West Greenland; courtesy of R. E. Beschel.] B: Allelopathic effects between vascular plants. Volatile terpenes released by the shrubs on the left (*Salvia leucophylla,* of the soft chaparral) are adsorbed on the particles of a clay soil and inhibit the germination and growth of plants of the annual grassland. During a dry period when the effect was most intense, the belt of bare ground surrounding the shrubs was 2 m wide, and partial inhibition of the grassland species extended about 6 m from the shrubs. [Photo taken near Santa Barbara, California; courtesy of C. M. Muller; see also Muller, 1966.]

(*Hieracium* and *Erigeron* species) produce chemical effects in the soil that destroy their own populations in one or a few years; they are then replaced by other species more tolerant of these chemical effects. Autotoxicity would not seem a profitable direction of evolution. Presumably, for some species of unstable communities, there has been greater selective advantage in the evolution of high concentrations of the secondary substances as defenses against enemies, than selective disadvantage from the effects of these substances on the short-lived populations of the species itself.

Defensive chemicals occur throughout the living world. Skunks are best known among vertebrate animals for chemical defense, but the skins of toads and salamanders contain concentrations of toxic materials. Many animals use venoms for capture of prey, and also for defense. Certain millepedes secrete organic cyanides. Terpenoids similar to those of plants are synthesized by a number of insects, and stink bugs (Pentatomidae) are defended by batteries of as many as 18 compounds. The monarch butterfly (*Danaus plexippus*) is celebrated as a protected species—as unpalatable in taste as it is striking in color—which is mimicked by another butterfly, the viceroy (*Limenitis archippus*). The monarch feeds as a caterpiller on plants of the milkweed family; the monarch and its relatives, and some other insects, have evolved the ability to tolerate the chemical defenses of milkweeds. The monarch butterfly has, in a further evolutionary step, turned this tolerance to advantage: it uses the repellent substances of the milkweed plant to make itself repellent to predators. It is the monarch's success in the biochemical interplay among species that makes it a protected model, which it has been to the advantage of the viceroy to mimic. The bombardier beetle (*Brachinus* spp.) is an insect "skunk." The beetles have a special two-chambered gland that secretes phenolic compounds and hydrogen peroxide. Mixture of the contents of the glands with enzymes produces a hot repellent spray that is released in a small explosion to discourage predators.

Bacteria, fungi, and lichens are engaged in their own chemical defenses. The antibiotics released by certain bacteria and fungi are the most widely known of such compounds because of their great medical value to man. Their value to man lies in their inhibition of bacterial growth; their value to the fungi and mycelial bacteria that secrete them doubtless lies in the same direction. They are chemical agents of conflict between species; their release into environment can, in some circumstances, give a fungus an advantage by killing or slowing the growth of other organisms that might compete with the fungus for food. These allelochemical interactions should be added to the complexity of transfer of substances in the soil already referred to. Lichens also produce defensive substances, some related to those of higher plants, some distinctive to lichens. Lichens in some cases produce bare zones around themselves by the release of chemicals that suppress other lichens; Figure 6.6B shows such an effect on a rock surface in the Arctic.

Plankton algae release into the water substantial fractions of the organic material they synthesize. It is because of this leakage from plant cells, together with animal excretion and the enzymatic effects of bacteria living on the surfaces of particles of dead organic matter, that the water of the oceans and lakes generally contains much more organic matter in solution than in living organisms and dead particles. Many aquatic plants and animals are able to absorb some of this dissolved material as food supplementary to their photosynthesis or ingestion. Uptake of foods, and also of vitamins, directly from the water is especially important to many of the unicellular organisms. These protists are both actively losing organic matter to the water and actively absorbing organic matter—primarily different substances. Even among the forms possessing chloroplasts many are dependent on external, dissolved organic matter; they live by mixtures of autotrophic and heterotrophic nutrition, mixtures that differ from species to species in the proportion of the heterotrophic needs and the identity of the organic compounds required from the water. There is thus active circulation of organic matter in the plankton that does not move through the classical food chains. It is also true that population dynamics of plankton are affected by the organic materials in the water that differently affect the growth of different species.

Along with nutrient ions, food, and vitamins the plankton cells are releasing wastes into the water. The organic wastes of one species may well be food for another. Some of the substances released may also be effective toxins. The common green alga *Chlorella* releases a compound, chlorellin, that has an inhibiting effect on the growth of other algal species, on the rate of grazing by plankton animals, and (if concentrations are high) on the growth of *Chlorella* itself. Other cases of chemical inhibition of one plant plankton species by another have been observed. The plant plankton species are consequently (along with the animals, bacteria, and fungi of the plankton) interrelated by a network of chemical exchange. Different positions in this biochemical network represent different positions in the community's function, hence differences in niche.

Allelochemic interactions are thus a major realm of species adaptations that are normally invisible to us. We might observe some evolutionary implications, with special reference to terrestrial communities. The evolutionary interplay of plants and insects (and other enemies of plants) probable proceeds as follows:

1. A plant species evolves chemical protection against animals; thus protected, the plants increase in abundance and with evolutionary time and geographic separations may differentiate into a group of species, and perhaps ultimately into a family.
2. Certain insect species evolve tolerance of the chemical protection; these insect species in turn may increase and diversify as consumers of the plants, in relative freedom from competition.

3. Consumption by the insects brings selection for intensification of the chemical protection in the plant, thus bringing also intensified selection for tolerance in the animals. Genetic feedback effects are occurring that are likely to bring about a balanced biochemical accommodation of plant and animal to each other. The plant is fed on by insects, but is not too heavily fed on.
4. The insect may, meanwhile, evolve to turn the plant's chemical protection to its own advantage. It may use the defensive chemicals as behavioral cues by which to locate the plants as food, and in some cases the plant's defensive chemicals may be used for defense by the insect.

These interactions between species may have further community-level implications:

5. Increase in the number of insect species dependent on the plant species up to some steady-state or maximum level. The number of insect species adapted to the chemistry of a plant species may be larger on dominant and geographically widespread plant species than on rare and local species.
6. Especially heavy loading of plant leaves with chemical defenses in some species of more arid climates—such as the dominants of the soft chaparral, the creosote bush desert, and tropical woodlands. Presumably, the slower-growing plants of these communities are less able to afford loss of leaves to herbivores than are the faster-growing plants of more favorable environments, and the former plants consequently spend a greater portion of their productivity on chemical defense.
7. Increasing divergence of the secondary chemistry of different plant species as selective pressures, exerted by different sets of insect species, lead to intensification of their chemical defenses in different directions.
8. Evolution in communities of chemical complexes of species, each comprising a plant species with its distinctive allelochemics, the animal, fungus, and bacterial species accommodated to this, and additional species (predators, parasites, and mimics) adapted to the species adapted to the plant. Such an assemblage can be considered a "component community," and a full community can be conceived as a "compound community" consisting of some number of such components.
9. Contribution of the component communities to high species diversity of the full community. The different components can include a number of sets of heterotroph species in parallel niches—species similar in mode of feeding and temporal and spatial position in the community, but differing in chemical adaptation to plant food. Most species relations to the component communities are loose. Even herbivorous insect and mycorrhizal fungus species seem mostly to feed on, or form symbiotic

relations with, at least a few plant species and not one only. The component communities are consequently only partly separate from one another. However, the more component communities there are, the more niches there are for species linking them together in various ways. Allelochemic interrelation is thus a major basis of community organization, niche differentiation, and the complexity of the community's niche space.

More generally we note that the environments of communities, the soil and water, contain diverse organic compounds of varied significance to populations; these include foods, vitamins and enzymes, wastes and decomposition products, antibiotics and allelopathics, repellents and attractants, and hormones and pheromones. The function of ecosystems involves a triple traffic of substances of three major groupings—inorganic nutrients, foods, and allelochemics—moving through varied routes in a chemical arabesque. The combined production, circulation, and utilization and effects of these substances represents a community-level metabolism, the complexity of which parallels that of the metabolism of organisms.

Soil Development and Classification

On land allelochemic substances carry the influence of the vegetation downward into the soil. Most soils are products of the interaction, under the influence of climate and topography, of a living community and a geological substrate; soils are blends of organic and inorganic materials. To produce a soil the rock or other geological substrate (which forms the *parent material* of the soil) must be weathered. The weathering occurs by varied combinations of physical effects (temperature change, and the expansion of ice that breaks rocks), chemical effects (change of minerals to different forms by exposure to water, oxygen, and weak acids of carbon dioxide and other compounds), and biological effects (splitting of cracked rocks by roots, and the contribution of carbon dioxide, and organic acids and chelating agents from plants). To the weathered inorganic materials are added varying proportions of organic remains, the litter and humus, that are mixed with the inorganic materials by water transport downward, by the death of roots in the soil, and often also by animal burrowing and, in cold climates, frost-heaving of the soil. As a result of these and other processes soils are of great complexity and variety. We cannot do justice to them here, but must use a small part of soil science that relates to our concerns with nutrient circulation.

The soil of a natural community is not an inert substrate, it is a part of the ecosystem. There is an interactive, complementary relation between the soils that support the community and affect its characteristics, and the community that develops and influences the character of the soil. Aspects of this complementary relationship have been described in connection with succession, detritus chains, and forest nutrient cycling. In the last of these we mentioned the role of acids derived from the community—carbonic and nitric acids, formed by the solution of carbon and nitrogen dioxides (from respiration and decay) in water, and organic acids, particularly phenolic compounds leached from plant surfaces and from decomposing plant litter. As we also observed, the soil contains in addition to dead organic matter living roots and fungi, and in many cases a rich biota of bacteria, protists, and animals. The soil is thus itself a living community, or subcommunity, much as is the above-ground part of the community.

Some soils develop on rock that is exposed at the earth's surface and that weathers in the soil as its parent material. Other soils develop on transported materials—the silt of a river floodplain, glacial till deposits, or wind-blown sand or loess. The nutrient elements in soils and communities are of three major sources. All communities receive nutrients from rainfall and the settling of particles from the atmosphere. Communities in which these atmospheric nutrient sources are almost the only ones are termed rain-supported or *ombrotrophic*. Many communities receive nutrients from weathering of the parent material. Some communities are currently receiving materials in the form of water-carried silt, wind-borne dust, or dissolved nutrients that flow downslope into the community within the soil or over the soil surface. We shall concern ourselves here with "typical" or "zonal" soils that are on relatively level terrain, and are subject at present to neither extensive downslope addition or removal of materials, nor substantial input of dust or silt.

Soils differ in age, degree of weathering and transport of materials, and extent of biological influences on them. One may consider first an ideal scheme of the stages of weathering and transport of materials downward in the soil. In a first stage weathering of a typical rock such as granite releases mineral elements, including silicon, calcium, magnesium, iron, and aluminum, and these combine into silicates—compounds of silicon with other elements—that predominate in the soil. With further weathering of the soil materials, iron is released and oxidized and colors the soil (yellow or red), while some of the silicates become clays. With further time the clays can be transported downward in water to a level in the subsoil where they accumulate. More resistant materials, including silica (silicon dioxide), certain clays, and oxides of iron and aluminum are then left behind in a soil that may be still rather rich in bases (calcium, potassium, and magnesium especially). With continued weathering the bases are leached downward, and some of the remaining clay minerals and silica may also be transported downward. The result of long-continued weathering and leaching of transportable materials downward is a soil strongly dominated by what remains, which may be primarily the oxides of iron and aluminum.

Departures from this ideal scheme are many, and result from differences in parent material, topographic position, and climate and vegetation.

Because soils are complex and intergrade with one another in many directions, their classification is difficult. The difficulties are comparable with those of classifying communities. Consider the problems of classification when there are many "objects" to be classified, each with a number of "characteristics" by which they may be classified, but the objects and the characteristics vary in a complex and multidimensional continuum, a "range of variation." With soils, as with communities (Chapter 4), two kinds of classification may be useful:

1. One may, on the one hand, seek to recognize in the range of variation some major types. These types are subjectively defined, weighing together a number of characteristics of the objects of classification, while emphasizing some characteristics that seem most significant in relation to others. The types are then points (or areas or volumes) of reference in the range of variation, defined by different combinations of significant characteristics. Because several characteristics are being considered, and these characteristics vary continuously and in partial independence of one another, the boundaries of the types may be vague. Because of the complexity of the range of variation, the types will generally not represent a full and inclusive classification of all the objects. The types may or may not be arranged into a hierarchy. This kind of classification, which we may call "typification," is not satisfactory if a formal, detailed classification of all the objects is needed. It may, however, express some major relationships in an effective way. Systems of biomes may be recognized as approaches of this sort, classifying communities into major types by balancing together characteristics of physiognomy, environment, and where possible animal communities.
2. On the other hand, a formal, hierarchical classification may be desired. This is to be based not on types but on formal classes, sets of objects that share certain defining characteristics. The classes now can have sharp boundaries, for each class has a definition that states that objects with certain characteristics belong to it, and objects without those characteristics do not. Furthermore, the classes can be arranged in a hierarchy; for a number of classes on a lower level can be defined (by their sharing of certain characteristics) as belonging to a larger class on a higher level. Classes on this second level can be grouped into still larger classes on a third level, and so on. Because the classes can be sharply defined and arranged in a hierarchy, all the objects can be classified in a system that expresses, in the classes on different levels, different degrees of relationship among the objects. The units can be given formal, standardized names to produce a taxonomy, a system of classification expressed in such names. The classification is thus definite, inclusive,

and orderly, even if the order is constructed by man by imposing a hierarchy on the range of variation. The most familiar hierarchical classification is the taxonomy of organisms from species up to phyla and kingdoms. Hierarchical classification is in this case aided by the fact that in some groups of organisms species are discontinuous with one another, and the fact that common evolutionary descent can be expressed in the classes. The system of Braun-Blanquet (Chapter 4) is a taxonomic classification applied to plant communities as objects, using diagnostic species as characteristics that define classes.

The approach to soils by typification was developed on the plains of Russia, where V. V. Dokuchaev, K. D. Glinka, and others founded soil science and recognized broad relationships of soil profile characteristics to climate and vegetation. The broadly defined units of this approach have been called "soil types" in Russia, but the international term for them is *great soil groups*. Ideally, the great soil groups are broadly defined types, recognized by profile characteristics of "zonal" soils—soils that develop on level or rolling terrain and that correspond in general to climatic climax or prevailing climax vegetation. In some major types of soils, development is strongly affected by the kind of parent material, or by conditions of drainage or water movement; and these soils may differ much from others that develop in the same climate. As soon as these non-zonal soils are accepted as great groups, the number of groups distinguished can become large. Given a large number of groups, the primary meaning of great groups —expressing only *major* soil types, and primarily those that can be related to climate, in the range of variation—can be lost sight of. Because the approach through great soil groups is one of typification, it cannot provide an inclusive classification and should not be expected to do so. Because different authors emphasize different characteristics and limit and name great groups differently, their classifications will differ at least in detail. Great soil groups, like biomes, may not offer much help in describing small-scale variation in ecosystems within a limited area. Because of the manners in which they are defined, however, great soil groups and biomes (or formations) will bear some relationship to one another, and both will be related to climate.

A practical system for local soil classification is also needed. It would be convenient if the system were designed as an inclusive taxonomy that could relate, on different levels, all the world's soils to one another with standardized names. A number of systems have developed in different countries, but the most ambitious hierarchy is that of the United States System of Soil Taxonomy (often still referred to as the "Seventh Approximation"), developed by the United States Department of Agriculture. The system begins with ten upper-level classes (termed orders), and the remaining levels of the hierarchy downward are suborders, "great groups," subgroups, families, and series. Classes are designated by a system of names that are outlandish but efficient.* Formal and generally quantitative definitions of the classes are given; these can thus be sharply bounded (subject to difficulties of measurement) and inclusive. To be definite, these boundaries must be based on a single defining characteristic separating a given class from its nearest neighbor(s) in the system. This single-characteristic definition is in contrast to the typification approach, and it has the consequence that closely related soils must often be placed in different higher-level classes. Many class definitions imply laboratory measurements on soils, year-around observation of soils, or measurements (such as mean annual soil temperature) reflecting climate; but some of these characteristics can be inferred from other evidences. Use of different defining characteristics on the different levels produces a classification that is more detailed and technical, and more useful for expressing smaller-scale soil differences than the great soil groups. The U. S. Soil Taxonomy is analogous, for soils, to the Braun-Blanquet classification of communities. Like the latter it is suited to local studies, yet also provides a framework for world-wide classification.

The Braun-Blanquet and the U. S. Soil taxonomies have similar limitations. Construction of the taxonomy requires the forcing of a hierarchy on objects related to one another in complex and multidimensional ways. It depends also on arbitrary (although carefully considered) choices of certain defining characteristics over others. Evolutionary relationships cannot guide the choices and give meaning to the units in the Braun-Blanquet and U. S. Soil taxonomies, as they can in the taxonomy of individual organisms. Although it is better than the great soil groups for some purposes, the U. S. Taxonomy seems relatively cumbersome and ineffective for soil geography. Its units can be used to discuss geographic relationships, but no single level of the system provides units as useful in this way as the great soil groups. If units on different levels are used, these are not

* The following are characterizations of the great soil groups discussed here in the terminology of the U.S. Soil Taxonomy (Soil Survey Staff 1960, Basile 1971). Equivalents are given in the form—Order-Suborder ("Great" groups). For the initiate the order and suborder are unneeded, for the group name contains stems indicating these. Thus Cryorthods denotes the group of the cold-climate (Cry-), true or common (-orth-, suborder), Spodosols (order, indicated by -ods). Taiga Podzols—Spodosols-Orthods (Cryorthods, Fragiorthods, Haplorthods). Forest Brownearths, defined here to include Brown Podzolic soils—Inceptisols-Ochrepts (Dystrochrepts) and Spodosols-Orthods (Entic Fragiorthods and Haplorthods); and Gray-Brown Podzolic soils—Alfisols-Udalfs (Fragiudalfs, Hapludalfs). Rainforest Latosols—Oxisols-Orthox (Haplorthox) and -Humox, Ultisols-Humults (Tropohumults). Chernozems—Mollisols-Borolls (Paleborolls, Argiborolls, Cryoborolls, Vermiborolls, Haploborolls), and Ustolls (Argiustolls, Vermustolls, Haplustolls). Sierozem and Desert soils—Aridisols-Orthids (Camborthids, Paleorthids, and Calciorthids), and -Argids (Durargids, Haplargids, Paleargids), and Solonchak—Aridisols-Orthids (Salorthids). Bog Peats—Histisols-Fibrists (Sphagnofibrists, Cryofibrists, Borofibrists), and -Hemists.

coordinate with one another as are the great soil groups. Furthermore, the manner in which the units are defined implies that some higher units will unite soils that should be separated in a geographic treatment, while both higher and lower units will in other cases divide what seem the most natural groupings of soils as these relate to climate and vegetation. The Braun-Blanquet system also is ineffective for world vegetation, and ecologists have come to accept the need for different classifications for local research (associations, dominance-types, and so on) and broader geographic treatment (formations or biomes).

The experience of ecologists with their classifications may suggest two further observations. The first is the possible confining influence of a formal hierarchy like that of Braun-Blanquet. No one classification can express adequately the relationships among the objects classified. In return for the benefits of a standardized classification, prices may be paid in commitment to certain criteria of classification and certain approaches to research. Some researchers tend to mistake their classification for their science, and to become intolerant of other approaches and uninterested in kinds of research for which other approaches are better suited. Despite the great contribution of the school of Braun-Blanquet, ecology would no doubt be poorer if only that school's approach to natural communities had been used. It may be that alternative classifications should be retained in soil science (even if in a role secondary to that of the U. S. Taxonomy) for their value in counterbalancing that system and offering different views of soil relationships. It seems in particular that the relation of the great soil groups to the U. S. system is complementary; soil science as a whole may gain from the use of these two approaches, differing in their perspectives and kinds of usefulness.

The second observation is the need for different names for different units of classification. Decades of dissension resulted in ecology when the term for a unit in one school was appropriated for a different unit by another school, which then insisted that its use of the term was the only correct use. A point of confusion of this sort in the U. S. Soil Taxonomy seems unfortunate. In one of history's most massive coinages of new terms, its authors chose to apply "great group" to a kind of unit that does not correspond to the great soil groups in concept, definition, function, or level, and to interpolate this term between family and order in their hierarchy. Confusion could be avoided by dropping "great" from "great group" in the U. S. Taxonomy. Great soil group in the international sense has clear priority, and many of the "great groups" (of which there are more than two hundred) of the U. S. Taxonomy are not very great. The international usage will be followed here as we consider a few great soil groups in relation to climate and the nutrient function of ecosystems. Their place in a world pattern of great soil groups is shown in Figure 6.7, which may be

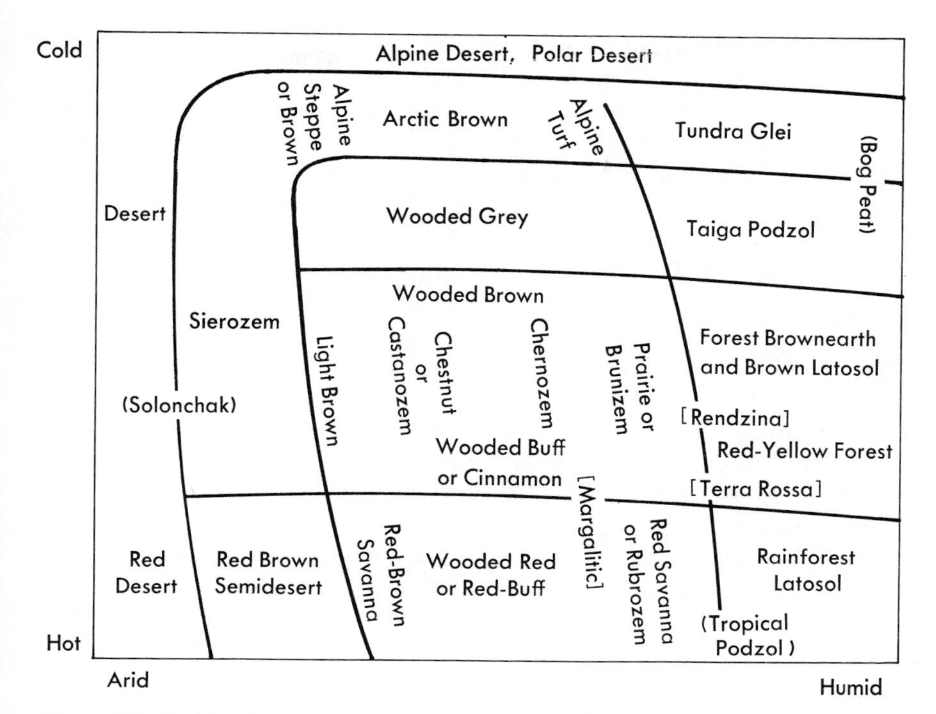

Figure 6.7. Great soil groups in relation to climate. Zonal great groups are shown by relative positions along climatic gradients of humidity on the horizontal axis, and temperature on the vertical. In intermediate climates both grass-dominated, and nongrass communities occur (Figure 4.10) and form different soils. Soils formed in grassland communities are written on the vertical; soils formed by woodlands and shrublands (or, in the Arctic, in tundra) are written on the horizontal for the same climates. Certain nonzonal great soil groups are also shown; those formed on limestone or in high-base situations are indicated in brackets, those formed with impeded drainage or special conditions of water movement in parentheses. Terminologies and numbers of great soil groups vary among authors. Forest brown earths as used here include more than one group in most classifications; brown latosols occur in tropical mountain forests.

compared with Figure 4.10. The statements on nutrients in vegetation and litter are based largely on the work of L. E. Rodin and N. I. Bazilevich.

Soils and Nutrients

1. Leached taiga soil, taiga podzol. In a northern spruce forest, in Canada or Finland, a soil develops in a cool and moderately humid climate. The decomposition of plant remains is slow, and the surface of the soil is covered by a mat of spruce needles, the mass of which may amount to

3 to 7 kg/m^2. This mat of acid, partly decomposed plant material, rather sharply separated from the underlying soil, is a *mor* litter layer. Gradual decomposition of the litter releases a continuing supply of organic acids that contribute to weathering the parent material and leaching materials downward in the soil. Beneath the mor, the upper soil is a leached layer in which silica predominates; this layer is gray or even white in color. This "bleached" soil material is characteristic of such soils and is the source of the Russian term for them—podzol (ash). Beneath the leached layer is a layer of deposit where iron-humus complexes and other materials carried downward by water accumulate. The humus and iron oxide may be cemented into a hardpan or an iron pan. In some podzols the hardpan is rocklike, and thick and strong enough to prevent plant roots from growing through to lower soil levels. Below the layer of deposit is the partly weathered parent material, and below this is the unweathered rock or other substrate. A podzol is a soil with a distinctive profile, or pattern of vertical differentiation (Figure 6.8). Podzols are traditionally thought typical of the taiga, but there is no real correspondence of great soil group and biome-type. Podzols occur in a wide range of climates and communities, and not all taiga is underlain by podzol.

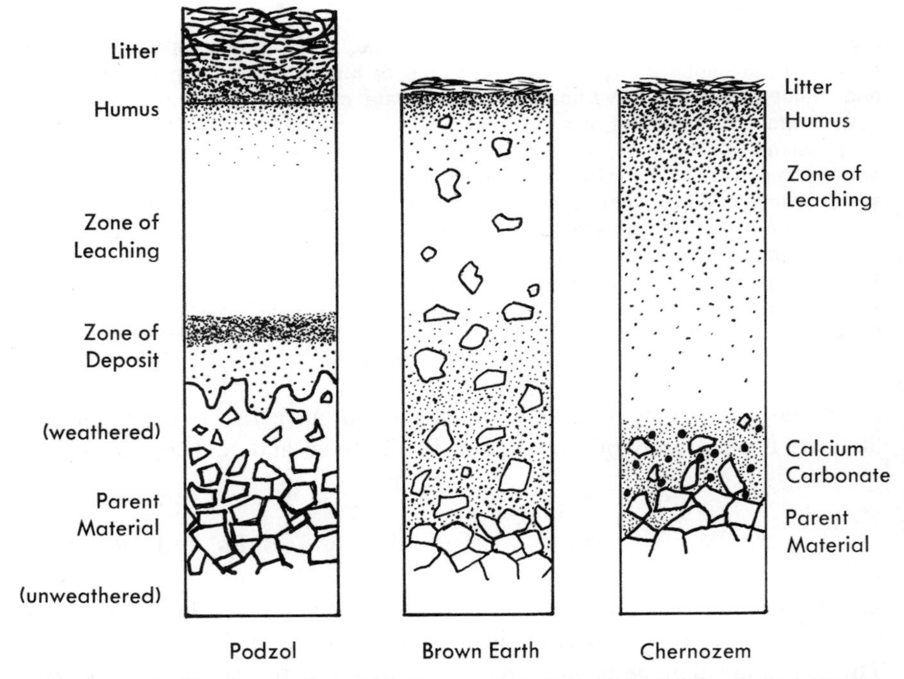

Figure 6.8. Profiles for three great soil groups. Left: A podzol, for a boreal spruce-fir forest. Middle: A brown earth for a temperate deciduous forest. Right: A chernozem for a temperate mixed grassland.

The leached layer is acid and low in nitrates and other plant nutrients; podzols are generally infertile soils. The spruce forest on podzol has a large part of the ecosystem's circulating nutrient stock in the plant biomass and litter, and a relatively small part in the soil. The content of nutrients in plant tissues is low compared with other forests; the over-all mineral ratio (ash content, excluding nitrogen, in total vegetation dry biomass including ash and nitrogen, above ground) is around 0.5 to 0.9 per cent; nitrogen content 0.2 to 0.4 per cent. Despite the abundance of silica in the soil, the silica content of the tissues is also low (2 to 5 per cent of the minerals) compared with some other communities. The spruce forest is an ecosystem with a large biomass and surface litter mass perched on a soil poor in nutrients, with slow cycling of nutrients from soil through plant tissues and delayed decay of litter back to soil, and with relative economy in use of nutrients in plant tissues. Some nutrients must be lost from the cycle by downward leaching in the soil. On some podzols these nutrients cannot be replaced by decomposition of parent material; for the net transport is downward, and the parent material may be sealed off from the community by the hardpan. Some communities on podzols are thus ombrotrophic, with the nutrients that are lost replaced largely or wholly from the atmosphere.

2. Brown forest soil or forest brown earth. In a deciduous forest in a somewhat warmer temperate climate the litter decomposes more rapidly, both because of higher temperature and because the deciduous leaves are thinner and less resistant to decay and animal consumption. The surface litter layer is consequently thin and grades continuously from leaves, through fragments, to particles, and humus colloids and solutes mixed with other soil materials. Such a litter layer, with effective mixing of the organic matter downward into the soil, is a *mull* layer. Whereas earthworms may be absent from the podzol, in the brown earth they may be abundant and important in litter breakdown and soil turnover. Although the upper soil levels are enriched in humus, they are leached of calcium and some other transportable materials; typically they are mildly acid and high in silica. Brown earths are thus affected (more weakly) by the same kind of leaching as podzols, and they are sometimes referred to as "podzolic" in consequence. Soil acidity decreases as the content of bases increases, downward to the level of deposit. The latter is not sharply defined and may merge gradually downward into unweathered parent material (Figure 6.8). Brown earths are in general more fertile than podzols; not only are levels of nitrates and other nutrients higher in the soil, but plant roots have unimpeded access to lower soil levels where nutrients are made available by weathering of the parent material (as well as by leaching).

The biomass of a mature deciduous forest is equal to, or greater than, that of a taiga forest; and this biomass is richer in nutrients (mineral ratio 1.0 to 1.3 per cent, nitrogen 0.3 to 0.5 per cent). The deciduous forest may thus have roughly twice the spruce forest's nutrient stock in plant tissues. The phosphorus content may be lower in the wood, but higher in the leaves of the deciduous forest; and since the leaves decay more rapidly, the phosphorus is more rapidly cycled. The litter mass is smaller, around 1 to 2 kg/m^2, but nutrient concentrations in it are higher than in the spruce forest. The deciduous forest thus functions with a richer nutrient economy, with a larger nutrient stock turned over more rapidly, and with a smaller fraction of the nutrients in plant tissue, a larger fraction in the soil. The community itself contributes to the maintenance of its higher soil nutrient levels. Nutrients leached downward or released by parent-material weathering can be taken from deeper soil levels by roots, used in the plants, and returned in litter to be released by decomposition on the surface of the soil. The community thus "pumps" nutrients from deeper to shallower soil levels.

3. Tropical red forest soil, rain forest latosol. A lowland tropical rain forest occurs on a soil that has been exposed for a very long time to intense weathering and leaching. Decomposition of organic material here is rapid, and the leaching is primarily by inorganic acids—the weak carbonic acid formed by CO_2 and the dilute nitric acid formed by NO_2 in solution in the water in the soil. The consequence (on parent materials that are not high in quartz) is in a sense opposite to that of the podzol; the silica has been transported downward, while the oxides of iron and aluminum remain along with immobile clay (kaolin). The red, leached layer of oxides and clay may be many meters deep, and the plant community may then have no means of nutrient replenishment from the parent material. The leached layer is at least mildly acid, and more than mildly impoverished in nutrients. Some tropical soils have accumulated iron that has been carried down into them by moving water and are strongly dominated by oxides of iron and aluminum. The soil material dominated by these, which hardens into a rocklike substance when exposed to air, is "laterite" in the strict sense. Confusion has resulted from use of the term laterite, along with latosol and ferallitic soil, for tropical soils with deep weathering and high iron and aluminum oxide content. Soils with related characteristics less strongly developed are often termed "lateritic," and these include many warm-temperate red and yellow forest soils of the southeastern United States.

Tropical rain forest biomass is comparable with that of temperate forests; the content of nutrients in the biomass appears to be of the same order as or higher than in temperate deciduous forest. An average mineral ratio for tropical rain forest is about 1.58 per cent, nitrogen content 0.57 per cent. A tropical rain forest produces a heavier fall of leaf-litter than a temperate forest; but this litter decomposes rapidly to leave a small surface litter mass (0.1 to 0.6 kg/m^2). The amount of organic matter dispersed through the deep leached layer in some of these soils is, however, large; it is much larger than would be recognized from the light colors of these soils. In the litter concentrations of aluminum and iron are higher, calcium conspicuously lower, and silicon apparently higher than in deciduous forest litter. High over-all nutrient content of the litter, combined with large litter fall and rapid decomposition, implies more rapid turnover of nutrients than in temperate forests. Average ratios of annual litter nutrient fall and release, to nutrient stock in biomass, are of the order of 14 per cent in tropical rain forest, 6 per cent in temperate deciduous forest, 4 per cent in taiga. The tropical rain forest thus has a relatively rich nutrient economy perched on a nutrient-poor substrate. Under constant, drenching rain the forest would be as effectively stripped of its nutrients as the open ocean plankton, if the forest did not function in a manner that retains them. It is probable that mycorrhizal fungi rapidly invade decomposing litter and recapture the nutrients for the plant-fungus partnership, and that nutrients leached from leaf surfaces are also effectively captured by the network of roots and fungal filaments. The rain forest thus has a "tight" nutrient circulation. Some nutrient loss must occur and is replaced by nutrients in rain; on a very deep latosol a tropical rain forest is ombrotrophic.

4. Temperate grassland soil, black earth or chernozem. In the American Middle West (and the Ukraine) grasslands occur in climates drier than those of the forests just described. The forest soils were affected by leaching, which occurs when (with precipitation exceeding evapotranspiration) a surplus of water moves downward through the soil. In a drier climate there is no such surplus. Water then tends, after penetrating into the soil to some depth during a rain, to be drawn back toward the surface by capillary action and root transport and lost from the soil by evapotranspiration from the soil and plant surfaces. Although some soluble nutrients (sodium and postassium) are carried downward, less soluble ones (especially calcium carbonate) remain relatively abundant in the soil. The soil is consequently mildly basic in reaction, and in these conditions the humic acids are neutralized and the clay minerals are relatively stable. Large amounts of organic matter (8 to 10 per cent of soil mass) are mixed into the soil by the death of the numerous fine roots of grasses that permeate the soil, and by animals. Organic content decreases downward, but a leached layer and layer of deposit are not distinct, as they are in many forest soils. There is, however, a lower soil level, in the transition to parent material, in which calcium carbonate nodules form; and some older grassland soils have zones of clay accumulation. A number of circumstances are favorable for chernozem soils. They are fertile because not leached, yet moist enough for moderately high productivity; the high organic content and calcium carbonate content give a loamy and granular structure; and the clay and organic particles contribute to fertility by their surfaces that can either hold nutrients or release them to plant roots.

The chernozems support a mixed prairie or steppe of both taller and shorter grasses, and they occur in the middle of the grassland belt between more humid tall-grass prairie and more arid short-grass plains. The above-ground biomass of the mixed prairie is only a fraction, about one tenth or one twentieth, that of a mature forest. The mass and nutrient content of the surface litter may exceed that of the above-ground living grasses, except after a fire. A high percentage (45 to 60 per cent) of the biomass reaches the litter each year, in contrast to around 1 per cent in a mature deciduous forest. Although the nutrient stock in plant tissues is far smaller, the annual return of mineral elements to the soil in litter may be higher than in a deciduous forest. Nutrient content is much higher per unit mass in grass and grassland litter than in forest trees, nitrogen content is around 1 per cent and the mineral ratio 2.5 to 3.5 per cent. Among these tissue minerals calcium is a smaller fraction than in temperate forests and silica a much higher fraction (one third to one half). The silica particles, released by litter decay, are also abundant in the soil. The grassland nutrient cycling is relatively "loose," with the stock of nutrients in the biomass small relative to that in the soil; and the nutrients are retained in the ecosystem more by the function of the soil than by their incorporation in plant tissues.

5. Gray desert soil, sierozem. In a desert climate the plant cover is sparse and much less effective in modifying the soil. Litter cover is incomplete or nearly absent, organic content of the soil low (less than 1 per cent), and weathering of the parent material slow. Largely unweathered materials may form the soil surface. The soils of a desert plain, lying below or between mountain ranges, are subject to a process that is the reverse of leaching. Soluble substances and other soil materials have been carried downward from the mountain slopes into the plain and accumulated in the soils there. Because of aridity, soil water moves upward and evaporation concentrates many of the soluble materials near the soil surface. These desert soils are consequently basic in reaction. Calcium carbonate, however, is generally accumulated at a level beneath the surface. In a gray desert soil, the layer of calcium carbonate deposit, that appeared as nodules in the chernozem, may become a continuous and rock-like layer (caliche).

A desert soil offers a relative abundance of some plant nutrients (except nitrogen, the content of which is generally low). The plant biomass ranges from as high as in a grassland in some semidesert scrubs, to very low. Because the desert is dominated by shrubs, the fraction of the nutrients in the biomass that turns over through the litter each year is lower than in a grassland, but higher than in a forest. The relative nutrient content of biomass and litter is high; the mineral ratio of the biomass is about 4 to 7 per cent, with a high content of sodium and chlorine in some species. Tissue concentrations of other, more critical nutrients (phosphorus, nitrogen, and potassium) are generally in the same ranges as in deciduous forest trees. To varying degrees the tissue concentrations of different nutrient elements are more labile—more widely varying in different species and in response to environmental difference—or more stabile—more consistent in a given kind of tissue in different plants and environments. The desert is an ecosystem of meager productivity and biomass provided with a superabundance of some nutrients. A small part of the abundant nutrients are circulated, even though tissues may be concentrated with them, while the circulation of other nutrients, especially nitrogen, appears to be relatively tight.

6. Salt desert soil, solonchak. The process of salt concentration described for a sierozem relates to that to be described for saline lakes in Chapter 7. Soil salinity increases from the edge of a desert plain at the base of the mountains to its center, which in some desert landscapes is occupied by a salt lake or a dry plain or salt flat. In this center the soils may be strongly saline, and in some such soils (solonchaks) the concentration of salts at the surface produces a whitish or gray crust on the soil. Some salt deserts have no vegetation, but other such deserts have a sparse cover of salt-adapted shrubs. Some of these shrubs have succulent, salt-flavored twigs, and others excrete salt to form crystals on the surfaces of their leaves. Mineral ratios of 10 to 14 per cent are reported for the plants in such deserts. Solonchaks, because their development depends on local drainage conditions, are not zonal soils. Although they are most extensive in desert basins, they can occur locally in desert, semidesert, and grassland climates from the Tropics to the Arctic.

7. Bog soil, peat. The blanket bogs of some cold, humid climates have been mentioned (biome-type 22). Bogs can form deep deposits of peat—compacted but relatively undecomposed plant remains that form an organic soil. In some cases the peat is on top of a leached silica layer, and the peat then suggests an exaggeration of the mor layer of a podzol. In other cases the peat is on relatively unweathered parent material, or on a permanently frozen layer, or in the bog succession described, on water. Peat soils occur in the most humid parts of both the tundra and taiga belts; but they are not thought typical of either biome, and they can occur in a wide range of climates. Because the roots of plants in the upper layers of a deep peat have no contact with weathering parent material, a bog may be ombrotrophic. The soil is acid and poor in nutrients; in response to sparse nutrient supply the vegetation is low in nutrient content. The bog and salt desert are in some respects in polar contrast to one another in their soils and nutrient economies.

Figure 6.7 shows these seven soils in the broader pattern of great soil groups in relation to climate. If we set complications aside, these soils may be related to one another along two of the ecoclines of Chapter 4. The first is the tropic-arctic series in humid climates (Figure 4.9D); and the series leads from tropical and subtropical latosols through warm-temperate red-yellow forest soils and mid-temperate brown earths to taiga podzols and tundra gleys and, as non-zonal extreme members, bog peat soils. Along this series the leaching process and effects on soil shift, from inorganic acid leaching with iron and aluminum oxides commonly left as residual material, toward humic acid leaching with silica commonly left as a residual material. Soil colors change from red (iron pigmented) latosols, through red and yellow lateritic soils, to brown earths, and gray (silica with some humus) podzols. Along the series the mass of the surface litter increases; the organic content of the soil itself decreases northward from brown earth to podzol and southward from brown earth to red-yellow forest soil, but may be high in rainforest latosol. The over-all nutrient content of the soil and soil pH increase from latosol to brown earth and decrease from the latter to podzol; soil nitrogen content generally increases northward but may be lower in the podzol. Nutrient stock in the community, and the fraction of that stock turned over per year, decrease northward; but relative dependence of the community on rain for its nutrient supply is higher toward the north and south extremes. Productivity and litter fall decrease northward, and the mineral content per unit mass of litter also decreases. There is consequently a striking decline in the return of nutrients to the soil by litter fall and root death from tropical forest to temperate deciduous forest to taiga; mean values are 128, 20, and 9 $g/m^2/yr$ for ash elements, and 26, 7, and 4 $g/m^2/yr$ for nitrogen. The relative concentrations of different nutrients in plant tissues show some average changes northward, from higher to lower content of silicon, aluminum, and iron, and from lower to higher content of calcium.

Along a humidity gradient in a temperate climate (Figure 4.9A), the series extends from forest brown earths through prairie soils, chernozems, and chestnut and light brown soils, to sierozems and desert soils, with solonchaks as non-zonal extreme members. The extent of weathering and the clay content of soils decrease along the series, or are at least lower in the desert soils. The behavior of water and nutrients shifts from predominant downward leaching in the forest, through relative balance of upward and downward movement in chernozem, to predominance of upward transport and evaporation, with solutes left near the surface, in desert soils. The over-all nutrient content of the soil and soil pH increase from forest to desert, but nitrogen content decreases from grassland to desert. Community biomass decreases along the series, but soil organic content and darkness of color increase from brown forest soils to the black chernozem, then decrease from the latter to the desert soils. From forest to desert the nutrient stock in the vegetation, and the fraction of the ecosystem's stock that is held in the vegetation, decrease; but the relative nutrient content per unit mass of vegetation increases. Average concentrations of different nutrients in plant tissues change, with increase in calcium, sodium, chlorine, and other elements. Productivity and litter fall decrease along the gradient, but nutrient content per unit mass of litter increases. Because of the combined effects of these two trends, the return of nutrients to the soil by litter fall and root death is highest in the middle of the gradient, on prairie and chernozem soils. Mean values for nutrient return to the soil in litter (above and below ground) are 20, 40, and 14 $g/m^2/yr$ for ash elements, and 7, 12, and 3.5 for nitrogen in deciduous forest, grassland, and semidesert. In their high soil organic content, high nutrient turnover, high tissue silica content, and other respects, the grasslands are not simply intermediate to forest and desert. There is a suggestion of the golden mean in the chernozems: These, in climates of intermediate conditions of both temperature and moisture, are some of the finest agricultural soils in the world.

Vegetation and Substrate

The last section has dealt with zonal or typical soils, developed in different regions and climates on fairly level topography from "normal" parent materials. To indicate the difficulty of generalization and the role of parent material, we mention that podzols with acid peat and distinctive communities occur on materials high in quartz (silica) in the Tropics. Many soils of mountains lack the kind of profile development described and remain thin and stony. There are also marked, permanent effects of kinds of rock or other parent material on soil development and characteristics. These interrelations of vegetation, parent material, and soil we shall illustrate with extreme cases, two of effects of substrate contrast on vegetation, and one of vegetation and leaching effects on substrate.

A significant difference in parent material in general implies some difference in vegetation and soil. Succession and soil development do tend to reduce the contrast in soil and vegetation on different rocks; but, as we have indicated in discussing succession and climax, such convergence is incomplete. The effects of parent-material differences on climax vegetation are of all degrees, from some that produce little or no recognizable difference, to others that produce striking contrasts in community structure and composition. The most widely observed of such contrasts are those between limestones and soils formed from acid rocks (such as granite or andesite). Because of these contrasts, separate great soil groups are recognized for soils formed from limestone in some climates (Figure 6.7). The effects of limestone are not consistent, however, in different climates or even on different limestones in a given climate. Zonal soils develop on many limestones; and impurities in a limestone—materials other than calcium (or magnesium) carbonate—may determine many of the characteristics of the soil that develops.

As limestone weathers, it releases abundant calcium carbonate that neutralizes and largely immobilizes the humic acids. Leaching effects are thereby much reduced. Many limestones, however, weather with cracks and channels through which water drains downward freely. Many limestone soils consequently remain shallow, stony, and relatively dry, though chernozems and other good soils develop on limestone in grassland climates. Plant communities on limestone are in many cases more xeric, or drought-adapted, than those on other rocks in the same area. Some contrasts of limestone versus non-limestone vegetation in the same climate in the United States are: oak-hickory forest versus oak-chestnut forest in the East, glades of junipers in grassland versus oak forest in the Middle West, mixed-grass prairie versus oak woodland on sandstone in the forest-prairie transition, and Chihuahuan desert versus taller Sonoran desert of mountain slopes in Arizona. The contrast in structure is greatest in intermediate moisture conditions, where limestone and other rocks support different communities among those (woodlands, shrublands, and grasslands) that occur between forests and deserts.

When communities of similar structure (both forests, or both deserts) are compared on limestone and acid rocks in the same area, many of the species in the two communities are different. Some species that occur on both are represented by genetically different ecotypes. The reasons for different species tolerances of limestone and acid soils are complex and incompletely understood. The soils in question differ in many chemical and physical characteristics; and root growth and nutrient uptake may be affected both by different nutrient availabilities, and by interactions among and relative toxicities of different elements. There is evidence that when species adapted to limestone are grown on acid soils, the aluminum and manganese in the latter may become toxic and the uptake of phosphorus may be limited by the aluminum. For species adapted to acid soils, the high levels of calcium and bicarbonate in limestone soils may limit uptake of potassium and iron. Despite differences in nutrient concentrations in the soil and their effects on plants, nutrient contents in plants of the same species can be quite similar on limestone and acid soils. Nutrient circulation in similar communities (such as different kinds of oak forest) on limestone and acid soils also may be similar in many respects; but in oak forests in Belgium studied by P. Duvigneaud, one on limestone had higher calcium and lower potassium content and turnover than one on an acid soil. The effects of limestone on community biomass and productivity are probably quite variable with climate, topographic position, and other circumstances. Belgian oak forests on different parent materials, including limestone, had similar productivities. Of a pair of samples on open southeast-facing slopes at middle elevations in the Santa Catalina Mountains, Arizona, the limestone supported open shrubland with grass (net primary productivity 185 $g/m^2/yr$, biomass 0.8 kg/m^2, both above-ground only), the granite supported pine-oak woodland (446 $g/m^2/yr$ and 11.4 kg/m^2). Rapid downward loss of water in this mountain-slope limestone is probably responsible for relative soil drought and low productivity.

Serpentines are rocks, often slick-surfaced and green in color, produced by metamorphosis from certain ultrabasic, intrusive rocks (peridotite, dunite) that look like very dark granites. Chemically these rocks differ widely from acid rocks such as granite, but in a direction quite different from that of limestones. Serpentines are in general very high in iron and magnesium, high in nickel, chromium, and cobalt, but low in calcium, potassium, sodium, and aluminum. The soils formed from serpentines are thus chemically distinctive, with a distorted representation of elements compared with other soils. The communities they support are, correspondingly, distinctive. The degree of the distinctiveness depends on climate, age of the land surface, size of the serpentine area and proximity to other areas, and chemical composition of the particular serpentine rock. Some contrasts of serpentine versus non-serpentine communities in the same climate are: tundra versus taiga in Quebec, pine woodland versus Douglas fir forest in Oregon, chaparral versus oak woodland in California, savanna and scrub versus tropical forest in Cuba and New Caledonia, tussock grassland versus southern beech forest in New Zealand. In some areas the serpentine lands are called "barrens" (Figure 6.9). In general the serpentine vegetation is more xeric in appearance, but not because of difference in moisture avail-

Figure 6.9. A serpentine barren, bare serpentine soil with scattered herbs surrounded by coniferous forests of *Abies lasiocarpa* and *Tsuga mertensiana* on normal sandstone soils, upper De Roux Creek, Wenatchee Mountains, Washington. [Courtesy of A. R. Kruckeberg.]

ability. Limiting nutrient conditions reduce vegetation structure in ways that simulate the effects of drought. The serpentine vegetation is doubtless of lower productivity, as well as biomass, but no measurements are available. The distorted soil nutrient conditions are more conspicuously reflected in the nutrient content of plants than is the case with limestone. In some species the total mineral content of leaves on serpentine is about two thirds that on other soils, with both calcium and potassium contents reduced by half while magnesium content is increased two to four times. It appears that magnesium is to some extent substituted for other bases in these serpentine plants. Some serpentine plants take nickel and chromium into their tissues to levels that would be highly toxic to other plants. Underlying the stunted appearance of serpentine vegetation is a deviant nutrient economy, compared with the communities of other soils.

The species composition of serpentine communities generally differs widely from that on nearby non-serpentine soils, and many of the species occurring on both do so as different ecotypes. Serpentine communities are noted among botanists for their concentrations of rare and narrowly endemic species—species largely or wholly confined to particular serpentine areas. On old land surfaces in warm climates serpentines bear some of the greatest concentrations of rare species in the world—such is true in the Coast Ranges and Klamath Region of southern Oregon and northern California, and to an even greater degree in New Caledonia. Experiments have shown that calcium deficiency limits the occurrence of some species on serpentine soils; toxicity of metals (magnesium, nickel, chromium), or unknown interactions among elements, probably limit others. The bay checkerspot was mentioned in Chapter 2 as an animal population of serpentine areas; it is easily observed that the animal, as well as plant, communities differ on serpentine soils from those on other soils nearby. One would expect the herbivorous insects to differ because of the different plant species, and the birds to differ when vegetation structure differs. The same is true of limestones compared with acid soils, and extensive areas of gypsum also have distinctive floras, faunas, and community nutrient relationships. The chemical contrasts should in all these cases imply marked differences in the organisms in the soil also, but such differences are unstudied.

Within an area soils differ because of effects of topographic position and drainage even if they have the same parent material. Down the slope of a hill from ridge top to valley floor, soils become increasingly moist, partly because of water drainage including the gradual movement of water downslope beneath the soil surface. The downslope flow of water, both on and beneath the soil surface, also transports some soil nutrients and soil particles, so that the soils toward the bottom of the slope tend to be deeper, to contain more fine particles, and to be more fertile. Soil characteristics change down the slope, and a soil gradient (catena) along the slope corresponds to the topographic moisture gradient and gradient of plant communities. Because of differences in topography and drainage, and often differences in parent material as well, the pattern of soils within a given area is usually quite complex. Soils representing one of the great soil groups may be zonal and prevalent in the area, but some of these zonal soils may differ considerably from one another, and most areas include also nonzonal soils. One of the detailed systems of classification, such as the U. S. Soil Taxonomy, is needed to describe a local pattern of soils and relate them to differences in environment and plant communities (or agricultural use of the soils). Our final example is an extreme case of local soil and vegetation difference developed on the same parent material.

In the coastal belt of Mendocino County, northern California, redwood forests are the prevailing climax. Along the coast are a series of ancient wave-cut terraces that were once beaches. These were lifted by coastal movement after they were formed, so that they are now a geological staircase from the present shore upward to an inner terrace 200 m above sea level and about one million years old. On the sloping edges and eroded parts of the terraces soil drainage is good, and redwood forests grow on forest brown earths without hardpans, with continuing nutrient release from parent-material weathering and good soil fertility. In the flat parts of the terraces water drainage is poor, and the soils are alternately water-logged in the rainy season and dust-dry in the summer. Under these conditions intense leaching occurs, with humic acids transporting nutrients downward to an iron or clay hardpan that in some of the soils is impermeable to plant roots and soil water. The soil above the hardpan is predominantly silica, white in color and flour-like in texture, and quite acid. The soil is thus an extreme podzol produced by leaching from the same parent material (a graywacke), under the same climate, as the soil of the redwood forests. The podzols with hardpan support a minuscule forest, or woodland, of pygmy cypresses and small pines mostly one to two meters tall, with low, shrubby undergrowth dominated by a blueberry and an endemic manzanita.

Redwood forest and pygmy forest occur within tens of meters of one another, but mixed evergreen forests (Douglas fir with sclerophyll trees) and pine forests occur between them. The vegetation forms a coenocline of an extraordinary degree of change within a limited distance (Figure 6.10). Community measurements decline steeply along the coenocline; estimates of some of these from redwood to pine forest to pygmy forest are: canopy height 63, 21, and 2 m, above-ground biomass $>$ 210, 40, and 2.4 kg/m^2, above-ground productivity 1325, 1190, and 300 g/m^2/yr, and surface litter mass 2.6, 4.7, and 0.5 kg/m^2. Most of the soil nutrients, in forms likely to be available to plants, decrease from redwood forest to pygmy forest, but aluminum is much higher in the pygmy forest soils and may make them toxic to some plants. Tissue concentrations of sodium in the pygmy forest express the large amounts of that element brought in the atmosphere from

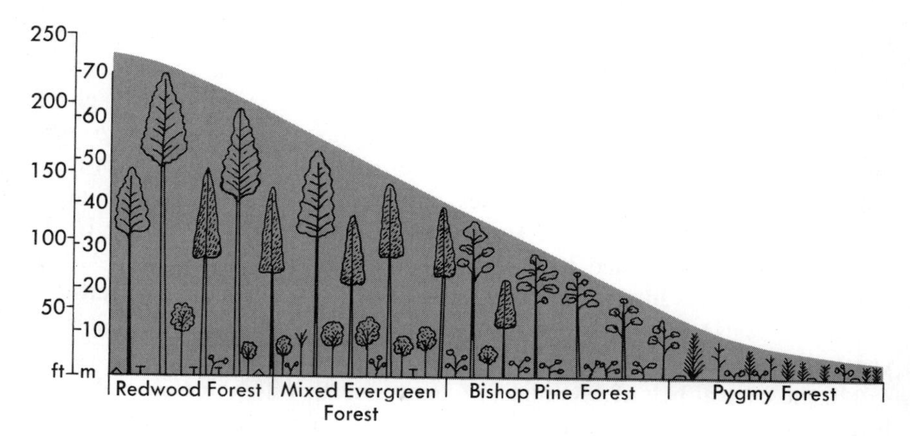

Figure 6.10. Mendocino terrace coenocline. Along a gradient of increasing soil leaching and nutrient impoverishment on coastal terraces in Mendocino County, California, forest composition and structure changes from giant forests of redwoods (*Sequoia sempervirens*) through intermediate types (*Pseudotsuga menziesii*—sclerophyll and *Pinus muricata* forests) to a pygmy forest of cypress (*Cupressus pygmaea*). [Jenny *et al.*, 1969; and unpublished data of W. E. Westman and R. H. Whittaker.]

the ocean nearby; they are several times as high as in a continental forest. Tissue concentrations of other nutrients are low, and manganese in particular is very low. The Mendocino terraces thus support both redwood forests with relatively enormous nutrient stocks in vegetation, and pygmy forests that are ombrotrophic and impoverished, stunted in stature and limited in nutrient economy by their extreme podzol. Other cases of intense soil leaching are known from varied climates, but that of the Mendocino terraces may be the most striking in its community expression in the world.

Watersheds

The chemical linkages of organisms and environments need, however, to be followed into broader contexts. A forest is part of a landscape; a plankton community is part of a water body. We shall expand our interest from forest stands to a mountain drainage basin or watershed as a landscape unit, and treat this as a larger ecosystem.

Small forested watersheds at the Hubbard Brook Experimental Forest, in the White Mountains of New Hampshire, have been studied in this way by F. H. Bormann, G. E. Likens, and others. The water leaving a watershed flows through a weir, a gauging device by which the volume of water flowing out of the watershed at different times can be measured. Chemical analysis of the water at different seasons makes possible calculation of the amounts of nutrient elements leaving the watershed in a year (output,

Table 6.4). The amount of rainfall also has been measured, and rainwater chemistry analyzed, around the year. Input in rainfall is indicated for several elements in Table 6.4. The amounts of the first four elements are not enough to balance the output, and there must be an additional source of nutrients for the ecosystem. This source is the weathering process, which gradually breaks down the parent material (granite and glacial deposits) in the soils of the watershed and releases soluble inorganic nutrients into the soil.

The weathering process is difficult to measure directly, but in the steady-state system the amounts released should approximately equal the difference values in Table 6.4. The Hubbard Brook forest is young and growing, and we cannot assume that many of the differences in Table 6.4 are in steady-state balance with weathering. If, however, we base our calculations on elements that are in least biological demand relative to supply (especially sodium), then we can use the differences to calculate weathering rate. Such calculations imply that the parent material is weathering at a rate of 70 $g/m^2/yr$. This rate is approximately equivalent to a lowering of the rock surface (by complete weathering) by a depth of 50 cm in the 13,000 years since glaciation, hence about 0.04 mm/yr. The rate at which the land surface is being lowered at Hubbard Brook is near an average rate for land surfaces. It is lower, however, than on some other mountain slopes, for the full forest cover of this watershed protects its soils from erosion.

In Table 6.4 inorganic nitrogen (in the forms of ammonium and nitrates) differs from all other elements in its negative difference. Nitrogen in forms available to plants may be added to the forest ecosystem by the fixation of atmospheric nitrogen by bacteria and blue-green algae, and also lost from the forest by denitrification by bacteria. There may thus be active

Table 6.4 Nutrient Budgets for Watersheds at Hubbard Brook. Values are kg/ha/yr (= $g/m^2/yr$ times 10). [Likens and Bormann, 1972.]

		Forested Watershed		*Cut Watershed*
	Precipitation Input	*Stream-Flow Output*	*Difference = Net Output*	*Difference = Net Output*
Calcium	2.6	11.7	9.1	77.9
Sodium	1.5	6.8	5.3	15.4
Magnesuim	0.7	2.8	2.1	15.6
Potassium	1.1	1.7	0.6	30.4
NH_4–nitrogen	2.1	0.3	−1.8	1.6
NO_3–nitrogen	3.7	2.0	−1.7	114.0
Sulfur	12.7	16.2	3.5	2.8
Silicon	trace	16.4	16.4	30.0
Aluminum	trace	1.8	1.8	20.7

cycling of nitrogen between the atmosphere and the forest ecosytem, with nitrogen entering the ecosystem by way of both precipitation and nitrogen fixation, and returning to the atmosphere by way of denitrification. There is also a much larger internal cycling of nitrogen. Nitrogen taken up from the soil by plants returns eventually to the soil in litter. Bacteria and fungi act on some of the litter to decompose proteins, its principal nitrogen-containing compounds, to release ammonium. Probably most of the ammonium is taken up directly by mycorrhizae and plants. Some of it may also be oxidized (nitrified) to nitrite and nitrate; but it appears that this oxidation is small in amount in the forest, perhaps partially inhibited by allelochemic effects on the soil. Despite all this internal transfer and transformation of nitrogen, the output (as ammonium, nitrates, and organic compounds) in the stream is small.

Observation of stream flow through the seasons gives further evidence on ecosystem function. The stream water is relatively constant in its concentrations of dissolved nutrient elements despite the changes in volume of stream water after storms and in different seasons. The water, most of which reaches the stream by subsurface movement in the soil, is largely stabilized in chemical composition by its interaction with the soil it flows through. There is somewhat more variability in the output of nutrients in particles suspended in the stream water. At most times the particle output is very small, but it is much higher when the stream is swollen and fast during storms and in the spring when the snow is melting. The capacity of moving water to carry particles and larger objects increases rapidly as the speed of the water increases. Despite the increase in particle output during storms and snow-melt, the year-around particle output is small compared with the dissolved output—2.5 $g/m^2/yr$ versus 14.0 $g/m^2/yr$, comparing all particulate and dissolved solids. Much of the output of particles is sand and other inorganic soil materials; the organic particle output is only about 1.0 $g/m^2/yr$. This organic matter (mainly particles and fragments from decaying leaves) represents net ecosystem production exported from the forest into the stream, where it may be used as food by stream organisms. Its amount is trivial in relation to the forest's productivity, but it is the major food source for the stream.

Productivity of the forest has been analyzed by some of the same procedures used in the Brookhaven forest. The net primary productivity of the forest was estimated as 760 $g/m^2/yr$ above-ground and 140 $g/m^2/yr$ below-ground for the period 1961–1965. Of this net production 217 $g/m^2/yr$ was added to the soil surface each year as woody litter, and 273 g/m^2 as leaf litter. The woody litter fall is smaller than the above-ground production of woody tissue (463 $g/m^2/yr$). The difference, 190 $g/m^2/yr$, represents net ecosytem production, growth in accumulating woody tissue, in this young forest, which was logged between 1909 and 1917. In 1965 the forest had a biomass of 13.3 kg/m^2 above ground, 2.8 kg/m^2 below ground; its above-ground mass is about twice that of the Brookhaven forest (Table 5.1). The biomass at Hubbard Brook would be expected to increase to about 35 kg/m^2 above-ground in the climax condition, if growth were to continue undisturbed for another two centuries. The annual output of organic material in the stream is only 0.5 per cent of the net ecosystem production above ground that is accumulating in the forest as woody tissue, and 0.01 per cent of the biomass already accumulated.

The increase in biomass implies increase in nutrient stocks held in the forest. Table 6.5 gives for several nutrient elements the above-ground pools of nutrients in woody tissues and in leaves, and the annual movement of nutrients into wood and bark, and to the soil by litter fall and leaching. Taking all ash elements together, the nutrient stock in above-ground woody tissues was increasing at about 1.4 per cent per year in 1961–5. For some elements the amounts going into woody tissue accumulation (about 1.2 times the amounts given, when below-ground woody growth is considered) are large in relation to the watershed budget values of Table 6.5. It is for this reason that sodium, an element of least accumulation in woody tissue relative to its release in weathering, is used in calculating weathering rates. A fraction of the nutrient pool, larger than that in the net ecosystem production, is returned to the soil each year as litter and by above-ground leaching. The leaching differs in some ways from that of the English oak

Table 6.5 Nutrients in the Tree Stratum at Hubbard Brook. Above-ground stocks of nutrient elements are given in g/m^2 for wood and bark, and for summer leaves. Nutrient movements are given in g/m^2/yr for wood and bark growth above ground, litter fall (including wood and bark), and leaching. Turnover rate is the ratio of leaching plus litter fall per year to above-ground stock of the same element. [Data of Whittaker *et al.*, 1974, Likens and Bormann, *Yale Univ. Forest Sch. Bull.*, 79, (1970); Gosz *et al.*, 1973, and Eaton *et al.*, 1973.]

	Wood and Bark					
	Stock	*Growth*	*Summer Leaves*	*Litter Fall*	*Leaching*	*Turnover, Rate/Yr*
Nitrogen	27.7	1.48	6.59	5.27	.992	0.178
Phosphorus	2.84	.154	.518	.36	.068	0.125
Sulfur	3.62	.165	.493	.55	2.10	0.633
Calcium	36.3	1.61	1.85	3.97	.673	0.121
Potassium	12.4	.561	2.84	1.77	3.01	0.310
Magnesium	3.11	.138	.458	.55	.200	0.208
Manganese	3.34	.154	.454	.98		0.30 *
Iron	.300	.011	.033	.040		0.14 *
Zinc	.447	.021	.033	.062		0.16 *
Sodium	.155	.0055	.0047	.010	.031	0.255
Copper	.038	.0017	.0027	.004		0.12 *
Carbon	5880	210	127	254	.506	0.042

* Estimated without leaching data.

woods (Table 6.2). Magnesium is less effectively leached at Hubbard Brook, and there is a net leaching of nitrogen there, in contrast to the net removal of nitrogen from rainwater in the oak woods. The amounts of leaching in Table 6.5 can be compared with the stream output of the same elements. Potassium is leached to the soil at a rate of 3 $g/m^2/yr$, but despite this leaching (plus potassum release in decomposition and weathering), only 0.06 $g/m^2/yr$ leaves the watershed.

The indications of nutrient retention by the forest suggest an experiment. In one of the Hubbard Brook watersheds, all trees were cut and all undergrowth plants were killed, leaving the dead remains of the trees and other plants on the ground. The effect of the cutting is shown in the right-hand column of Table 6.4. The outflow of almost all the nutrients measured (except ammonium and sulfate) increased strikingly. The increase was most dramatic in the output of one key nutrient, nitrate-nitrogen, which rose from 0.12 $g/m^2/yr$ from the uncut watershed to more than 10 $g/m^2/yr$ from the cut watershed. The increase in output of nitrates, but not of ammonium, results from disruption of the forest's nitrogen cycle. In the cut watershed protein decomposition releases ammonium; but in the absence of the plants almost all of this may be oxidized by bacteria to nitrites and nitrates. The resulting large amounts of nitrate are leached from the soil into the stream, and nitrate as an anion helps carry metallic cations (calcium, magnesium, potassium, sodium, and so on) in combination with it from the soil into the stream and out of the watershed as indicated in Table 6.4. Along with this whole-watershed leaching of nutrients, the cutting increased the outflow of water by 30 to 40 per cent above what would occur with the forest present, because water was no longer being transpired by plants. Output of particulate organic matter and inorganic soil particles also increased eleven-fold above that leaving the uncut watershed, even though erosion was minimized by the way the forest was cut—without construction of log trails or removal of wood. When bare soils are exposed on mountain slopes by logging, loss of organic matter and soil from the watershed by erosion may be much greater. The uncut forest in contrast has a self-regulating, self-protective function. Much of the rainwater penetrates into the soil and leaves it in a slow and continuing manner by uptake and transpiration by plants, and by subsurface flow within the soil downhill and into the stream. The forest thus tends to hold its stock of nutrients, and its soil and organic material, against loss by leaching and erosion.

The cut and uncut watersheds are parts of a larger landscape including other small watersheds, the streams draining them, the lake into which some of them flow, and a larger stream leaving the area. Some effects of the cutting extend downstream. Increased nutrient inflow, plus exposure to sunlight, produced a blooming of algae in the stream in the cut watershed; and the nutrient enrichment was transmitted to the larger stream into which it flows. The three ecosystems—forest, stream, and lake—are linked by nutrient flow. Figure 6.11 illustrates the carbon flow for the three, from a forest watershed, to a tributary stream with its meager productivity, to the lake. Inorganic nutrients also flow through these cycles, so that it is quite possible for a potassium atom to circulate from soil through forest tree and back to the soil, and move out of the watershed and down the stream to circulate through plankton organisms of the lake, in the same summer. If it leaves the lake by way of the stream, the potassium atom may be on its way to the Merrimac River of Thoreau and the Atlantic Ocean off New England.

The rates at which different nutrient elements leave a forested watershed are affected by the manners in which these elements are circulated between forest and soil—especially the degree to which a given element is bound into organic matter and held in tight circulation. No matter how tight the circulation, however, some loss of nutrients in water moving under and over the soil surface into the stream is inevitable. The level of production of the forest must be to some degree affected, and in the long run limited, by the rate of nutrient input. Production of the lake is in turn affected by the nutrient flow from its watershed, for this flow replaces nutrients it must lose to stream output and to its sediment. There are further points of interest: (1) Life on land is to a degree dependent on the weathering process to replace nutrients that must be lost from the soil by transport in moving water. (2) Rainwater is "pure," but not pure like distilled water. Life on land is also to a degree dependent on the inorganic nutrients in precipitation, which become part of the function of natural and agricultural communities.

Biogeochemistry and the Oceans

The question of where the nutrients in precipitation come from is an important one. They come, in large part, from the sea. Over the 70 per cent of the earth's surface that is ocean, winds and wave spray mix droplets of sea water into the lower atmosphere; the droplets evaporate to leave particles of salts in the air, and turbulence and convection mix these upward in the atmosphere. The particles may later, over the land, serve as the condensation centers for snow, or may be dissolved into raindrops, or may settle; in any case they become part of the nutrient content of precipitation.

Thus it is likely that a potassium atom that reaches the Hubbard Brook forest in spring snow has arrived in air from over the ocean; and that potassium washed from the forest into the stream is on its way back to the ocean. It is possible for an atom to do a year's tour of duty in the nutrient circulation of a forest or wheatfield, between years in which it participates in the water chemistry and nutrient function of the marine plankton ecosystem. The reader will recognize that the land and the sea are coupled

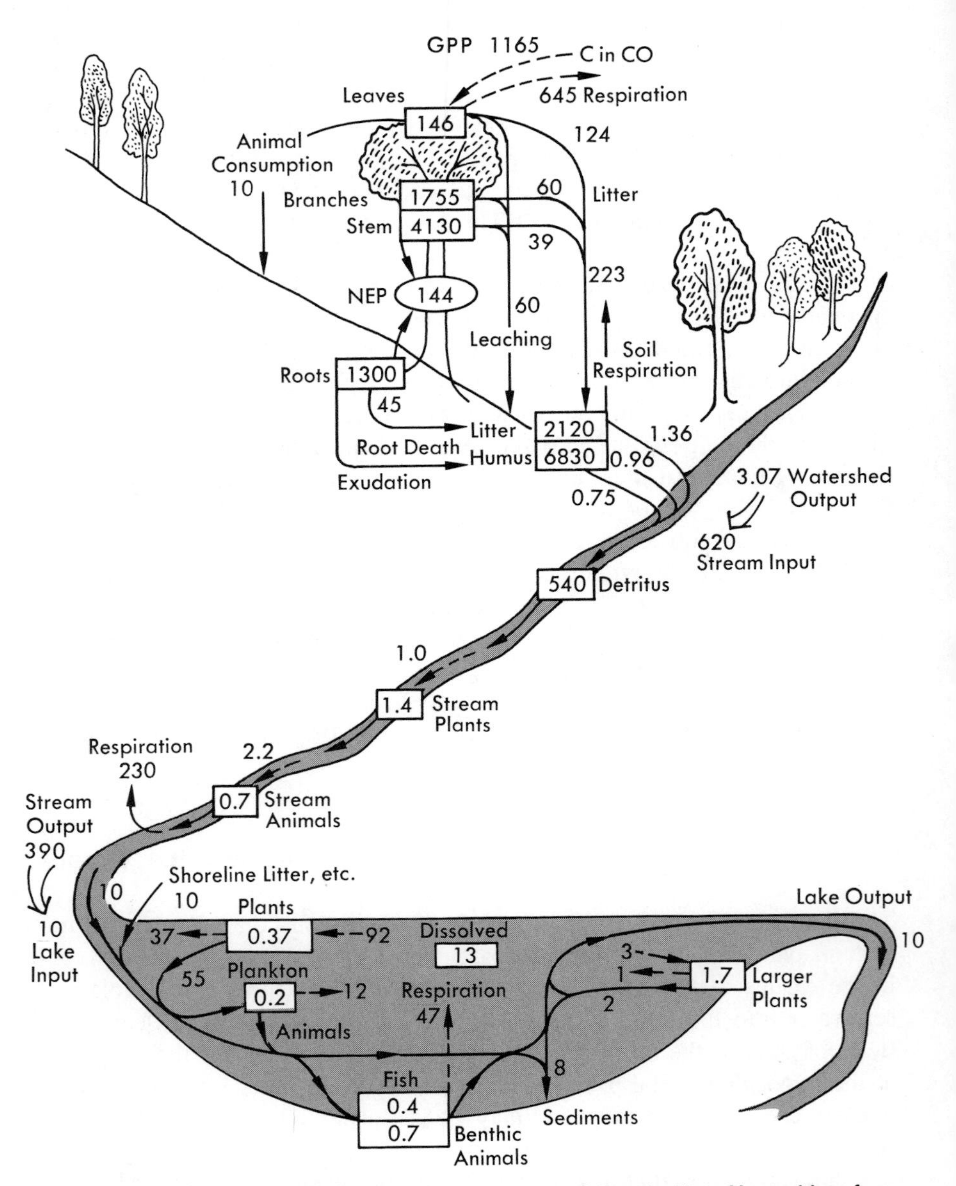

Figure 6.11. Carbon flow in a landscape, at Hubbard Brook, New Hampshire, from a forest cycle, on the top, through a small stream (Bear Brook), to a lake (Mirror Lake). Numbers in boxes are biomass pools in grams of carbon per meter square; numbers on arrows are transfers in $gC/m^2/yr$ (multiply by 2.2 to convert carbon values to dry matter). Plant respiration values are estimates. The hollow arrows indicate conversion to different area bases when carbon moves from one ecosystem to another—from watershed (130 ha) to stream (0.59 ha) to lake (85.1 ha). Similar streams, not Bear Brook, drain into Mirror Lake. [Whittaker *et al.*, 1974; Gosz *et al.*, 1973; Eaton *et al.*, 1973; Fisher and Likens, 1972; Jordan and Likens, *Verh. Int. Ver. Limnol.* **19** (in press); and unpublished data of G. E. Likens and R. H. Whittaker.]

in nutrient circulation, and that the expanding context in which nutrient circulation should be studied leads ultimately to the world ecosystem. The study of worldwide chemical circulation, concentration, deposit, and release is the subject of biogeochemistry.

The proportions of elements in rainwater, Table 6.4, do not simply correspond to their proportions in sea water. Calcium and nitrogen, for example, are several times as abundant (considered not in absolute amounts but by ratios to sodium) in rainwater as in sea water. Our statement of the marine origin of the nutrients in precipitation must be modified for these and other elements. The proportion of calcium in precipitation increases from the coast inland in some areas. Some of the calcium in the precipitation comes from the ocean, but much comes from dust from land surfaces, mixed upward into the atmosphere in one place and brought down again (either as particles or in solution) in another area. A primary source of the calcium is the dust from soils in areas of limestone rocks, for limestone is predominantly $CaCO_3$. It is not the case that this calcium is simply terrestrial in origin, however, for limestone is a rock deposited on the ocean bottom and brought to the surface on land millions of years later by geological processes. There are thus two major routes by which calcium circulates from sea to land: a shorter-term route by sea-spray into the atmosphere; and a larger and longer-term route of deposit, formation of rock, elevation of rock, exposure, weathering, and dust. Major features of the biogeochemical cycle for calcium are illustrated in Figure 6.12.

The contribution of the land to the sea is considerable. It is estimated that 2.73×10^9 t of dissolved materials are brought to the seas each year by rivers. Fresh water is less dense than sea water because of the dissolved salts in the latter. Nutrients brought to the sea in rivers consequently tend to be mixed into the surface waters near coasts, where these nutrients may contribute to the productivity of coastal waters. Some of the nutrients in rivers are also caught and used in the productivity of estuaries, the mouths of rivers draining into the sea, where river water and sea water meet and are mixed by tidal flow. Along with dissolved nutrients the rivers carry organic particles, and clay, silt, and sand into the sea at a rate of approximately 9.3×10^9 t/yr before extensive disturbance of the land surface by man, probably about 24×10^9 t/yr currently. Other particles from the land enter the sea: till from glaciers and icebergs, wind-blown sand from desert coasts, and air-borne dust carried aloft over the continents and brought down in precipitation on the ocean. Some of these particles contribute nutrients to the sea. They form also an important part of the various sediments, although their share in the sediments (and the particle sizes involved) tends to decrease from the coasts outward to deeper waters.

Nutrients mixed into shallow coastal waters may be carried out to the open sea in currents, or by slow drift of surface waters away from the coast. They may be carried in time—periods of weeks or months—into the

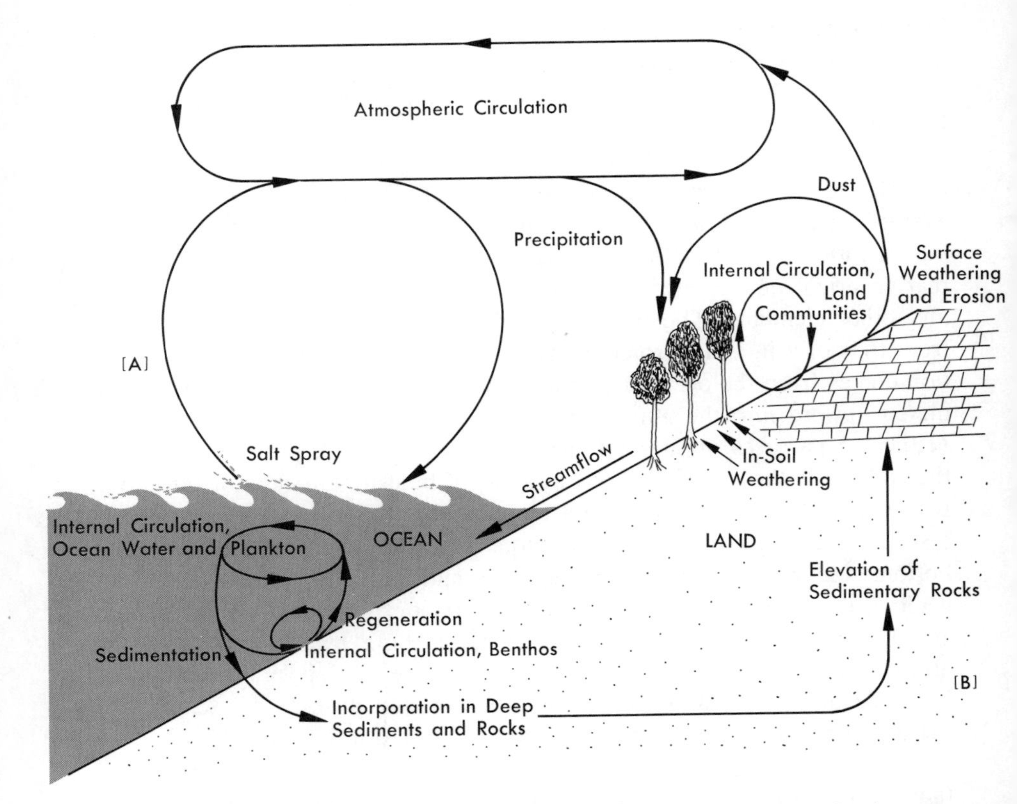

Figure 6.12. Major features of biogeochemical cycles for calcium and other elements. Circulations within ecosystems, on land and in the sea, are linked into the global circulation through cycles. **A:** Shorter-term cycles from the ocean surface into the atmosphere by salt spray, into terrestrial ecosystems by precipitation, and back to the ocean in stream water. **B:** Longer-term cycles from ocean waters into deep sediments and rocks, to exposure on land surfaces after the elevation of these rocks, and back to the ocean by varied routes involving in-soil weathering, dust, stream flow, or atmospheric circulation and precipitation.

central open waters of the oceans. In the major ocean basins (excepting the Southern or Antarctic Ocean) the surface waters flow in great whorls or gyres, turning clockwise in the Northern and counterclockwise in the Southern Hemisphere. The water toward the center of an ocean basin, and the center of its gyre, is stable and stratified; warmer surface water is separated by a thermocline from the cold waters of the depths. In certain areas of the oceans, particularly in the Arctic and Antarctic, cold surface water sinks and moves toward the equator beneath other, warmer water. The sinking of this water is balanced by the rise of water from the depths elsewhere. In addition to the surface currents other currents flow in other directions in intermediate and deeper waters. Surface and deeper currents flow between the ocean basins, and connect these also with the smaller and partly enclosed basins of various gulfs and seas. The oceans and their seas are thus a great circulatory system, interconnected by currents the complexity of which is not shown by diagrams of surface currents.

Water movements in the oceans are extensive but, on a human time scale, unhurried. The surface currents move mostly at speeds below 30 km/day (0.8 mi/hr), though the fastest major current speeds northward along the East African Coast during the southwest monsoon at more than 60 km/day (1.56 mi/hr). The deeper waters mostly move less as distinct currents than as slow drifts of great masses. The speeds of these drifts are not easily clocked, but the decay of radiocarbon, ^{14}C, to ^{12}C provides a means of estimating the age of water and inferring time between different parts of an ocean. Applied to the greatest of the deep currents, the Pacific Deep Water below 2500 m, the radiocarbon measurement gives a speed of 0.05 cm/sec or 0.043 km/day, and a time of about 600 years from the Antarctic Ocean to the Pacific at 30° north latitude. Relative stability versus mobility of water masses can be expressed also as residence times. Given a river inflow of 27×10^{12} m^3/yr, a mean residence time for ocean water of 50,000 years is implied (leaving out of consideration the evaporation from and precipitation back onto the ocean surface). Within the ocean itself, residence times for major water masses and areas can be estimated, based on volumes and rates of inflow and outflow in currents. In one simplified model, residence times for surface water masses are about 10 years (North Atlantic and South Atlantic) and 25 years (Pacific and Indian Ocean Surface Water). Suggested residence times for deep waters are 600 years for Atlantic Deep Water and 1300 for Pacific and Indian Deep Water. Intermediate to these are residence times in the Arctic (45 years) and the Antarctic (100 years) Oceans, in which turbulent exchange between surface and depths occurs both as upwelling and as the sinking of water in areas of convergence.

The routes followed by nutrients through the maze of surface and deeper currents of the oceans are complex and varied; but we can construct as an example an imaginary course for a potassium atom: from the coastal waters of New England into the northern edge of the Gulf Stream, northeastward in this current across the North Atlantic to Iceland and then, in mixture with colder Arctic waters, downward from the surface to become part of the Atlantic Deep Water traveling southward, across the equator in this deep water to the Antarctic Ocean and back to the surface in the Antarctic upwelling areas, eastward in the West Wind Drift current past the Indian Ocean and across the Pacific to South America, northward in the Peru Current along that coast to curve westward into the Equatorial Current of the South Pacific, westward in this current and the great South Pacific gyre to curve south past New Zealand to the Antarctic convergence, downward in this to deeper waters moving northward beneath the surface

Pacific circulations, across the equator to the Bering Sea in the northern North Pacific, through the Bering Strait with the small flow into the Arctic Ocean and through the under-ice circulation of this to flow outward along the east coast of Greenland and southward to the Labrador Current and the waters off the Maritime Provinces and New England. Here the potassium atom might be carried back onto the continent in a northeasterly storm. Such a journey, without serious delays en route, might take 1000 years.

During this journey the potassium atom has a long opportunity not only to participate in the nutrient function of the plankton, but to be carried downward to the sediments. Throughout the oceans nutrients move downward in settling organisms and particles; the more productive the surface waters, the greater the loss of nutrients by settling is likely to be. Many of the nutrients carried out of surface waters in settling detritus are released by decomposition in water at middle depths. Some of the detritus reaches the ocean bottom, however, where it may serve as food for benthic organisms. Either directly (in the case of particles of skeletal material) or indirectly after use in the detritus chains of the benthos, nutrients may be incorporated permanently into the sediments. The sediments, the marine muds or oozes, are varying mixtures of land-derived particles, skeletons of marine organisms, and modest amounts of organic matter. The rates of sediment accumulation in the ocean are smaller, on the average, than the rates of land surface erosion. The average growth of sediments in the ocean is about 0.01 mm/year; but the rates decrease from values higher than this in coastal waters, to around 0.005–0.01 mm/yr in foraminiferan (*Globigerina*) ooze, to less than 0.005 mm/year in some of the deeper sediments with diatom or red clay ooze.

Because of the loss by settling, concentrations of elements in sea water are by no means determined simply by their solubilities and abundances in weathering products and stream water. Concentrations are determined by the need of organisms for substances and the consequent rates of removal and deposit in sediments, in relation to the input by streams and solubility. Assuming that the ocean is in steady state for these nutrients, then inputs and outputs are equal, and residence times (see Table 6.3) can be calculated by dividing the size of the pool of a nutrient in the ocean by either the net input rate or the rate of loss to the sediment. Table 6.6 gives mean ocean water concentrations, compared with mean river water and marine plant concentrations, and residence times for some major elements in the sea. Nutrient contents in marine plants are quite different from those in land plants. Given the abundance of many elements in sea water, and the advantage of osmotic balance with sea water, marine algae take up salts in amounts that give mineral ratios much higher than those of most land plants. Amounts in different kinds of algae depart widely from the values in Table 6.6; diatoms, for example, have around 200,000 ppm of silica.

Table 6.6 Elements in Sea Water, River Water, and Marine Plants. Average concentrations are given in parts per million of sea water, compared with concentrations in river water and the dry matter of marine plants (brown algae). [Goldberg 1963, Livingstone 1963, Bowen 1966.]

	Concentrations			*Residence Time*
	Sea Water ppm	*River Water ppm*	*Marine Plants ppm*	*Sea Water years*
Sodium	10,500	6.3	33,000	2.6×10^8
Magnesium	1,350	4.1	5,200	4.5×10^7
Calcium	400	15	11,500	8×10^6
Potassium	380	2.3	52,000	11×10^6
Strontium	8	0.08	1,400	19×10^6
Iron	0.01	0.67	700	1.4×10^2
Manganese	0.002	0.012	53	1.4×10^3
Silicon	3	6.5	1500	8×10^3
Carbon	28	11	345,000	
Chlorine	19,000	7.8	4,700	
Sulfur	885	3.7	12,000	
Bromine	65	0.021	740	
Boron	4.6	0.013	120	
Fluorine	1.3	0.09	4.5	
Nitrogen	0.5	0.23	15,000	2.5×10^3
Phosphorus	0.07	0.005	2,800	

There are some striking concentrations of particular elements into particular species. Some of these concentrations, and the sulfuric acid in the vacuoles of the alga *Desmarestia,* are probably defenses against animal consumers.

Some elements are strongly concentrated into marine organisms, but other elements are abundant in sea water because they are in small demand by organisms relative to their high solubility and input rates. These elements, which include sodium, magnesium, potassium, calcium, chlorine, and sulfur have long residence times expressed in millions of years (Table 6.6). Other elements with large inputs to the ocean occur there in much lower concentrations because of the demand for these substances by organisms and consequent loss to the bottom sediments; these include silicon, nitrogen, manganese, and phosphorus, with residence times measured in thousands of years. Like soil chemistry, sea water chemistry is so edited and amended by organisms as to obscure the composition of the original text—the geological input by weathering.

Many of the transfer rates for biogeochemical cycling are difficult to measure. One that is of interest to man, and departs from the pattern of transfers in Figure 6.12, is the return of phosphate to the land surface in the dung, or guano, of marine birds feeding on fish in the sea and returning to nest on islands or coastal headlands. The rate of phosphorus transfer to the land by guano and fisheries is estimated at 1.0×10^5 t/yr, and is small, but not insignificant in relation to river transport of phosphorus to the sea, estimated at 1.4×10^7 t/yr. The greatest guano deposits are along the coast of Peru; these and other phosphate deposits are being mined by man for use in fertilizer as part of current techniques of intensive agriculture. The use of phosphorus in fertilizer is now above 7×10^6 t/yr, hence more than 50 per cent of the natural movement of phosphorus into the sea. Some of this fertilizer phosphorus must return from farm fields into streams and into the sea; by this and other means man is accelerating the flow of phosphate nutrient into some coastal waters.

Figure 6.13 gives estimates of world pools and rates for another nutrient element of major concern, nitrogen. Nitrogen is abundantly present in the environment, forming about four fifths of the atmosphere. This abundance of nitrogen is not simply available to higher organisms; apparently no higher plant is able by itself to take in and use atmospheric nitrogen in its metabolism. A scattering of lower, procaryotic organisms are able to do so. Some blue-green algae fix significant amounts of nitrogen in the oceans, lakes, and the soil. Symbiotic bacteria in the root nodules of legumes (and also species of alder, buckbrush, and a number of other genera that are not legumes) fix atmospheric nitrogen. A pair of widespread genera of bacteria of the oceans and soil, *Nitrosomonas* and *Nitrobacter,* convert ammonium into nitrite, and nitrite into nitrate, as occurred extensively in the cut watershed at Hubbard Brook. These bacteria primarily use the energy of dead organic matter to convert ammonium from dead organic matter (protein residues and so on) into nitrates. Their role is thus not fixation of atmospheric nitrogen but participation in the circulation of nitrogen within ecosystems. Other, denitrifying bacteria break down nitrates and release molecular nitrogen back to the atmosphere. Other routes of nitrogen fixation exist, including other bacteria and the action of lightning in the atmosphere (from which some part of the nitrogen in the precipitation of Table 6.4 is derived). Man is increasingly active in fixing nitrogen both intentionally by industrial processes, and unintentionally in internal combustion engines, particularly those of automobiles. The greater part of the nitrates in precipitation in many areas is now from sources related to man, although such sources are not included in Figure 6.13. Residence times for nitrogen are of the order of 3×10^8 years for atmospheric N_2, 2500 years for nitrogen in the sea (treating nitrogen of nitrates and organic compounds as a single pool), generally less than a year for nitrites and nitrates in the soil.

Some points of interest on the nitrogen cycle and biogeochemistry are:

1. The existence of an invisible dependence of the higher organisms upon a few of the lower. The higher, eucaryotic organisms are largely dependent on certain bacteria and blue-green algae for nitrogen fixation and

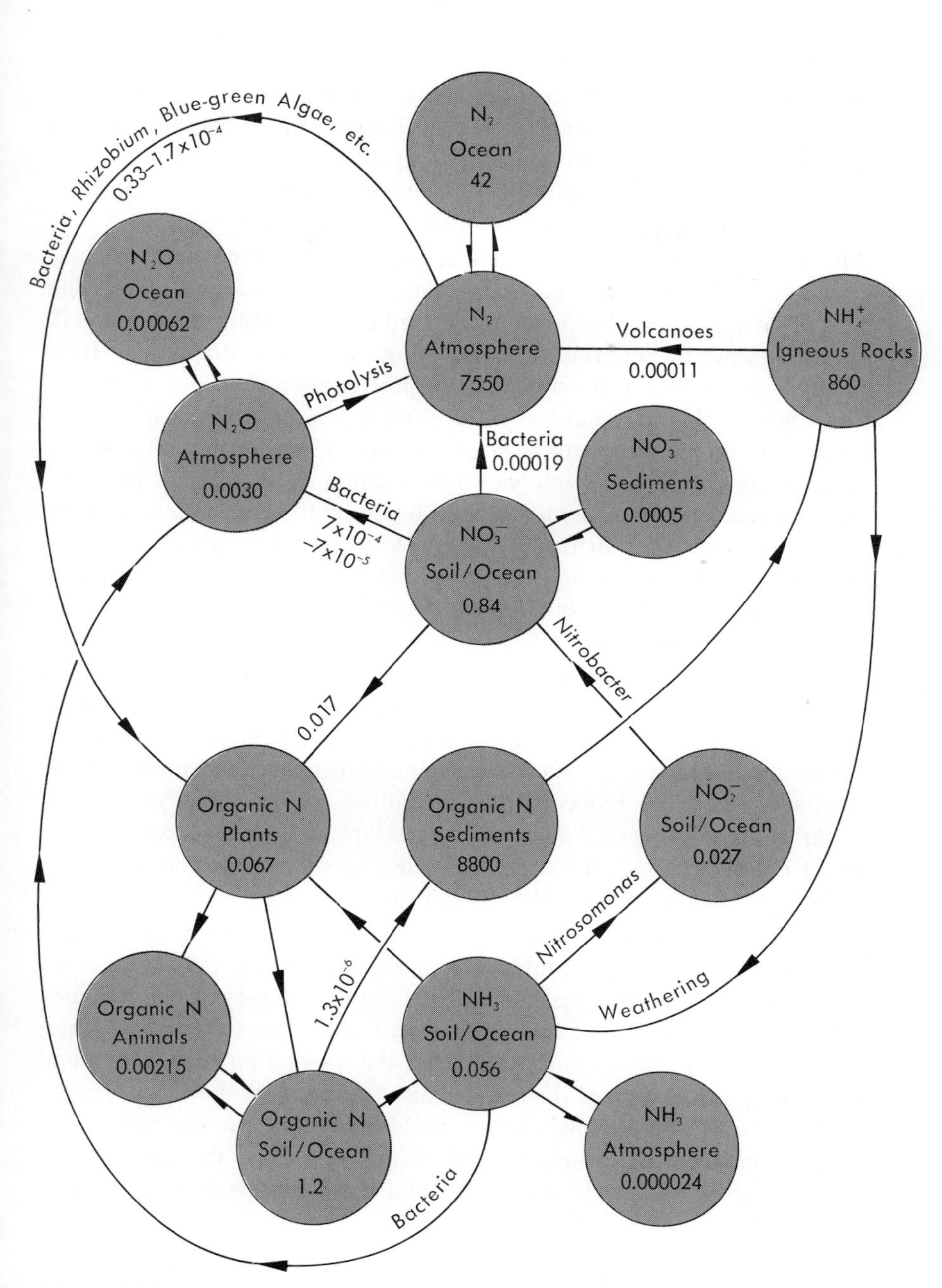

Figure 6.13. Biogeochemical cycle for nitrogen. Numbers in circles are amounts of nitrogen in pools, in kilograms per square meter of the earth's surface. Numbers on arrows are transfer rates in kilograms of nitrogen per square meter per year. [Bowen, 1966; see also Delwiche, *Scient. Amer.*, **223:**137 (1970).]

nitrogen circulation, without which the function of natural communities would slow down to a much lower level. Of nitrogen fixed by natural means, about 90 per cent is fixed by organisms, about 10 per cent by lightning. Man's total application of agricultural fertilizer is about 10 per cent of the amount of nitrogen fixed by microorganisms.

2. The existence of steady states on different levels of nitrogen circulation, as of other biogeochemical circulations. In a steady-state forest the rate of nitrogen input by precipitation and organic nitrogen fixation will balance output by denitrification and ground-water and stream loss. Rate of input to the ocean by stream flow, precipitation, and nitrogen fixation together should equal loss by denitrification and deposit in bottom sediments. In the biosphere at large, the loss by sedimentation and denitrification should equal the gain from fixation by organisms, weathering, volcanic release of ammonia, and lightning fixation. It is not really known what changes in this steady-state system of interlinked steady-state systems are occurring. Man, however, by increased fertilization of agricultural fields and in other ways is increasing the movement of nutrients into water bodies.
3. The significance of these cycles for the dissolved substances in the sea, as well as those of soil waters and streams. It is familiar that life evolved through most of its history in the sea, and that land and fresh-water organisms bear the imprint of sea-water chemistry in their chemistry of inorganic ions. The blood of man resembles a dilute form of the sea water to which his earlier ancestors were adapted. It may be less evident, but it is also true that the chemical characteristics of the sea, soil, and air are profoundly influenced by organisms.
4. The complexity of the biogeochemical circulation, and the biogeochemical linkages of land, sea, atmosphere, soil, fresh water, and organisms. Each element or substance has its own pattern of biogeochemical circulation, differing at least in quantitative details from those of all other elements. All together these cycles, and the moving air and water that are an important part of their mechanism, imply that the local ecosystems of the world are woven into a world-ecosystem, the biosphere (or, considering that the biosphere is its living part only, the *ecosphere*). Man is part of the world ecosystem, and its environment is man's environment.

SUMMARY

The function of ecosystems involves a kind of metabolism—complex patterns of transfer, transformation, use, and accumulation of inorganic and organic materials. Some aspects of this metabolism can be studied by releasing radioactive tracers, such as radiophosphorus, into aquaria or lakes and following their movements. Radiophosphorus circulates very rapidly between water and plankton, moves more slowly into shore plants and animals, and gradually accumulates in bottom sediment. When phosphate fertilizer is added to a lake, there is a temporary increase in productivity, following which the concentration of phosphate in the water returns to what it was before fertilizer introduction. Nutrient transfers link all parts of the ecosystem to one another, and the amount of the nutrient in the water is determined not just by input, but by the over-all steady-state function of the ecosystem. In a forest ecosystem nutrients from the soil enter plants by way of mycorrhizal fungi and roots and are distributed to the various tissues of the plant. The largest share of nutrients go to leaves and other short-lived tissues that return nutrients to the soil within a short time, to complete the cycle. Nutrients are also being washed, or leached, from plant leaves to the soil and downward within the soil. Organic substances too are washed from leaf surfaces to the soil, and some of these substances are inhibitory to other plants. The chemical inhibition of plants by other plants is only one case of allelochemic influences, chemical effects of one species on another. The most widespread of such effects are the uses of chemicals by organisms to defend themselves against their enemies. Three broad groups of substances—inorganic nutrients, foods, and allelochemics—participate in community metabolism.

Effects of the community are exerted downward into the soil. Carbon dioxide and organic acids from the community contribute to the soil weathering and leaching; humus particles are added to the soil, and roots and animals affect it. Climate, and the kind of community that develops in response to climate, together influence the direction of soil development and the characteristics of mature soil. Some of the major kinds of soil, or great soil groups, can be related to climate in a manner roughly paralleling that of biomes. Along the climatic gradient northward in forest climates soils show changes in the character of leaching and kind of residual material left near the soil surface, and sizes of nutrient pools and rates of nutrient circulation through the vegetation decrease northward. Along the climatic gradient from forest to desert in temperate latitudes, concentrations of nutrients (other than nitrogen) in the soil and in plant tissues tend to increase, but amounts in circulation between soil and vegetation decrease. The interrelation of soil and community is shown also by the manner in which the rocks from which soils are formed can, along with climate, determine nutrient function of ecosystems and thereby the characteristics of soils and communities. The effects of communities on the soil, on the other hand, are shown by the manner in which nutrients in soil water and stream water are stabilized in a forested watershed, and the much faster movement of the nutrients into the stream and out of the watershed when the forest is cut.

Substances are transferred between, as well as within, local ecosystems.

Nutrient circulations of land and water ecosystems are coupled by the transfer of nutrients from land to the sea in rivers, and from the sea to the land by precipitation, and a longer route into the ocean sediments and onto land by the elevation and exposure of these as rocks. Ecosystems of the world are thus linked together in biogeochemical cycles, patterns of transfer and concentration of substances. Chemical characteristics of the atmosphere, the soil, and the ocean waters are determined or strongly influenced by the activities of organisms. To the biosphere, as the community of all organisms of the world, corresponds a world ecosystem or ecosphere, as a functional system comprising the earth's air, soils, surface waters, and organisms.

References

Phosphorus in Aquaria and Water Bodies

CONFER, J. L. 1972. Interrelations among plankton, attached algae, and the phosphorous cycle in artificial open systems. *Ecological Monographs* **42**:1–23.

*HAYES, F. R., J. A. MCCARTER, M. L. CAMERON and D. A. LIVINGSTON. 1952. On the kinetics of phosphorus exchange in lakes. *Journal of Ecology* **40**: 202–216.

HAYES, F. R. and J. E. PHILLIPS. 1958. Lake water and sediment. IV. Radiophosphorus equilibrium with mud, plants, and bacteria under oxidized and reduced conditions. *Limnology and Oceanography* **3**:459–475.

†HUTCHINSON, G. EVELYN. 1957. *A Treatise on Limnology*. Vol. I. Geography, Physics, and Chemistry. New York: Wiley. xiv + 1015 pp.

*HUTCHINSON, G. E. and V. T. BOWEN. 1950. Limnological studies in Connecticut. IX. A quantitative radiochemical study of the phosphorus cycle in Linsley Pond. *Ecology* **31**:194–203.

IMBODEN, D. M. 1974. Phosphorous model of lake eutrophication. *Limnology and Oceanography* **19**:297–304.

MCCARTER, J. A., F. R. HAYES, L. H. JODREY and M. L. CAMERON. 1952. Movement of materials in the hypolimnion of a lake as studied by the addition of radioactive phosphorus. *Canadian Journal of Zoology* **30**:128–133.

*LEAN, D. R. S. 1973. Phosphorus dynamics in lake water. *Science* **179**:678–680.

MORTIMER, C. H. 1941–2. The exchange of dissolved substances between mud and water in lakes. *Journal of Ecology* **29**:280–329, **30**:147–201.

RIGLER, F. H. 1956. A tracer study of the phosphorus cycle in lake water. *Ecology* **37**:550–562.

RUTTNER, FRANZ. 1963. Fundamentals of Limnology. 3rd ed. Univ. Toronto. xvi + 295 pp.

Schindler, D. W., F. A. J. Armstrong, S. K. Holmgren and G. J. Brunskill. 1971. Eutrophication of Lake 227, Experimental Lakes Area, northwestern Ontario, by addition of phosphorus and nitrate. *Journal of the Fisheries Research Board of Canada* **28**:1763–1782.

Watt, W. D. and F. R. Hayes. 1963. Tracer study of the phosphorus cycle in sea water. *Limnology and Oceanography* **8**:276–285.

Whittaker, R. H. 1961. Experiments with radiophosphorus tracer in aquarium microcosms. *Ecological Monographs* **31**:157–188.

Nutrients in Land Communities

*Art, H. W., F. H. Bormann, G. K. Voigt and G. M. Woodwell. 1974. Barrier island forest ecosystem: role of meteorologic nutrient inputs. *Science* **184**:60–62.

*Chapman, S. B. 1967. Nutrient budgets for a dry heath ecosystem in the South of England. *Journal of Ecology* **55**:677–689.

Duvigneaud, Paul, editor. 1971. *Productivity of Forest Ecosystems:* Proceedings of the Brussels Symposium 1969. Paris: Unesco. 707 pp.

Koelling, M. R. and C. L. Kucera. 1965. Dry matter losses and mineral leaching in bluestem standing crop and litter. *Ecology* **46**:529–532.

Monk, C. D. 1966. An ecological significance of evergreenness. *Ecology* **47**:504–505.

Ovington, J. D. 1965. Organic production, turnover and mineral cycling in woodlands. *Biological Reviews* **40**:295–336.

Pomeroy, L. R. 1970. The strategy of mineral cycling. *Annual Review of Ecology and Systematics* **1**:171–190.

Reichle, David E., editor 1970. *Analysis of Temperate Forest Ecosystems.* New York: Springer. xii + 304 pp.

†Rodin, L. E. and N. I. Bazilevich. 1968. *Production and Mineral Cycling in Terrestrial Vegetation.* Edinburgh: Oliver & Boyd. v + 288 pp.

Tamm, C. O. 1951. Seasonal variation in composition of birch leaves. *Physiologia Plantarum* **4**:461–469.

Thomas, W. A. 1969. Accumulation and cycling of calcium by dogwood trees. *Ecological Monographs* **39**:101–120.

Tyler, G., C. Gullstrand, K. Å. Holmquist and A.-M. Kjellstrand. 1973. Primary production and distribution of organic matter and metal elements in two heath ecosystems. *Journal of Ecology* **61**:251–268.

Allelochemics

Brower, L. P. 1969. Ecological chemistry. *Scientific American* **220**(2):22–29.

Chambers, Kenton L., editor. 1970. *Biochemical Coevolution.* Oregon State University Annual Biology Colloquia **29**, x + 117 pp.

Ehrlich, P. R. and P. H. Raven. 1965. Butterflies and plants: a study in coevolution. *Evolution* **18**:586–608.

*EHRLICH, P. R. and P. H. RAVEN. 1967. Butterflies and plants. *Scientific American* **216**(6):104–113.

*EISNER, T. and J. MEINWALD. 1966. Defensive secretions of arthropods. *Science* **153**:1341–1350.

FRAENKEL, G. S. 1959. The raison d'être of secondary plant substances. *Science* **129**:1466–1470.

HARBORNE, J. B., editor. 1972. *Phytochemical Ecology:* Proceedings of the Phytochemical Society Symposium, Surrey, 1971. London: Academic. xiv + 272 pp.

HARTMAN, R. T. 1960. Algae and the metabolites of natural waters. In *The Ecology of Algae,* ed. C. A. Tryon, Jr. and R. T. Hartman. Pymatuning Laboratory. Univ. Pittsburgh, Special Publication **2**:38–55.

JANZEN, D. H. 1968. Host plants as islands in evolutionary and contemporary time. *American Naturalist* **102**:592–595.

JANZEN, D. H. 1974. Tropical blackwater rivers, animals, and mast fruiting by the Dipterocarpaceae. *Biotropica* **6**:69–103.

MULLER, C. H. 1966. The role of chemical inhibition (allelopathy) in vegetational composition. *Bulletin of the Torrey Botanical Club* **93**:332–351.

*MULLER, C. H., R. B. HANAWALT and J. B. MCPHERSON. 1968. Allelopathic control of herb growth in the fire cycle of California chaparral. *Bulletin of the Torrey Botanical Club* **95**:225–231.

PIMENTEL, D. 1968. Population regulation and genetic feedback. *Science* **159**: 1432–1437.

ROOT, R. B. 1973. Organization of a plant-arthropod association in simple and diverse habitats: the fauna of collards (*Brassica oleracea*). *Ecological Monographs* **43**:95–124.

†SONDHEIMER, ERNEST and J. B. SIMEONE, editors. 1970. *Chemical Ecology.* New York: Academic. xv + 336 pp.

SOUTHWOOD, T. R. E. 1961. The number of species associated with various trees. *Journal of Animal Ecology* **30**:1–8.

†WHITTAKER, R. H. and P. P. FEENY. 1971. Allelochemics: chemical interactions between species. *Science* **171**:757–770.

Soil Development and Classification

BRADY, NYLE C. 1974. *The Nature and Properties of Soils.* 8th ed. New York: Macmillan. xvi + 639 pp.

BUOL, STANLEY W., F. D. HOLE and R. J. MCCRACKEN. 1973. *Soil Genesis and Classification.* Ames: Iowa State Univ. x + 360 pp.

CLINE, M. G. 1949. Basic principles of soil classification. *Soil Science* **67**:81–91.

CLINE, M. G. 1963. Logic of the new system of soil classification. *Soil Science* **96**:17–22.

CROWTHER, E. M. 1953. The sceptical soil chemist. *Journal of Soil Science* **4**:107–122.

CURTIS, C. D. 1970. Differences between lateritic and podzolic weathering. *Geochimica Cosmochimica Acta* **34**:1351–1353.

DUCHAFOUR, PHILIPPE. 1965. *Précis de pédologie.* 2nd ed. Paris: Masson. 481 pp.

GERASIMOV, I. P. and YE. N. IVANOVA. 1958. Comparison of three scientific trends in resolving general questions of soil classification. *Soviet Soil Science* **1958(11)**:1190–1205.

HUNT, CHARLES B. 1972. *Geology of Soils: their Evolution, Classification, and Uses.* San Francisco: Freeman. xii + 344 pp.

JENNY, HANS. 1941. *Factors of Soil Formation: A System of Quantitative Pedology.* New York: McGraw-Hill. xii + 281 pp.

KUBIËNA, WALTER L. 1948. *Entwicklungslehre des Bodens.* Wien: Springer. xi + 215 pp.

LEGGET, ROBERT F., editor. 1965. *Soils in Canada; Geological, Pedological, and Engineering Studies.* Rev. ed. Univ. Toronto. x + 240 pp.

SIMONSON, R. W. 1962. Soil classification in the United States. *Science* **137**: 1027–1034.

SOIL SURVEY STAFF. 1960. *Soil Classification, a Comprehensive System—7th Approximation.* Washington: U. S. Department of Agriculture. v + 265 pp.

THOMPSON, LOUIS M. and F. R. TOEH. 1973. *Soils and Soil Fertility.* New York: McGraw-Hill. x + 495 pp.

WEBSTER, R. 1968. Fundamental objections to the 7th Approximation. *Journal of Soil Science* **19**:354–366.

Soil Geography

ARKLEY, R. J. 1967. Climates of some great soil groups in the western United States. *Soil Science* **103**:389–400.

BASILE, ROBERT M. 1971. *A Geography of Soils.* Dubuque, Iowa: Brown. vii + 152 pp.

*BRIDGES, E. M. 1970. *World Soils.* London: Cambridge Univ. 89 pp.

BUNTING, BRIAN T. 1967. *The Geography of Soil.* Chicago: Aldine. xii + 213 pp.

EYRE, S. R. 1968. *Vegetation and Soils: A World Picture.* 2nd ed. London: Arnold. xvi + 328 pp.

FITZPATRICK, EWART A. 1971. *Pedology: A Systematic Approach to Soil Science.* Edinburgh: Oliver & Boyd. xvi + 306 pp.

GERASIMOV, I. P. and M. A. GLAZOVSKAYA. 1965. *Fundamentals of Soil Science and Soil Geography.* Jerusalem: Israel Program for Scientific Translations. ix + 382 pp.

GLINKA, K. D. 1927. *The Great Soil Groups of the World and their Development.* Ann Arbor: Edwards. 235 pp.

KUBIËNA, WALTER L. 1953. *The Soils of Europe.* London: Murby. 318 pp.

PAPADAKIS, JUAN. 1969. *Soils of the World.* Amsterdam: Elsevier. xiv + 208 pp.

RODIN and BAZILEVICH 1968.

TAVERNIER, R. and G. D. SMITH. 1957. The concept of Braunerde (Brown Forest soil) in Europe and the United States. *Advances in Agronomy* **9**:217–289.

TEDROW, J. C. F. and H. HARRIES. 1960. Tundra soil in relation to vegetation, permafrost and glaciation. *Oikos* **11**:237–249.

VOLOBUEV, V. R. 1964. *Ecology of Soils.* Jerusalem: Israel Program for Scientific Translations. ii + 260 pp.

Vegetation and Substrate

ANDERSON, G. D. and D. J. HERLOCKER. 1973. Soil factors affecting the distribution of the vegetation types and their utilization by wild animals in Ngorongoro Crater, Tanzania. *Journal of Ecology* **61**:627–651.

BILLINGS, W. D. 1950. Vegetation and plant growth as affected by chemically altered rocks in the western Great Basin. *Ecology* **31**:62–74.

CLINE, M. G. 1953. Major kinds of profiles and their relationships in New York. *Proceedings of the Soil Science Society of America* **17**:123–127.

DUVIGNEAUD, P. et S. DENAEYER-DE SMET. 1971. Cycle des éléments biogènes dans les écosystèmes forestiers d'Europe, pp. 527–542 in *Productivity of Forest Ecosystems,* ed. P. Duvigneaud. Paris: Unesco.

GOODLAND, R. and R. DOLLARD. 1973. The Brazilian cerrado vegetation: a fertility gradient. *Journal of Ecology* **61**:219–224.

JENNY, H., R. J. ARKLEY and A. M. SCHULTZ. 1969. The pygmy forest-podsol ecosystem and its dune associates of the Mendocino coast. *Madroño* **20**:60–74.

LYFORD, W. H. 1974. Narrow soils and intricate soil patterns in southern New England. *Geoderma* **11**:195–208.

KRAUSE, W. 1958. Andere Bodenspezialisten. *Encyclopedia of Plant Physiology* **4**:755–806.

MASON, H. L. 1946. The edaphic factor in narrow endemism. *Madroño* **8**:209–226, 241–257.

NYE, P. H. 1954. Some soil-forming processes in the humid Tropics. I. A field study of a catena in the West African forest. *Journal of Soil Science* **5**:7–21.

PLATT, R. B. 1951. An ecological study of the mid-Appalachian shale barrens and of the plants endemic to them. *Ecological Monographs* **21**:269–300.

QUARTERMAN, E. 1950. Major plant communities of Tennessee cedar glades. *Ecology* **31**:234–254.

RUNE, O. 1953. Plant life on serpentines and related rocks in the North of Sweden. *Acta Phytogeographica Suecica* **31**:1–139.

WESTMAN, W. E. 1975. Edaphic climax pattern of the pygmy forest region of California. *Ecological Monographs* (in press).

*WHITTAKER, R. H., R. B. WALKER and A. R. KRUCKEBERG. 1954. The Ecology of Serpentine Soils. *Ecology* **35**:258–288.

*BORMANN, F. H., G. E. LIKENS. 1970. The nutrient cycles of an ecosystem. *Scientific American* **223**(4):92–101.

BORMANN, F. H., G. E. LIKENS and J. S. EATON. 1969. Biotic regulation of particulate and solution losses from a forest ecosystem. *BioScience* **19**:600–610.

BORMANN, F. H., G. E. LIKENS, D. W. FISHER and R. S. PIERCE. 1968. Nutrient loss accelerated by clear-cutting of a forest ecosystem. *Science* **159**:882–884.

EATON, J. S., G. E. LIKENS and F. H. BORMANN. 1973. Throughfall and stemflow chemistry in a northern hardwood forest. *Journal of Ecology* **61**:495–508.

FISHER, S. F. and G. E. LIKENS. 1972. Stream ecosystem: organic energy budget. *BioScience* **22**:33–35.

GOSZ, J. R., G. E. LIKENS and F. H. BORMANN. 1973. Nutrient release from decomposing leaf and branch litter in the Hubbard Brook Forest, New Hampshire. *Ecological Monographs* **43**:173–191.

HOBBIE, J. E. and G. E. LIKENS. 1973. Output of phosphorus, dissolved organic carbon, and fine particulate carbon from Hubbard Brook watersheds. *Limnology and Oceanography* **18**:734–742.

JOHNSON, N. M., G. E. LIKENS, F. H. BORMANN and R. S. PIERCE. 1968. Rate of chemical weathering of silicate mineral in New Hampshire. *Geochimica Cosmochimica Acta* **32**:531–545.

*LIKENS, G. E. and F. H. BORMANN. 1972. Nutrient cycling in ecosystems. In *Ecosystem Structure and Function,* ed. John H. Wiens. Oregon State University Annual Biology Colloquia **31**:25–67.

LIKENS, G. E., F. H. BORMANN, N. M. JOHNSON, D. W. FISHER and R. S. PIERCE. 1970. Effects of forest cutting and herbicide treatment on nutrient budgets in the Hubbard Brook Watershed ecosystem. *Ecological Monographs* **40**:23–47.

LIKENS, G. E., F. H. BORMANN, N. M. JOHNSON and R. S. PIERCE. 1967. The calcium, magnesium, potassium, and sodium budgets for a small forested ecosystem. *Ecology* **48**:772–785.

MARKS, P. L. and F. H. BORMANN. 1972. Revegetation following forest cutting: mechanisms for return to steady-state nutrient cycling. *Science* **176**:914–915.

WHITTAKER, R. H., F. H. BORMANN, G. E. LIKENS and T. G. SICCAMA. 1974. The Hubbard Brook ecosystem study: forest biomass and production. *Ecological Monographs* **44**:233–254.

Biogeochemistry and the Oceans

†BOWEN, H. J. M. 1966. *Trace Elements in Biochemistry*. London: Academic. ix + 241 pp.

CARSON, RACHEL L. 1951. *The Sea Around Us*. New York: Oxford Univ. vii + 230 pp.

DUXBURY, ALYN C. 1971. *The Earth and its Oceans*. Reading, Mass.: Addison-Wesley, xv + 381 pp.

†GOLDBERG, E. D. 1963. The oceans as a chemical system. In *The Sea,* ed. M. N. Hill, Vol. **2**:3–25. London: Interscience.

HUTCHINSON, G. E. 1950. The Biogeochemistry of Vertebrate Excretion. *Bulletin of the American Museum of Natural History*. New York: **96,** xviii + 554 pp.

*HUTCHINSON, G. E. and others. 1970. The Biosphere. *Scientific American* **223**(3):44–208.

LIVINGSTONE, D. A. 1963. Chemical composition of rivers and lakes. *U. S. Geological Survey Professional Paper* **440–G,** vii + 64 pp.

*MACINTYRE, F. 1970. Why the sea is salt. *Scientific American* **223**(5):104–115.

MACINTYRE, F. 1974. The top millimeter of the ocean. *Scientific American* **230**(5):62–77.

REDFIELD, A. C. 1958. The biological control of chemical factors in the environment. *American Scientist* **46**:205–221.

REVELLE, R., and others. 1969. The Ocean. *Scientific American* **221**(3):54–234.

SILLÉN, L. G. 1967. The ocean as a chemical system. *Science* **156**:1189–1197.

TAIT, RONALD V. 1968. *Elements of Marine Ecology: An Introductory Course.* London: Butterworths. 272 pp.

TUREKIAN, KARL K. 1968. *Oceans.* Englewood Cliffs, N. J.: Prentice-Hall. viii + 120 pp.

7

Pollution

The surface of the earth is an arena of moving currents of air and water and moving organisms. The ways substances move in the environment and are concentrated from environment by organisms are part of the background for understanding pollution problems. Like all open systems, man's industry must discharge wastes and by-products to environment. These substances become foreign materials, toxins, or excess nutrients that circulate in ecosystems and may distort their function, may be concentrated into organisms to their detriment or man's, and may be transported by wind and water to appear in unexpected places. The loading of local ecosystems and environments, and thereby of the world ecosphere, with detrimental materials is pollution. Several aspects of pollution are of concern to us: the significance of such different pollutants as radioisotopes, pesticides, heavy metals, and combustion products; the behavior of these in terrestrial, freshwater, and marine environments; and the kinds of effects on communities that pollution and other stresses produce.

Radioisotope Contamination

Experiences with radioisotopes have provided valuable instruction in the meaning of pollution processes. During World War II water from the Columbia River was used to cool plutonium-producing reactors at the Hanford Works in eastern Washington. The intense neutron flux in the reactors made certain elements in the water radioactive. Retention of the water in basins for a period during which much of its radioactivity decayed, dilution of the water in the large volume of the river, and careful monitoring of the radioactivity of river water and organisms were means of ensuring that no hazard to human health developed. The levels of radiophosphorus and other isotopes in the Columbia River organisms were not hazardous, but they were sufficient to emphasize the necessity of control measures to prevent hazard. Decreasing levels of radioactivity in water and organisms could be traced down the Columbia River and into the Pacific Ocean.

Other radioactive materials escaped from the smokestacks of Hanford and other atomic plants. Contamination by this means of the atmosphere and vegetation has required study of the concentration of radioiodine and other isotopes in terrestrial vertebrate animals. Tests of atomic weapons up until the test limitation treaty of 1963 released increasing amounts of radioactive materials into the atmosphere. These materials were carried into the upper atmosphere and circulated around the earth, and produced measurable increases in radioactivity in precipitation over the whole of the earth's surface. One weapon test showered radioactive dust on a Japanese fishing ship and spread unexpected amounts of radioactivity to Pacific islands, the ocean waters, and tuna and other fish. In general the amount of fallout from weapon tests has not been sufficient to constitute a short-term physiological hazard to man, or so far as is known to other organisms, outside the restricted areas in which the tests were carried out.

Exposure of organisms to radioactive materials implies exposure of their protoplasm (especially the nuclei of their cells) to the adverse effects of ionizing radiation. Radioactive materials in the environment are consequently toxic pollutants. Ionizing radiation produces interrelated effects on three time scales: (1) intense, and usually acute or short-term, exposure producing acute radiation sickness, (2) lower-level, and often chronic, exposures producing in the individuals exposed delayed consequences, which may include cancer and relatively subtle effects in accelerated aging, and (3) effects, which can occur at any level of exposure, on the genetic material of the reproductive cells—effects that are expressed as unfavorable mutations in descendants of the organism exposed. All three effects may occur primarily from damaging hits by ionizing radiation on or near the genetic material in nuclei—the chromosomes. The sensitivity of organisms to radiation effects is in consequence affected by the sizes of their chromosomes. Sensitivity varies widely among kinds of organisms: man and other vertebrates are sensitive, arthropods and many plants are less sensitive, microorganisms relatively insensitive.

Throughout the world naturally radioactive elements and cosmic rays produce low levels of natural or background radioactivity. This natural ionizing radiation is responsible for part of the natural mutation rate of organisms, although hits by ionizing radiation in general account for a minority of mutations. An organism has a complement of genes that has resulted from long-term selection. The genes selected for are favorable both individually and in their interrelations in the whole genetic system of the organism and species. The great majority of mutations are necessarily deleterious, even if some few mutations are advantageous and their selection is the basis of evolution. Increasing the level of environmental radiation has the inescapable consequence of increasing the numbers of unfavorable mutant genes, genes whose effects on human beings range from the mildly disadvantageous to the tragic. It is this less obvious genetic effect of radiation, which has no threshold and occurs at levels far below those of direct, physiological radiation damage, that has been of most concern. The treaty limiting above-ground testing of atomic weapons by the major atomic powers (although by no means a complete and permanent solution to the problems of radiation release into the environment) is one of civilized man's greatest achievements in limiting the destructive implications of technology.

The broadcasting of radioisotopes into the air, water, and soil has brought concern both with isotopes of elements of known biological significance (tritium or hydrogen-3, carbon-14, phosphorus-32, sulfur-35, potassium-40, calcium-45, manganese-54, iron-59, zinc-65, iodine-131, and so on) and those of a number of elements exotic to the chemistry of life (rubidium-86, strontium-89 and 90 and yttrium-90, zirconium-95 and niobium-95, cesium-134 and 137, cerium-144, ruthenium-106, thorium-232, plutonium-239). Some of these have been of special concern as pollutants. Radioiodine is of concern to man because of its concentration along food chains from low levels in the environment to relatively high levels in the thyroid gland. Radiostrontium is of concern because of its similarity in chemical behavior to calcium. Strontium-90 is concentrated from low levels in the environment to higher levels in human bones; the concentration ratios are highest in children who are adding calcium to bone more rapidly than adults and who receive strontium in milk as well as in other foods. Cesium-137 is taken up by plants along with potassium. One of the least expected hazards to human beings involved food-chain concentration of cesium from dust from atomic explosions, to soil and lichens in the arctic tundra, to reindeer feeding on these lichens, to Eskimos eating the reindeer.

If the radioisotope and its nonradioactive relative are at stable levels in an aquatic environment, and certain other simplifying assumptions are accepted, then we can express environmental radioactivity for an isotope as equation (1), in Table 7.1. The concentration of a substance from environment into an organism's tissues can be expressed as a *concentration ratio* (equation 2). In the case of radioactive substances, the concentration ratios for the stable, naturally occurring isotope or element (R) and for the radioactive isotope (R') are not identical because of the effect of radioactive decay on the latter. There may, furthermore, be a small but measurable difference in the chemical behavior of the radioisotope and the

Table 7.1 Radioisotope Concentration into Organisms

For radioactivity of an isotope in an aquatic environment,

$$D_e = C_e I_e A_i \qquad (1)$$

(D_e, the activity density of the water, its radioactivity measured in curies per gram of water) = (C_e, concentration of the nonradioactive isotope or element per gram of water) × (I_e, isotopic proportion or ratio by numbers of atoms of the radioisotope to the nonradioactive equivalent) × (A_i, the specific activity of the isotope in curies per gram).

The concentration ratio, R, from environment into an organism,

$$R = C_o/C_e \qquad (2)$$

C_o is the concentration of the element in grams per gram of the organism (by dry weight, or in some treatments by live weight).

The concentration ratio for the radioisotope is then R', and D_o is an equilibrium activity density for the organism,

$$D_o = R'D_e = C_o I_e A_i e^{-\lambda t} \qquad (3)$$

in which λ is the constant for radioactive decay of the isotope and t is a biological time for the isotope.

Radioactivity in the organism approaches this equilibrium level as,

$$D_t = D_o(1 - e^{-(k + \lambda)t}) \qquad (4)$$

D_t is the activity density of the organism at time t; k is a constant for turnover of the substance in the organism.

nonradioactive equivalent in uptake and metabolism. Equation (3), for equilibrium radioactivity of the organism, has not been complicated to allow for this difference. The constant of radioactive decay for the isotope is λ; t is the time of the isotope in biological transfer and metabolism and includes (a) mean residence time for an atom of the isotope in the pool of atoms of that isotope within the organism, and (b) food-chain time, time spent by the isotope moving along a food chain from environment to food source if the isotope is taken up in food rather than directly from environment. The manner in which the organism's radioactivity approaches equilibrium after its exposure to a radioisotope in its environment is given by equation (4). The equation, of the same essential form as equation (3) in Table 6.3, treats k as a constant, analogous to λ, for rate of loss by turnover and excretion of atoms from the pool of the element in the tissues of the organism, related to time (a) above.

The concentration of a radioisotope in an organism (activity density, D_o, in curies per gram) is necessarily greater with a higher isotopic proportion ratio (I_e). This ratio is increased by either increased abundance of the pollutant isotope, or scarcity of its nonradioactive relative, or by both. Activity density must also increase with a high concentration ratio. As observed in the last chapter, activity concentration ratios of one and two million times for radiophosphorus can result from the high elemental concentration ratio (R) when nonradioactive phosphate is scarce in the water. Ratios for many other elements are measured in thousands or tens of thousands. High decay constant (λ) values express rapid loss of radioactivity by the radioisotope and imply that the concentration ratio for this isotope (R') will be markedly lower than that for the element (R). A number of radioisotopes of short half lives that are abundant products of reactors and atomic weapons are of minor concern in the environment under most circumstances. Concentration of one of these isotopes will be lower in animals if the time (t) includes food-chain time (but some of these isotopes are directly absorbed from fresh water by animals). In some cases, however, concentration of the inactive relative of the radioisotope (C_o) increases up a food chain from plants to animals, and from herbivorous animals to carnivorous ones. In this case concentration ratios for the isotope and radioactivity of tissues may increase up a food chain. Such may be the case despite the fact that the approach to equilibrium, as expressed in equation (4), is slower in larger organisms.

In general, sensitivity to radioactivity also increases up the food chain. For higher animals the crucial concentration ratios are those that apply not to the whole organism, but to particular tissues in which a given isotope is concentrated—for example, the thyroid gland for iodine, and bone for strontium. There is a tendency, despite wide differences in species on a given trophic level, for the effects of radiopollution to hit hardest those organisms that are highest in food-chain level and in evolutionary level.

Man is low in neither sense, and would suffer the consequences of extensive release of radioactive materials accordingly.

Radioactivity, like fire, is a potentially destructive servant of man. Rapid expansion of radioactivity tamed and caged, in the form of nuclear power plants, is planned to permit continued expansion of industrial economies while petroleum supplies decrease. It has been proposed that hundreds of nuclear reactors should be built around the world by the year 2000 at a rate of expansion (with a doubling time of 2 to 3 years) that for a massive industry is almost without precedent. The more tightly the release of radioactive materials from nuclear reactors and plants processing nuclear fuels is controlled, the greater the cost of that control. In general the escape of radioactive substances is not wholly prevented; it is minimized and held to levels that for an individual nuclear plant are not (barring accident) detrimental. Some release of radioactive materials occurs, but the environmental concentrations of these—primarily gaseous tritium and krypton-85 in present plants—can be kept very low. The individual nuclear reactor does not seem dangerous in this respect, and present projections of environmental krypton-85 and tritium to the year 2000 are not in themselves alarming. Nevertheless, significant increase in environmental radioactivity is likely to result from the cumulative effects of many nuclear reactors and processing plants, some in countries where technological skills are limited.

Present nuclear energy reactors are "burner" reactors in the sense that they consume their fuel (uranium-235) without replacement, to release heat that can be converted to electrical energy. If the great expansion of nuclear energy production were to be based only on burner reactors, the world supply of uranium-235 could be exhausted in a fairly short time. It is hoped, therefore, that large-scale "breeder" reactors will become technologically feasible. A breeder reactor, while using uranium-235 for fuel, at the same time converts a larger amount of the much more common nonfissionable uranium-238 into fuel (plutonium-239), or converts thorium-232 into fuel (uranium-233). In principle, breeder reactors might make possible the use of most of the available uranium and thorium for fuel, instead of only the limited amount of uranium-235. Breeder reactors could thus greatly increase the nuclear energy available to man as well as, unfortunately, the amount of fissionable and potentially explosive plutonium in the world. The record to the present of burner nuclear energy reactors in the United States is unimpressive as regards sustained energy production and frequency of accidents or breakdowns. (Some of the latter have involved melting of the cores of fuel of reactors with the damage contained, however, within their protective shields.) Further reliance on technologically more demanding breeder reactors could increase not only energy but problems: the cumulative release of some radioactive substances, the burden of great quantities of intensely radioactive wastes to be transported and placed in permanent containment somewhere on the earth, and the escape of plutonium to illicit hands. The multiplication of nuclear reactors may also impose heavy demands on water supplies for cooling (at the same time other demands on the same water are increasing) and extensive heat-pollution on water bodies. For societies committed to exponentially increasing energy use, development of and reliance on nuclear power has seemed an inescapable decision. If the societies were not thus committed, the decision would seem a dangerous choice.

Current radioactivity in the environment is not a serious concern. Experiences with radioisotope contamination to date suggest, however, some points of interest (that relate to the observations on tracers in microcosms):

1. The effectiveness of world-wide transport in wind and water, and the unity of the ecosphere in consequence of this transport.
2. The effectiveness of organisms and food chains in concentrating substances and the manner in which ecosphere transfer and concentration by organisms collaborate to produce those hazards that develop.
3. The variety and complexity of transfer and concentration routes and patterns, the variation in detail despite similarity in principle that affects different isotopes and elements.
4. The partial unpredictableness of the patterns and hazards. Although relevant principles of biogeochemical circulation and ecosystem function are understood, detailed effects are not simply predictable and hazards may be discovered only when already well advanced. Unwelcome surprises are normal among pollution effects.

Pesticides and Heavy Metals

Since World War II there also has been a rapid increase in release into the environment of poisons that kill organisms unwelcome to man. Among these are chemicals of most varied character aimed at insects and mites, nematodes and fungi, fish, rodents and mammalian predators, and algae and higher plants. The most massive uses of poisons are against man's principal rivals for the consumption of agricultural crops—insect pests—, and the chemicals used are generally referred to as pesticides. In the diversity of chemical agents used two families of compounds are of most concern.

The organophosphate compounds include parathion, malathion, chlorthion, phosdrin, thimet, TEPP, and others. These poisons are inhibitors of the enzyme cholinesterase, important in the transmission of nerve impulses; in particular they block the normal function of synapses. These substances are among the most intensively effective poisons known; and their use in most cases requires stringent controls to avoid human exposure. Sensitivity to their effects on nervous function is shared by arthropod targets and vertebrates, including man; parathion in particular has gained from cases of poisoning a reputation for lethality to man. Aquatic organisms seem in general less affected than terrestrial ones. Organophosphates can produce extensive mortality of organisms other than the target pests in areas where they are applied, and they can disturb the interactions on which biological control (by predators and parasites of the pest species) may have been based. Unlike the chlorinated hydrocarbons, however, the organophosphates are relatively unstable both in environments and in tissues; and they do not in general spread widely through environments or accumulate in environments or in tissues. The same is true of another group of compounds that act as cholinesterase inhibitors and are now in extensive use, the carbamates (carbaryl or sevin, and so on).

These are of much less serious concern in pollution effects than the organochlorines or chlorinated hydrocarbons. Organochlorines as chemical weapons were not first used by man; a number of such compounds are produced by fungi and are effective antibiotics. Man, however, is using large quantities of some compounds of this group (DDT, chlordane, dieldrin, endrin, aldrin, heptachlor, toxaphene, lindane, mirex, and so on). These poisons combine intense toxicity to a wide range of organisms with relative chemical stability. They consequently accumulate in environments and along food chains, are distributed by wind and water in the ecosphere, and persist in foods that reach man. Their effects on physiological processes may be varied, but the best-known substance, DDT, acts on the central nervous system to produce tremors and convulsions. DDT is accumulated in fatty tissues, and may be withdrawn from them into the blood stream with lethal effects when fat reserves are used by the organism. Aquatic, as well as terrestrial, organisms are highly sensitive to chlorinated hydrocarbons. Vulnerableness to chlorinated hydrocarbons is shared by invertebrate and vertebrate animals; birds and fish are highly vulnerable, mammals relatively less so.

The ecological behavior of these substances resembles that of radioisotopes in a number of ways. Many of the relationships discussed in the previous section apply, although one cannot simply relate pesticides to nontoxic relatives whose ecologic and metabolic routes they follow. The ecological significance of pesticides is strongly affected by their half lives of decay into inactive compounds. The organophosphates with short chemical half lives are of less ecological significance than the chlorinated hydrocarbons with long ones. As certain radioisotopes decay to yield another radioisotope, so some of the chlorinated hydrocarbons change chemically to other hydrocarbons, also toxic (aldrin–dieldrin, DDT–DDD–DDE). Like radioisotopes, pesticides are actively taken up through living membranes from the environment, or from the contents of digestive tracts. Concentration ratios of many thousand times from environment into organisms result and are part of the basis of their effectiveness as poisons used with purpose. Like radioisotopes the persistent pesticides are transferred along food chains. In some cases, though not all, concentration ratios increase along food chains to the tissues of vertebrate predators. Equilibria in animal tissues are affected by such factors as a compound's stability, food-chain relations if uptake is indirect, and turnover time in and metabolic treatment by the animal. Like the radioisotopes, the more stable pesticides are transported long distances at low concentrations in wind and water, to appear in biologically significant concentrations at points distant from those of their release (Figure 7.1).

DDT is one of the oldest and most widely used of these poisons, and the one for which unintended dispersal and biological effects are best known. Some of the earliest observations of effects of chlorinated hydrocarbons involved the death of robins and other birds in cities where trees had been sprayed with DDT in the effort to control Dutch elm disease. Bird populations in a number of cities were temporarily reduced by 30 to 90 per cent following elm spraying. Mortality was by no means limited to robins but involved, with varying death rates, most or all of the bird species

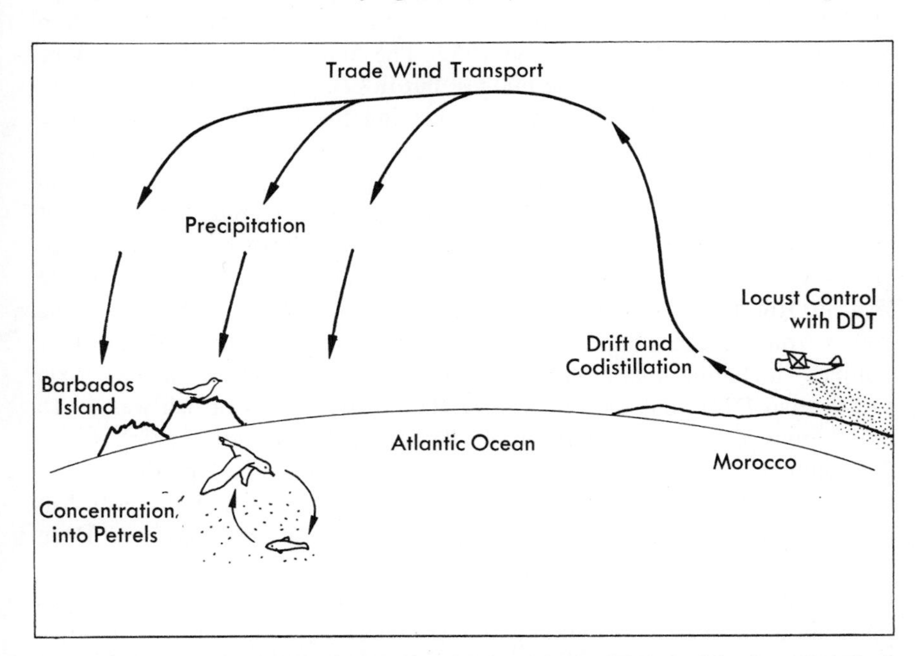

Figure 7.1. Example of DDT transport in the ecosphere. DDT used for locust control in Morocco, North Africa, is carried into the atmosphere. It is then transported across the Atlantic Ocean in westerly trade winds and brought down again in precipitation. DDT reaching the surface waters around Barbados Island is concentrated along food chains from water into plankton, fish, and the Bermuda petrel, a marine bird nesting on the island, the population of which is in decline because of the DDT it receives. [Risebrough *et al., Science,* **159:**1233 (1968); Wurster and Wingate, *Science,* **159:**979 (1968); Rudd, Fig. 7 in Murdoch, 1971.]

nesting and feeding in the sprayed area. The death rates of robins were highest following rains, when earthworms are most abundant as food on the surface of the ground. The effects were based on a food-chain concentration process. DDT reaches the soil by the settling of the spray, in falling leaves (if these have been sprayed in summer) and probably by rain wash from leaf and bark surfaces to the soil. Earthworms may concentrate the DDT by factors of more than ten times from their food (decaying leaves and other soil materials) into their tissues. Some tissues of the robin concentrate the DDT further; levels of DDT and DDE with lethal effects on the nervous system of robins are easily reached following ordinary intensities of spraying. The deaths of birds that feed in the tree foliage rather than on the ground (warblers, orioles, and so on) indicate poisoning through other kinds of food chains.

Clear Lake is a popular recreational center that has had excellent sport fishing, about one hundred miles north of San Francisco. A midge (*Chaoborus*) that is related to mosquitoes but does not bite is so abundant at the lake as to be a nuisance. The chlorinated hydrocarbon DDD was applied to the lake at concentrations of one part DDD to 50 to 70 million parts of water in 1949, 1954, and 1957. The first and second applications killed about 99 per cent of the midges, but in each case the midge population recovered rapidly in the years following the DDD application. The DDD further caused extensive deaths of and tissue contamination in other invertebrate animals in the lake, and consequently in the birds and fish feeding on these. A colony of one thousand western grebes that bred at the lake disappeared from it, and extensive deaths have been observed in grebes visiting the lake in winter, after breeding in other areas. A DDD level of 1,600 ppm (parts per million) was found in visceral fat of these grebes from the lake; high levels were recorded in some hundreds of other samples, including plankton, fish, frogs, and birds. DDD is less toxic to fish than DDT, but the range of DDD in fatty tissues of fish was 40 to 2,500 ppm. Concentration in edible flesh of fish was not so high (1 to 200 ppm, predominantly above 20), but it was higher in predatory fish favored as human food than in herbivorous species. Flesh of most fish exceeded the maximum tolerance level of 7 ppm set by the Food and Drug Administration for DDD residues in marketed foods. The third application of DDD produced only a limited and brief reduction of the midge population (suggesting evolution of resistance), and the effort at control by DDD was abandoned as unsuccessful. Pesticide residues in, and damage to, the lake persisted.

DDT and DDD are thus among the poisons subject to progressive concentration or "biological magnification" up food chains. In Clear Lake concentration ratios for DDD were about 265 times from water into plankton, 500 times into small fishes, up to 80,000 times into the fat of grebes, and 75,000 times into the fat of predatory fishes. In a study of a

Long Island saltmarsh, where the current concentration of DDT in water was only .05 ppb (parts per U. S. billion, 10^9), concentrations in organisms were 0.04 ppm in plankton, 0.16 ppm in shrimp, 0.30 ppm in marsh insects, 1.2 to 2.1 ppm in predatory fish, 3 to 18 ppm in birds of various diets, and 23 to 26 ppm in mergansers and cormorants (birds predatory on fish). Concentrations would be still higher in a predatory bird feeding on other birds, such as the peregrine falcon, now extinct in the eastern United States because of the concentration of DDT into its tissues. Both the rapid and intense concentration of DDT–DDD into plankton cells, and the progressive concentrations up animal food chains result from the affinity of these hydrocarbons for fats. The progressive concentration implies that these poisons (and some others) tend to decapitate food chains: to knock the top member off the food-chain sequence.

DDT has now achieved worldwide distribution in the biosphere. Although it is of low solubility in water, DDT is carried by moving water into streams and lakes, and from these into the oceans. DDT is also volatilized into the atmosphere and transported long distances as a low-level atmospheric pollutant. In addition the talc dust that is used for aerial spraying with DDT has apparently now become worldwide in the manner of radioactive dusts; it has been found as the principal dust component in air samples from over Pacific islands. Some marine bird and fish individuals on the Antarctic Coast, most remote in the world from agricultural areas, have low levels of DDT and DDE in their tissues, as do many animals of the open oceans. Pesticide residues occur very widely in terrestrial vertebrates. Pesticide poisoning has produced widespread population declines in some species of predatory birds in the United States and in Europe. Populations have declined both in species that rely on terrestrial food chains (peregrine falcon, European sparrow hawk) and in those with aquatic food chains (osprey, bald eagle, pelican). Extinctions of natural populations can be gradual effects of lower levels of tissue contamination with pesticides than those producing tremors and death in experiments. Population decline occurs by less obvious processes: effects on the endocrine system that interfere with reproductive behavior in adult birds, result in thin-shelled and frequently broken eggs, and reduce survival of embryos and young birds. There may also be increases in death rates of adults at times of hunger stress, when DDT is released from fat into the blood. In fish, effects on behavior that reduce reproduction and survival may result from concentrations of DDT that are too low to kill by direct, physiological effects.

Despite the ecological consequences of pesticides that are now widely recognized, their use in many areas has increased at accelerating rates. Certain chlorinated hydrocarbons, the polychlorinated biphenyls, have been widely used in the plastics industry and are now widespread contaminants in environment and the tissues of birds and other organisms. Use of DDT in the United States has declined since 1964, and there has been some shift of emphasis in use of pesticides away from the chlorinated hydrocarbons and toward the less persistent organophosphates and carbamates. Tissue loads of DDT–DDE are decreasing in some bird species, and some bird populations are showing signs of recovery. In the world as a whole, production and use of the chlorinated hydrocarbons as a group is still increasing. Further accumulation in the environment and wildlife must be expected. Adequate data on the accumulation rate and residence time of pesticides in the oceans are not yet available. Their present occurrence in the ocean, combined with the continuing accumulation from streams draining into the ocean and precipitation from the air over the ocean, should imply population decline and extinction for vulnerable marine species. Such effects should appear first on the continental shelves, and in marine birds and in fish that are top carnivores—the species most used as food by man. Apart from mortality, pesticide contamination may render fish unsuitable for human food. Coho salmon from Lake Michigan were found (in 1969) to contain DDT at 4 to 6 times the tolerance standard for sale and human consumption.

Chlorinated hydrocarbons have been of great value in agriculture. They often provide some years of dependable, persistent, and relatively complete control of insect pests with less expense, effort, and biological skill and understanding than are needed in the use of resistant strains of crops, natural enemies of pests, other chemical agents, and integrated control using more than one of these. In many cases, however, a pesticide application destroys the population of a predator or parasite (that controls a pest population) more effectively than it destroys the pest population itself. Increased likelihood of a damaging outbreak of the pest results, requiring continued and increasing pesticide use to maintain control. In some cases control of the pest by pesticide releases the population of another, more resistant pest, requiring additional control measures.

In the earlier days of California a prosperous industry in oranges and other citrus crops was centered in a spacious and beautiful valley half-rimmed with mountains, the Los Angeles Basin. By 1877 this industry was seriously threatened by an introduced pest, the cottony cushion scale (*Icerya purchasi*). Natural enemies of the scale insect were brought to California from Australia; and the scale insect was effectively controlled by these, particularly a ladybug, the vedalia beetle (*Rodalia cardinalis*). In the 1950's use of DDT and other pesticides in southern California led not to improved control but to new outbreaks of the pest, for pesticides were more toxic to the beetles than to the scale insects. Control was in this case re-established by abandoning the use of DDT and relying on the beetle. In a number of areas control of insect pests with pesticides has led to outbreaks of mites, resistant to the poisons, as new and serious pests.

The survival of more resistant individuals in a pest population treated with pesticide can produce rapid evolution of a more resistant population.

Increased intensity of application or continual introduction of new poisons may be needed to maintain control. Insect pests of cotton in the southeastern United States were for a time successfully controlled by two chlorinated hydrocarbons—DDT and benzene hexachloride. The security of this control was broken in 1955 by the appearance of resistant strains of cotton's worst enemy, the boll weevil; and before long widespread resistance had evolved in other cotton pests. Replacement of the hydrocarbons by organophosphorus compounds and carbamates produced temporary improvement in control, followed by further evolution of resistance. By the end of the 1963 season almost all the major cotton pest species included local populations with resistance to one or more of the chlorinated hydrocarbons.

Certain metals share with the chlorinated hydrocarbons an unwelcome combination of properties—toxicity and persistence. As elements these metals are wholly indestructible by chemical processes and must persist in the environment until they are removed from it, primarily by deposit in sediments. Arsenic compounds were widely used to control insect pests before the time of organophosphate and chlorinated hydrocarbon pesticides. Considerable arsenic accumulated in soils in some areas, but the arsenic is not sufficiently mobile in the environment to have become a widespread environmental problem. Mercury compounds have been used to treat seeds to protect them against fungi, and mercury is used in various industrial processes. Mercury is a poison that affects the nervous system, and local cases of human mercury poisoning have occurred. Mercury is also relatively mobile in environment, and mercury compounds are transported in water and deposited in sediments in lakes and the sea. Permanent deposit in the sediments would be appropriate for these compounds, but they are subject instead to conversion to methyl compounds of mercury that are both highly toxic and mobile. Dimethyl mercury is volatile and can escape from the water body into the air for further movement in the environment. If the water is acid, the monomethyl mercury formed does not escape but remains in the water body and may be transferred through food chains. Marked increases in mercury content of fish and aquatic birds have been observed in Sweden. In some areas fish have been blacklisted as unsafe for human consumption because of their mercury content. Consumption of fish from Lake Erie and some other water bodies in the northeastern United States is discouraged because of mercury pollution. A wide-ranging marine predator, the swordfish, has been designated as unsafe for food in the United States. Man's addition of mercury to the oceans by river transport is estimated as roughly equal to the natural river input of mercury to the oceans (both about 5000 t/yr). This amount seems trivial in relation to the total amount of mercury in the oceans, but the mercury is being added to the smaller mercury pools in surface waters near coasts, from which many of the fish used by man are taken.

Lead, like mercury, is a heavy metal that affects the nervous system and has produced acute poisoning of human beings. Lead is also released to, and is mobile in, environment. The principal source is the tetra-ethyl lead used in gasoline in amounts (3.5×10^5 t/yr) that much exceed the natural rate of river transport of lead to the ocean (1.5×10^5 t/yr). The amount is again very small in relation to the whole-ocean pool of lead; but the effect is on surface waters. Lead concentrations in these in the Northern Hemisphere have risen from 0.01–0.02 to 0.07 ppb during the use of tetra-ethyl lead since 1924. Lead is thus a low-level contaminant of environment on a global scale. Much of the spread of lead contamination around the world occurs by transport in the atmosphere. The accumulation of lead in snow of different ages in the Greenland ice cap (Figure 7.2) provides an effective record of this transport, and the kind of accelerating accumula-

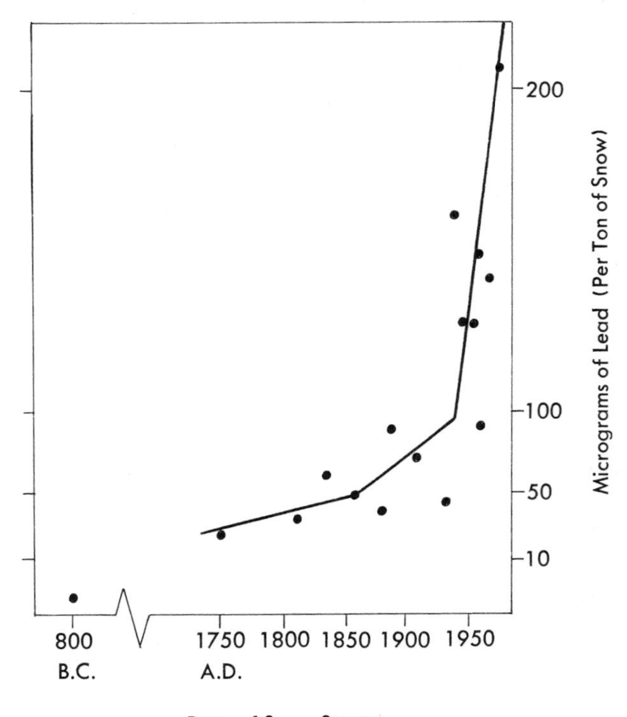

Figure 7.2. Lead accumulation in the ecosphere. Lead released into the air, primarily in the exhaust from automobiles, is transported around the world in the atmosphere and brought down in precipitation. The concentration of lead in snow of different ages in the Greenland icecap provides a record of the historic increase in lead content of the ecosphere resulting from industrial uses of lead. The scale for lead concentrations is logarithmic. [Patterson and Salvia, *Environment,* **10**:66 (1968); MacIntyre and Holmes, Fig. 5 in Murdoch, 1971.]

tion that many pollutants show. Lead in the atmosphere of American cities with heavy automobile traffic has reached average levels (1.5 to 2.5 $\mu g/m^3$) about fifty times those in rural air, and probably several thousand times that in the air before the time of the automobile. The effects of chronic exposure on human beings are uncertain and difficult to separate from those of other urban stresses and the intake of lead in food. Current steps to reduce the release of both lead and mercury into environment are welcome precautions. Among other metals that could cause environmental problems are copper, nickel, zinc, silver if widely used in cloud seeding, and some less familiar—cadmium, vanadium, beryllium, uranium, and plutonium, of which the last (because of its effectiveness in inducing cancer) is often characterized as perhaps the most deadly substance known to man.

Despite laws on heavy metal content, some lead and mercury reach human beings in food. The average level of lead in human blood in the United States is 0.25 ppm; recognizable poisoning generally results above 0.8 ppm. Pesticide contents of food also are subject to legal limits and controls. It is necessarily the case, however, that persistent pesticides circulate back from agricultural use through food, and probably also in water and the atmosphere, to man himself. All of us carry small amounts of these poisons in our tissues. So far as is known there is no present reason for concern for human health in general. Tissue levels in man are (if cases of intense exposure in agricultural areas are excluded) much below those that produce short-term toxic, physiological effects. Grounds for long-range confidence about effects on human health are less secure. The effects of chronic exposure to low levels of varied mixtures of pesticides in tissues are unknown. Use of DDT and some other chlorinated hydrocarbons is now effectively restricted in the United States and some other countries. These national limits should slow, but not prevent, continued global accumulation of chlorinated hydrocarbons.

Three observations on pollution effects on man—including those of air pollution to be discussed next—may be mentioned in conclusion:

1. The most important effects on human health are not acute and dramatic cases but are subtle and primarily statistical in their expression. Many of the effects appear as increased frequency of the diseases—circulatory, respiratory, and cancerous—of age and decline in man. Complex correlations link these diseases with metals in environment, air pollutants, and other stresses and with one another, while some of the environmental stresses also are correlated with one another. Effects of a given pollution exposure are inconsistent between individuals; effects vary with other aspects of health and apparently unpredictable vulnerableness to different chemicals. Effects may consequently be difficult to recognize. The greater the number of pollution stresses to which man

is exposed, the more difficult determining the importance of any one of these stresses becomes.

2. Different pollution stresses—including some that are quite different in character—can intensify one another in unexpected ways. For example, uranium miners in Colorado were found subject to a fourfold increase in cancer frequency compared with the general population (Table 7.2). Smoking also increases cancer frequency, by about ten times in this sample. For the miners who were also smokers, the increase in cancer frequency was not the sum of these effects but their product, hence a fortyfold increase in cancer frequency.
3. Lags in the expression of pollution effects can make recognizing and preventing them difficult. The lags apply to many pollution effects, but can be illustrated with one particularly clear case. Asbestos, which is widely used for insulation, fire protection, automobile brakes, and other purposes, is a fibrous form of some of the rocks we have discussed as serpentine. Workers in a plant processing asbestos, or in shipyards or construction using it, inhale the microscopic fibers without noticeable short-term effect. These workers develop, however, after lag periods of twenty to thirty years, abnormally high frequencies of cancer including a distinctive and fatal cancer of the lining of the chest cavity caused by the asbestos fibers. The fibers are easily airborne; and cancers develop, after long delays, in some people who have worked in shipyard areas or lived in asbestos mining areas, but had no direct contact with asbestos.

Air Pollution

The wealth of energy used, which is the real basis of the industrial wealth of the advanced nations, comes largely from the burning of fossil fuels—coal and petroleum products. Worldwide, water power provides only 0.03 per cent of industrial energy, though it provides a larger fraction of electric power production. Like all energy conversions, burning is an imperfect process. Different combustion processes release varying amounts of un-

Table 7.2 Uranium Mining, Smoking, and Cancer. Observed cancer numbers versus expected cancer numbers (according to normal occurrence in U. S. population) in 3414 uranium miners in Colorado. [Lundin *et al., Health Physics* 16:571 (1967).]

Person-Years of Observation	*Cancers Expected*	*Cancers Observed*	
9,047	0.5	2	In non-smokers
26,392	15.5	60	In smokers

oxidized and incompletely oxidized hydrocarbons into the atmosphere along with toxic inorganic oxides—CO, NO, NO_2, and SO_2. Photochemical reactions in the atmosphere in sunlight produce other carbon and nitrogen compounds, and ozone (O_3).

As in other pollution processes, effects that are both local and general, and both acute and chronic, are observed. The toxic products of combustion from a city or industrial area are normally mixed upward into a large volume of air, without a hazard to man resulting. When air is held beneath an atmospheric inversion, particularly if it is also confined within a valley or basin ringed by mountains, hazardous local pollution may result. An early acute air pollution crisis in the United States occurred at Donora, Pennsylvania. The town and a local concentration of industry are situated in a horseshoe-shaped valley. In October, 1948, in a period when an inversion prevented the dispersal of industrial fumes, these fumes accumulated until some 40 per cent of the population were made ill and twenty people killed.

A major chronic air pollution problem now exists in the Los Angeles Basin, California, and is produced by the combination of a large city and its suburbs with very heavy automobile traffic, frequent inversions, and the mountains around the valley inland from the Coast. The inversions and the mountains together tend to hold pollutants in the basin. Combustion products accumulate in the air and are acted on by sunlight to produce a distinctive smog that is both directly unpleasant, causing eye irritation and mild nausea, and a longer-range health hazard, as indicated in the frequency of emphysema and other respiratory ailments. Automobiles are believed to be the source of more than half the combustion pollutants in the United States, and in the Los Angeles Basin 80 to 90 per cent. Industrial pollution in the Los Angeles Basin has been controlled and efforts have been made toward reducing the release of pollutants by automobiles, but the Los Angeles smog has continued to increase in depth and extent. The smog can sometimes be seen cascading over the mountains inland from Los Angeles; visible smog from the Los Angeles area now spreads into all the valleys surrounding the Basin and at times extends, with densities decreasing with distance, across the deserts into Nevada and Arizona. Los Angeles smog has killed extensive areas of pine forests in the mountains above the city and apparently is also affecting the deserts, some plants of which are very sensitive to ozone. Smog effects have driven the citrus industry and other crops out of the Los Angeles Basin to other California valley areas, to which the expanding smog is now pursuing them. Reduction of yield and smog-marking of leaves are now observed in other areas. Ozone and PAN (peroxyacyl nitrate) are believed to be the principal ingredients of Los Angeles smog that affect plants and produce characteristic markings on their leaves.

New York City is not ringed by mountains but is part of a large area of industry, population, and automobile traffic. The city has chronic air pollution, which is less conspicuously unpleasant than that of Los Angeles, but has apparently greater effect on human health. Sulfur dioxide may be the most important single air pollutant in New York, as in other areas without strong photochemical effects on smog. A period of acute pollution effects occurred from November 12 to 22, 1953, when a stagnant air mass covered the industrialized northeastern states. In addition to the direct irritation of the smog in New York City, later statistical analysis indicated an increase in the death rate in the city during and following the pollution period amounting to between 175 and 260 deaths beyond those that would be normal for the period. Other periods of fairly acute pollution have occurred to the present, despite progress in reducing some forms of pollutant release in the city. During periods of slow air movement the mantle of pollution now covers the whole of the larger metropolitan area—from Boston to Washington—of which New York City is part. Effects of pollution on vegetation are less evident than in California, but some plants are affected. Experiments comparing plants grown in the normal air of the metropolitan area and in air purified by charcoal filtering have shown damage and reduction of growth in some of the former.

Air pollution effects on human health are, as we have indicated, statistical in their expression. Comparing smaller and larger cities, the death rate from all forms of cancer is about twice as high (0.002/yr versus 0.001/yr) in New York and Boston as in El Paso, Texas and Charlotte, North Carolina. In major urban areas the death rates for arteriosclerosis and heart disease (0.0029 to 0.0043/yr) are two to four times as high as in smaller cities. Higher rates of death and sickness in urban areas apply to a wide range of maladies, including respiratory disease (chronic bronchitis and emphysema) and cancer of both the respiratory and the digestive tracts, as well as heart disease. Chronic exposure to carbon monoxide at levels that do not produce immediate effects may contribute, along with sulfur dioxide and other pollutants, to the unfavorable statistics of urban life.

Pollution in the Tokyo metropolitan area is apparently more severe than any in the United States. Pollution comparable with that of Los Angeles and New York now affects major urban and industrial areas throughout the Northern Hemisphere; air pollution in urban areas is present, although less advanced, in the Southern Hemisphere. Atmospheric pollution is thus in transition from a local problem to a general phenomenon of the ecosphere. Lower-level pollution is now often of subcontinental extent in the United States. When a slow-moving high-pressure system is centered over the eastern states, the circulating air accumulates pollutants from the cities and highways of many states. Visible, mild or moderate pollution then covers most of the eastern United States and intensifies into heavy pollution over the major urban areas of the Atlantic Coast and the Lake

States. Small cities are then subject to marked air pollution, for their own pollutants are added onto those accumulated in the air before it reaches them. Visible pollution of subcontinental extent occurs also in the western states, when stable air masses in the intermountain region accumulate pollutants from both coastal and inland areas.

Contributions from man's combustion to the world atmosphere now include 2.9×10^8 t/yr of carbon monoxide, 14.7×10^7 t/yr of sulfur dioxide, 5.3×10^7 t/yr of nitrogen oxides, and 2.3×10^7 t/yr of smoke particles. Pollutants are removed from the atmosphere by chemical changes, by settling as particles, and in solution in rain. Despite this removal, as rates of input of pollutants into the atmosphere increase, the magnitudes of the pools of pollutants in the atmosphere must increase. Air moves around the earth in periods of fifteen to twenty-five days in mid latitudes. Residence times of some pollutants are shorter than this—sulfur dioxide (one to four days) and nitrogen oxides (three or four days)—, but that of carbon monoxide may be longer (probably one to two months). Most pollutants are washed out of air crossing an ocean, but some pollutants can persist in air moving from one continent to another. To some extent an intercontinental reinforcement of pollution effects is now possible; pollutants from New York are added onto pollutants from around the world, including New York. Even as multiple local pollution sources increase the general pollution level, so the general pollution level increases the intensity of local pollution effects.

Air pollution is thus becoming a part of biogeochemistry, as we may illustrate for two phenomena—acid rain and carbon dioxide increase. Rain normally has some carbon dioxide and other substances in solution in it and is mildly acid (pH around 5.5). Industrial and automobile pollution add to the air sulfur dioxide and nitrogen oxides; and these in solution in rain become, directly or with oxidation, sulfurous and sulfuric, nitrous and nitric acids. Because of these, rain over much of the United States is now distinctly more acid than before the spread of industry. The pattern of acidity—generally decreasing from the Pacific Coast industrial areas across the intermountain region to the Rocky Mountains, and increasing from the Rockies eastward to the Atlantic Coast and especially the northeastern states—corresponds to the pattern of pollution visible from the air. Rainfall in the northeastern states is now predominantly between 3 and 4 in pH, and is at times more acid than pH 3. As the northeastern United States receives pollutants from a wider area, so Sweden receives them from Great Britain and Germany; and a similar increase in rainfall acidity has been observed in Sweden. Changes in the floras and faunas of fresh-water bodies are resulting, and reductions in rates of forest growth have been suggested in both Sweden and New England. The basis of the latter effect is not established, but might involve increased leaching of nutrients both from plant leaves and from some soils. Substantial long-term effects on land communities are possible, but not yet predictable. The works of man, too, are vulnerable. Figure 7.3 illustrates another now widespread process, a kind of leaching of part of the West's cultural heritage.

The effect of pollution on biogeochemical cycling is most studied for the largest single product of combustion, carbon dioxide. The biogeochemical cycle for carbon is illustrated in Figure 7.4. One transfer rate, from fossil fuels to atmospheric CO_2, is enormously increased by man's combustion. A rate value now about 25 $g/m^2/yr$ implies the release of 13×10^9 metric tons (3×10^{13} lbs) of CO_2 into the atmosphere per year beyond the amount released by respiration and other routes before the time of industrial man. Major shifts in rate values imply shifts in pool magnitudes in such a system. The relatively small pool of CO_2 in the atmosphere is in exchange equilibrium with the much larger pool of CO_2 and carbonates in the ocean. The oceanic CO_2 and carbonates act as a buffering system that reduces the effect of combustion in raising atmospheric CO_2 levels. Nevertheless, the mean CO_2 content of the atmosphere increased from 290 ppm in 1900 to 320 ppm by volume in 1970.

CO_2 contributes to the greenhouse effect of the atmosphere. This effect

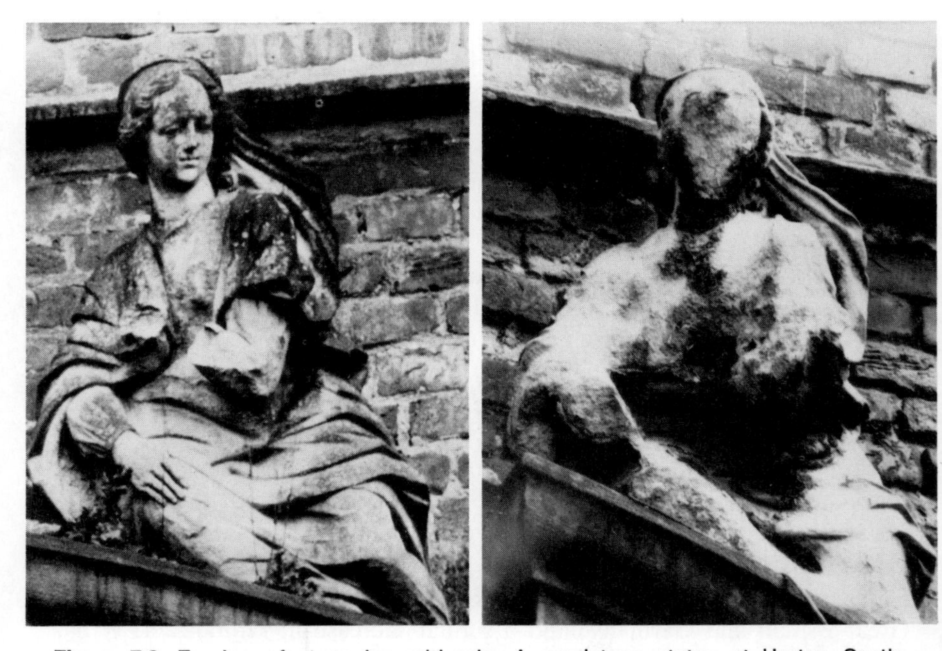

Figure 7.3. Erosion of stone by acid rain. A sandstone statue at Herten Castle, near Recklinghausen, Westphalia, Germany, as eroded by air pollution in the industrial Ruhr. Left: 1908, with light to moderate damage. Right: 1969, with almost complete destruction. The castle was built in 1702. [Courtesy of E. M. Winkler, from *Stone: Properties, Durability in Man's Environment.* Springer-Verlag, Vienna-New York (1973).]

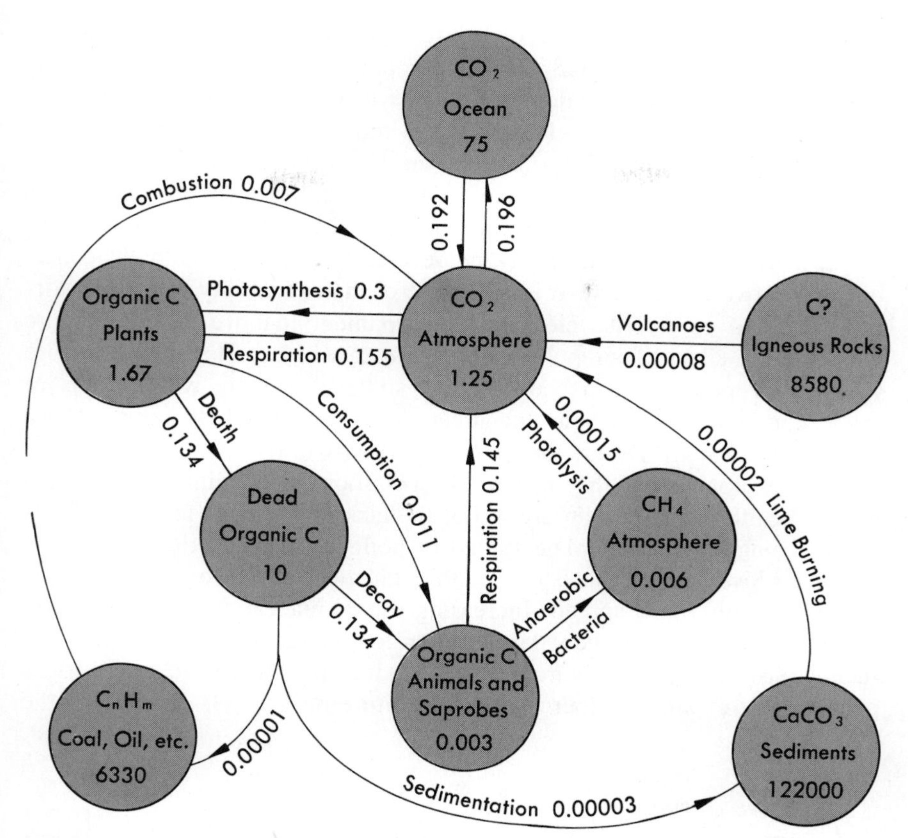

Figure 7.4. Biogeochemical cycle for carbon. Numbers in circles are amounts of carbon in pools, in kilograms per square meter of the earth's surface; numbers on arrows are transfer rates in kilograms of carbon per square meter per year. [Modified from H. J. M. Bowen, *Trace Elements in Biochemistry* (1966); see also Reiners in *Brookhaven Symp. Biol.*, **24**:368 (1973).]

is the capacity of the atmosphere to transmit heat energy from the sun to the earth's surface, but to absorb heat energy in different wave lengths radiating from the earth's surface and to return some of this heat by re-radiation back to the earth's surface. Increased CO_2 content of the atmosphere may thus imply the strengthening of the greenhouse effect and the warming of the earth's surface. A modest effect of CO_2 increase on mean temperatures of the earth's climates has probably occurred. Mean air temperatures for the world increased by about 0.6°C from 1880 to 1940; but the trend then reversed with cooling by about 0.2°C by 1960. The recent cooling may have resulted from the increased content of particles in the atmosphere; but it is not possible to assess with confidence the relative contributions of carbon dioxide, volcanic dust, particles released by industry, dust from arid lands and those bared by man, and unknown factors.

There is consequently no secure basis for projecting long-range implications. It has been suggested that higher temperatures might, by melting polar ice caps, increase the depth of ocean waters until coastal cities around the world were submerged. It now seems more likely that the rapidly increasing concentration of particles in the atmosphere will, because these particles reflect sunlight, continue to lower temperatures at the earth's surface in the future. Neither the carbon dioxide release nor the increase of particles in the air can easily be reversed now if the climatic consequences prove detrimental. It is indicative of man's inability to predict adequately or to control his effects on environment that we cannot be sure what the net effect of air pollution on climate will be. We cannot be sure whether air pollution can contribute to shifts of climate with serious or tragic consequences, such as the great drought in the belt across Africa south of the Sahara Desert that began in 1968–70.

It does not appear that general air pollution has significant effects on human health outside urban areas, but the lack of evidence for such effects is not strong reassurance. The spread of pollution itself to the countryside between cities implies that some of the statistical effects of that pollution on health should also spread. Increasing use of nuclear fuels in the future has been planned, but very large expansion in the combustion of fossil fuels is also expected (Figure 7.5). It is clear that there are prices to be paid for this expansion. Increasing expenditure on controls can reduce the

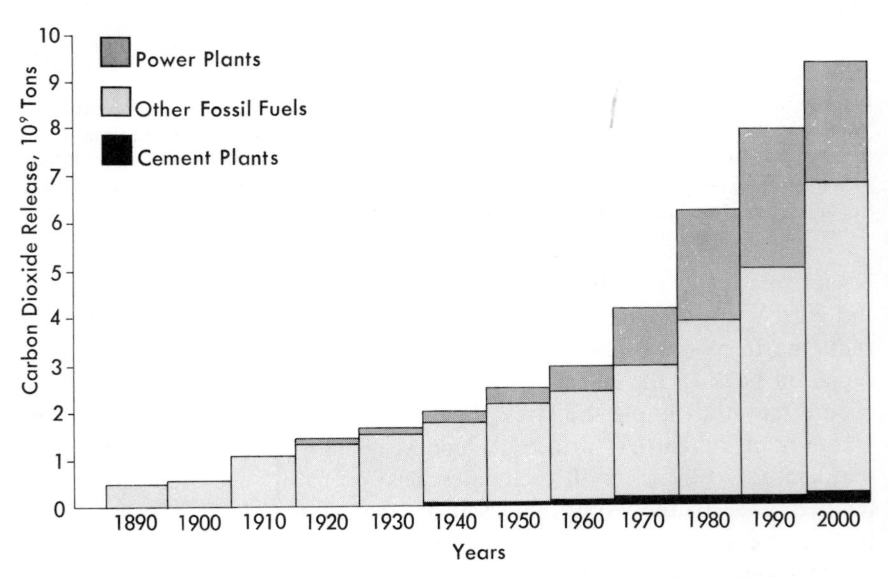

Figure 7.5. Release of carbon dioxide by industry and automobiles. United States totals in billions of tons per year. [Rohrman *et al., Science,* **156:**931 (1967).] Despite increasing use of nuclear power, projections indicate rapidly increasing use of fossil fuels also through the remainder of the century.

rate of increase, but not eliminate, pollutant release from combustion. Exponentially increasing energy production can be expected to produce exponentially increasing problems, whether the energy is from combustion, or nuclear reactors, or both. It may be hoped that some limitation of industrial growth, and controls on release, will together contribute to reducing the intensity of air pollution. Controls are made difficult because nations accept as imperative growth in energy release for the sake of growth in wealth and power, and because of the multiplicity of pollution sources. There is, however, one respect in which it is easier to reduce air pollution than other forms. Because of the short residence times of pollutants in the atmosphere, reduction of pollutant release can produce prompt reduction of pollution levels. This fast response is in contrast to the slow decrease in longer-lived pollutants, such as persistent pesticides and heavy metals, in soil and water when man reduces the rate at which they are released.

Lake Production and Eutrophication

It is possible to arrange many lakes along a gradient from *oligotrophic* lakes, with low nutrient content in the water and low productivity, to *eutrophic* lakes, with high nutrient levels and productivity. Because lakes differ from one another in so many different ways, the gradient is a simplification. It serves, however, to relate some characteristics of lakes in ways that are appropriate to our present concerns. Table 7.3 summarizes some

Table 7.3 Some Characteristics of Types of Lakes. Typical ranges of characteristics for different types of lakes are given. Some lakes do not fit into this scheme, and additional types of lakes (sterile or ultra-oligotrophic, and polluted or hypertrophic but not saline) can be recognized. [Likens in H. Lieth and R. H. Whittaker, *The Primary Production of the Biosphere,* Springer, New York (1975).]

	Oligo-trophic	*Meso-trophic*	*Eutrophic*	*Saline*	*Dys-trophic*
Net primary productivity, g/m^2/yr	15–50	50–150	150–500	500–2500	10–100
Phytoplankton biomass, mg/m^3	20–200	200–600	600–10,000	1000–20,000	20–400
Total organic matter, ppm	1–5	2–10	10–100	20–200	20–100
Chlorophyll a, ppb	0.3–3	2–15	10–500	50–1000	0.01–2.0
Light penetration, m	20–120	5–40	3–20	2–10	1–5
Total phosphorus, ppb	<1–5	5–10	10–30	30–100	1–10
Inorganic nitrogen, ppb	<1–200	200–400	300–650	400–5000	1–200
Total inorganic solutes, ppm	2–20	10–200	100–500	1000–100,000	5–100

[ppm = mg/liter, ppb = μg/liter or mg/m^3; light penetration is the estimated depth to which 1 per cent of sunlight penetrates at midday.]

characteristics of lakes that range from the oligotrophic through the intermediate or mesotrophic to the eutrophic. This gradient can be extended beyond eutrophic lakes to the saline lakes of arid climates. Dystrophic or bog lakes are special cases, often characterized by low inorganic but high organic content of the water, and by limitation of light penetration by the dark organic matter in the water. The other four types of lakes from oligotrophic to saline show trends in many of their characteristics (some of which are less strongly correlated than the table may suggest), but especially in the characteristic of key concern to us—productivity.

A number of factors affect the nutrient circulation and productivity of lakes. Effects of these factors are interrelated, but the following paragraphs should be understood as statements in the form: If other characteristics of two lakes are closely similar, then the effect on productivity of a difference between them in this factor would be. . . .

1. Fertility of the drainage basin. Of two lakes, one receiving inflowing water from an area of infertile rocks (quartzite or granite mountains, say), the other from an area of fertile soils (as of a rich farming area on limestone), the latter should be the more productive. Continuing input of greater quantities of nutrients maintains larger pools of circulating nutrients as a basis of productivity.
2. Lake depth and slope of shore. Comparing two lakes, one deep with steep, rocky shores, the other shallow with sloping shores of mud and sand, the latter should be the more productive. In the first most of the bottom is out of reach of sunlight, and plants cannot grow on the bottom and contribute to the lake's production. The fact that the lake is deep may imply also that thermal stratification prevents effective movement of nutrients from the depths of the lake into the lighted surface waters during much of the warm season. The steep, rocky shores imply that shallows and the growth of plants along the shore are limited. In the shallow lake extensive growth of shore and bottom plants adds their productivity to that of the plankton. It is thus to be expected that on the average the productivity of shallow lakes will be higher—not merely in terms of productivity per unit of water volume, but in production per unit of the lake's surface area.
3. Form of shore line. Closely related to the preceding is the form of the shore line. Of two lakes with similar size, depth, and slope of the shores, the lake with a long and irregular shore line with many inlets should have greater productivity than one with a short shore line. The longer shore line implies a greater area of production by the shore plants relative to the area of the lake. High productivity is thus favored both by low ratio of depth to area and by high ratio of shore length to area.
4. Temperature. Of two similar lakes the one in the warmer climate has

the longer season of biological activity and greater nutrient turnover through the year, and should have the higher productivity. Warmer temperature may also imply more rapid weathering of soil parent material in the watershed, and therefore a larger supply of nutrients to the lake. Temperature is one basis for the contrast between some oligotrophic mountain lakes of very low productivity and crystal water, and some eutrophic lowland lakes of high productivity and more turbid waters.

5. Water turnover. Climatic humidity and manner of water inflow and outflow indirectly affect productivity. If a lake is virtually a wide part of a river, with inflow and outflow large relative to its volume, plankton productivity may be limited by the short residence time of water in the lake. A lake with a smaller inflow and outflow may receive smaller amounts of nutrients in its inflow, but use more of these for plankton and shore-plant productivity and retain a greater share of them as a circulating stock within the lake basin. The manner of water turnover in lakes is affected also by climatic humidity. In a humid climate the amount of water leaving the lake by evaporation from the surface is generally small compared with that leaving by stream (and underground) outflow. In a drier climate, the fraction of loss by evaporation from the surface is larger, and in an arid climate many lakes have no stream outflow, losing their water only by evaporation. Such distillation of lake water into the atmosphere leaves the dissolved nutrients behind. There is thus a greater tendency for nutrients to accumulate in many lakes of drier climates; and the manner in which lakes change with time, or age, is different in humid and arid climates.
6. Lake age. Lake age, especially as it affects the filling of the basin, thus influences lake productivity. Most lakes are relatively short-lived; many were formed by glaciers or changes in river courses and are only a few thousand years old. (There are some notable exceptions in large, deep, old, biologically distinctive lakes.) A lake ages in time as sediments are brought in and, along with organic materials from the lake itself, accumulate on the lake's bottom. There are different routes by which lakes age according to climate (and the character of their basins).

a. In cool and humid climates many smaller lakes in fully enclosed basins, without stream outflow, age as bogs by growth of a floating mat of vegetation and accumulation of peat filling the basin. We have described the process as an example of succession. The aging process is primarily one of accumulation of organic materials, not inorganic nutrients; and dystrophic or bog lakes are characteristically low in nutrient content of the water and in productivity (Figure 4.11).

b. In arid regions many lakes age in a widely different way, as saline bodies. With water leaving primarily or solely by evaporation, the salts brought in accumulate and the lake becomes progressively more saline as it ages. As salinity (expressed as total inorganic solutes in parts per million) increases, the lake's biological character changes profoundly. At about 300 to 500 ppm, while the lake is not noticeably saline, the typical freshwater plankton is replaced by other species. With salinity above 1000 ppm, biotas fully distinctive for saline bodies occur; freshwater fish fail to survive above 7000 to 15000 ppm. With salinity in tens of thousands parts per million, the number of species is severely limited. Great Salt Lake in Utah is such a body, with salinity over 200,000 ppm, and a fauna limited to the brine shrimp (*Artemia salina*), and syrphid fly larvae (*Ephedra*). Despite decreasing species diversity, productivity tends to increase with increasing salinity. Great Salt Lake is highly productive, and the dead shrimps and fly larvae at times form massive accumulations on its shores. Even the Dead Sea of Israel, with salinity about 226,000 ppm, has a productive plant plankton, although no animals. Great Salt Lake is, however, indicative of the fate of many saline lakes, given time. The lake is much reduced in area since postglacial time and is still shrinking, exposing as it does so vast tracts of salt desert. The end stage for such a lake is a playa or salt flat. Some salt flats receive water and have shallow, highly saline, temporary lakes, but are occupied more of the time by mirages inviting the unwary than by water.

c. Between these two types are the many lakes that age in ways we think more normal. The lake basin gradually fills, and the lake becomes shallower and its shores less steep. The lake may progress in time from the deep, steep-shored form of many oligotrophic lakes, to the shallow, sloping-shored form of typical eutrophic lakes (Figure 7.6). Thus the lake ages, and the rate of this aging under natural conditions is determined by the input of sediments and the size and form of the basin, and by the fertility and condition of the watershed. Given time, the progression may lead from a shallow lake to a marsh or swamp and thence to dry land. Some lakes age from relatively oligotrophic youth through eutrophic maturity to senescence and disappearance.

Lakes age primarily by the filling of their basins with inorganic and organic deposits. Increasing fertility and productivity of a lake, or development from an oligotrophic toward a eutrophic condition, is termed *eutrophication*. The aging of a lake may lead to eutrophication, but the two processes need not run in parallel. If aging and eutrophication are used to characterize lake development, much as we speak of succession on land, the complexity and variety in the ways particular lakes develop should not be lost sight of. Productivities of lakes can both increase and decrease with age, particularly as the conditions of their watersheds change. Disturbance of the vegetation can permit more rapid movement of nutrients into the lake, increasing its productivity. As the land vegetation later recovers, the

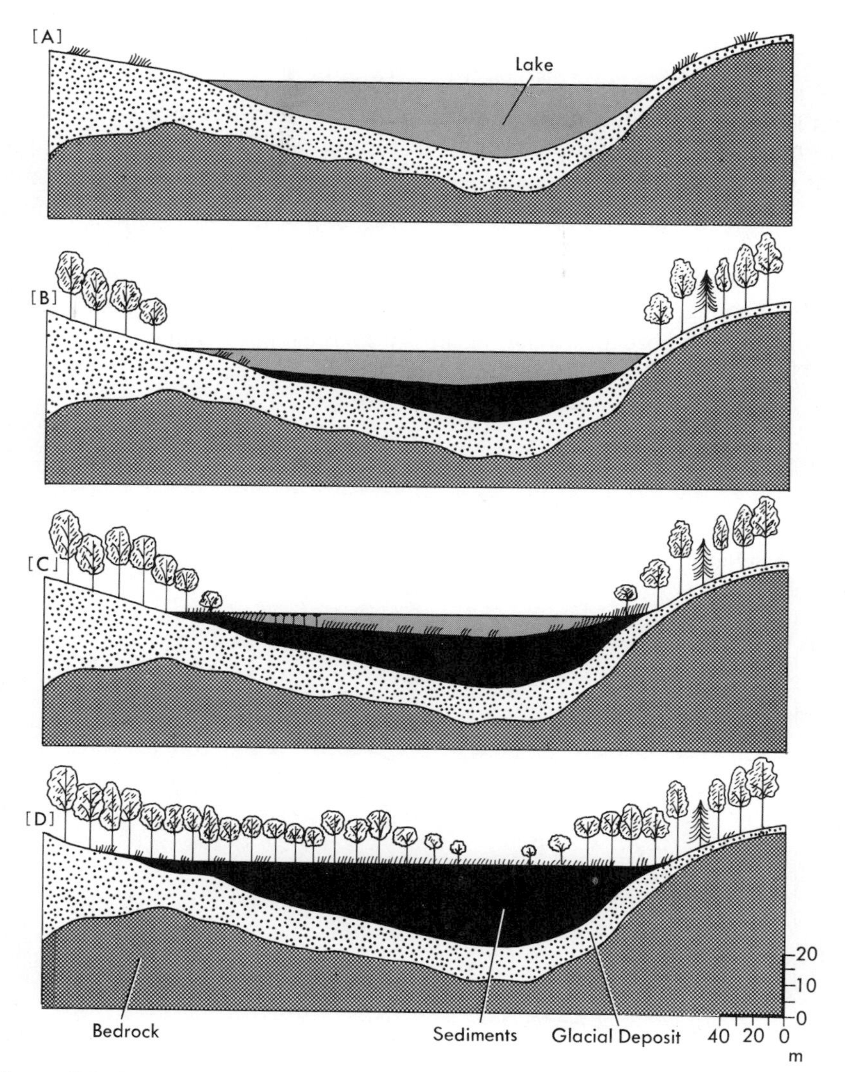

Figure 7.6. Aging of a small temperate lake.

A: Shortly after the retreat of the glacier, about 12,000 yrs ago. A layer of glacial deposit, or till, is left covering the bedrock; and a ridge or dam of this till is responsible for forming the lake. The lake is highly oligotrophic. Vegetation is beginning to reoccupy the land surface.

B: The lake at present [unpublished data of R. Mohler and G. E. Likens for Mirror Lake, New Hampshire]. The lake basin is about half filled with mud sediments but is still oligotrophic and has only a meager growth of higher aquatic plants in shallow water. The lake has eroded a shore on the left (south) side; its watershed is now forested.

C: A few thousand years from now. The lake has become shallow and eutrophic; its shores are gradual mud slopes with marshes, and growth of higher aquatic plants is extensive. Productivity is high.

D: The lake aged into land vegetation. Continued accumulation of sediments has filled the lake basin. Marsh vegetation has advanced from its shores (in **C**) to what was the center of the lake, and forests have occupied the older shore sediments. Given additional centuries, the full area of the lake may be occupied by forests.

lake's productivity may decrease. Many lakes remain oligotrophic for very long periods, with their nutrient stocks and productivities in a steady state determined by nutrient input, as this in turn is regulated by the characteristics of the watershed. For many lakes real eutrophication may occur only late in their life cycles, when filling of the basin has made the lake quite shallow. Figure 7.7 illustrates, in a diagrammatic way, the three routes of aging as they involve the accumulation of organic and inorganic materials. The figure omits complexities, intermediates, and deviant routes to show only the broadest relationships of lake aging to climate.

The productivity of lakes, like that of the oceans, is governed most

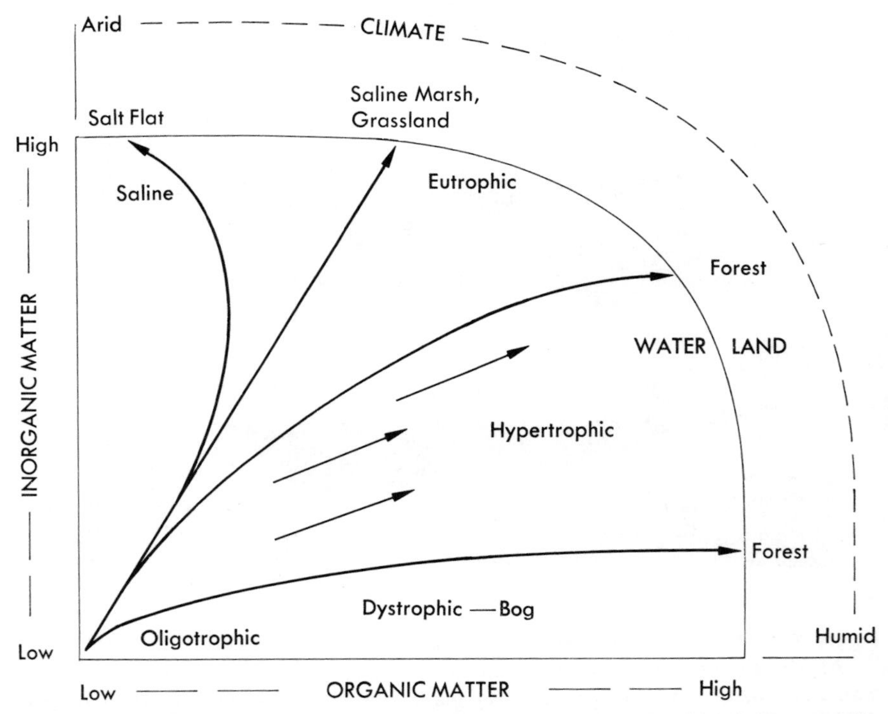

Figure 7.7. Routes of lake aging. Lakes can age in different ways, influenced by climate, with different patterns of accumulation of organic and inorganic materials. Routes indicated are (a) dystrophic aging or bog succession (Figure 4.11) with low inorganic nutrient availability and extensive accumulation of organic matter as peat; (b) "normal" aging from an oligotrophic to a eutrophic condition; the lake may continue to age to a land community dominated by trees in a more humid climate (Figure 7.6) or grasses in a less humid climate; (c) saline aging with extensive accumulation of inorganic salts leading to a salt flat or salt desert. The oblique arrows indicate unnatural or forced eutrophication, or lake hypertrophy. The figure represents relative content of organic and inorganic matter in lake waters, not actual amounts; saline lakes are highly eutrophic with high organic content, but with inorganic materials predominant.

strongly by nutrients. In broad geographic comparisons lake productivity is significantly correlated with latitude, and consequently with temperature and input of sunlight energy. It is thus tempting to think that sunlight energy might control productivity. In the Far North the limited sunlight and the shortness of the summer do in fact limit the potential productivity of lakes. However, arctic lakes are in general young and low in nutrients, as well as cold, short in open season, and unproductive. Nutrient release in the arctic soils is slow, and few nutrients are moving into lakes while both lakes and the surrounding soils are frozen solid. The productivities of arctic lakes can be much increased by fertilization. Even in the Arctic nutrients have important, direct effects on the productivity of lakes. Many tropical lakes are old and nutrient-rich, as well as warm and highly productive (although the very deep, old tropical lakes are not highly productive). Between the arctic and tropic extremes, a number of influences, of which temperature is one, bear on lake productivity by way of nutrient flow. It is not uncommon to find eutrophic and oligotrophic lakes in the same area subject to the same climatic temperature and sunlight energy, but with productivities differing several fold because of factors other than these. The correlation of lake productivity with latitude and sunlight energy reaching the water is an example of an ecological correlation that is largely (although not wholly) coincidental or secondary. Differences among lakes in productivity are not in general directly caused by differences in sunlight energy available for photosynthesis.

Man has his own uses for the shores of lakes and of the streams that drain into them. From settlements, cities, and industries water that has been used or affected by man drains into water bodies. Some reaches of streams are turned into sewage channels, or into aquatic deserts by toxic products of industry; and lakes receive pollutants both through streams and from human settlements and industries on their shores. Although some industrial wastes are toxic, the more widespread problem is pollution with organic wastes and nutrients that are not inherently toxic. Large amounts of nutrients, however, cause artificial eutrophication of lakes. Especially when combined with the input of sewage or other organic material, large amounts of nutrients do not produce a favorable increase in the natural productivity of the lake, but a distorted, unbalanced, and unfavorable increase in productivity. The nutrients deflect the normal aging process, as indicated in Figure 7.7, in a way that is often termed *cultural eutrophication*. The prefix "eu-" implies good nutrient conditions, however, and unnatural over-fertilization of a lake to its detriment might well be termed lake "hypertrophy."

The power of these processes to change the quality of lakes is illustrated by the recent history of a large American lake, Lake Erie. Cultural eutrophication is expressed in the curve of dissolved solids in the lake water, Figure 7.8. The water itself has become increasingly malodorous,

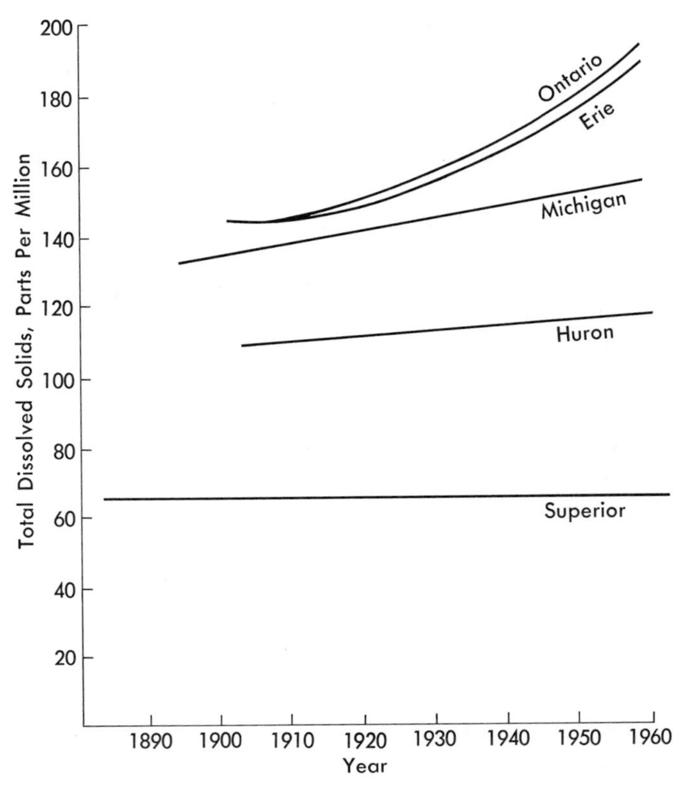

Figure 7.8. Cultural eutrophication of Lake Erie and other Great Lakes. Eutrophication is reflected in the total dissolved solids in the water; Lakes Erie and Ontario have shown rapid eutrophication effects in the last 35 yrs, Lakes Michigan and Huron less striking ones. Lake Superior is more strongly oligotrophic than the other lakes and has not yet been enough affected by man to show cultural eutrophication. [Beeton, 1965.]

unpalatable, and unattractive for swimming. Windrows of decaying algae and millions of dead fish are at times cast up on the shore. The natural bottom communities of animals are transformed in some areas and replaced by forms tolerant of pollution. Desirable commercial food fish formerly harvested from the lake—20 million pounds of cisco, 15 of blue pike, 2 of whitefish, and so on—have largely disappeared to be replaced by less desirable rough fish. The lake drains through Niagara Falls, and to the visual beauty and aural impressiveness of the great falls is now added a third, less welcome sensation, the scent of pollution. Through Niagara Falls the water of Lake Erie reaches Lake Ontario, now also subject to cultural eutrophication. Effects are observed also in the larger Lake Michigan, especially at the southern end influenced by the Chicago metropolitan area.

In lakes where the input of pollutants can be controlled, the abnormal eutrophication can be reversed. Lake Washington has the city of Seattle, Washington, on its western shore and has been increasingly surrounded by smaller cities of the Seattle metropolitan area. Research on the lake by W. T. Edmondson has provided an unusually detailed record of this lake's response to nutrient input. As the population grew, an increasing number of sewage plants released processed sewage into the lake. Although it was treated, the sewage contained nutrients, especially phosphorus, that increased the fertility and growth of algae in the lake; and by 1955 growths of blue-green algae warned of development toward serious eutrophication. Public concern was aroused and support gained for construction of new means of treating the sewage and releasing it into the deep water of Puget Sound at some distance from the city. Lake Washington, relieved of its overload of fertilizer, has recovered rapidly since 1968; concentrations of phosphates and densities of algal cells have decreased, and water transparency has increased. The transfer of nutrients from the lake to Puget Sound is not considered to have a serious effect on the latter for the present.

It will be difficult and expensive to reverse the increasing and unfavorable eutrophication of Lake Erie and the other Great Lakes, with their massive and diverse pollution inputs from an area of major and increasing concentration of population and industry. It is hoped that cooperation among the cities, and states and provinces, surrounding these lakes will make some improvement in them possible. Control is difficult for the thousands of smaller water bodies across the country now subject to increasing cultural eutrophication.

"Nutrients" have been referred to as affecting lake productivity without specifying the elements involved. It is likely that phosphate-phosphorus is most significant, over all, in both natural and cultural eutrophication. Nitrogen probably is also important, and man is greatly increasing the flow of fixed nitrogen in the ecosphere. In some streams and ground-water, as well as in lakes, nitrate concentrations are at levels potentially detrimental to human health. Potassium, iron, and sulfur are in short supply in some lakes, and in some oligotrophic lakes molybdenum, zinc, and manganese have been found to be limiting. Increased inputs of some of these could contribute to eutrophication of some lakes. Deficiency of manganese, as well as excesses of other nutrients, may contribute to the replacement of diatoms by blue-green algae in cultural eutrophication. Carbon dioxide, which easily diffuses into lake water from the air, appears not to be an important factor in determining productivity. The primary importance of phosphorus emphasizes an inconvenient circumstance: the rapid, recent increase in the use of phosphates in detergents. The earlier detergents that replaced soaps were resistant to biological decomposition and caused foaming and toxic effects in the water bodies they reached. These detergents were replaced with others that were degradable—subject to breakdown in environment by bacteria or chemical processes. Their breakdown releases phosphates that are now a major contribution to the nutrients producing cultural eutrophication. In Canada the phosphates in detergents have been replaced with NTA (nitrilotriacetate), which is regarded as less detrimental to water bodies.

Cultural eutrophication is one of those processes that become increasingly rapid and widespread with the increase of human populations. As the population and industry of the United States increase, these influences bear simultaneously on lakes: (1) increasing input of sewage from towns and cities or from recreational homes surrounding the lakes, and boats, (2) increasing input of industrial wastes, some toxic, some occurring as organic wastes, some including inorganic nutrients, (3) increasing domestic wastes other than sewage, notably detergents, (4) increasingly intensive farming with heavy use of fertilizers, some of which move into streams and lakes as nutrients supporting eutrophication, and (5) increasing input of nitrates from the atmosphere, where content of nitrogen oxides is rising from pollution by automobile engines and other sources. The first two sources of eutrophication are soluble in principle, but the expense of such solution is so great as to render it difficult if not unattainable on a widespread basis. The last three are illustrations of technology's left hand—the adverse environmental effects of technology that is favorable in its own sphere.

Man and the Seas

Except for some that flow into inland basins, all the world's rivers drain into the oceans. The oceans are thus the end-place, the sink, for most of the materials from the land transported by rivers—inorganic nutrients and eroded soil, detritus and sewage, industrial wastes and toxins. The oceans are salt (3.4 per cent or 34,000 ppm) for the same reason as saline lakes: fresh water with dilute salts flows in by rivers and leaves by evaporation without the salts, which accumulate. The oceans are thus rather like a giant saline lake, with a volume (1.37×10^{18} m^3) perhaps 6000 times that of all the inland lakes, fresh and saline, of the world together.

Size and function as a sink are the ocean's defense and danger. In earlier times man could comfortably assume that any waste released into the oceans was lost in their immensity. The effect of two billion human beings on the seas was truly superficial. They occupied the shores, sailed the surface, and contributed miscellaneous pieces of wood and metal to the bottom deposits, without changing the oceans themselves in a measurable way. It is different, in the 1970's with a much larger human population and an enormous and rapidly expanding industry. The mark of man is now upon the seas in a number of ways.

There is, first, the occupation and alteration of estuaries, the river mouths, inlets, and bays where fresh and salt water are mixed by tidal water movement. Many estuaries include saltmarsh and mudflats and are highly productive ecosystems. Some saltmarshes are among the most productive communities known, and the detritus they release is food not only for the marsh and mudflat, but for the coastal waters of the oceans. Estuaries are thus an important part of the food base for the coastal waters where (along with upwelling areas) the major part of the world's fisheries are concentrated. They are a base for those fisheries in a further way, for many of the major commercial fish species spawn, or feed during part of their life cycles in estuaries. Species dependent on estuaries make up about 85 per cent of the commercial fisheries catch of the United States, about 97 per cent of the catch in the Gulf of Mexico. Dredging and industrial occupation and pollution of estuaries, and destruction of saltmarshes for housing and other development, are thus indirect attacks on these fisheries. The same is true of the thermal pollution or heat loading which occurs when estuarine waters are used to cool nuclear reactors. The increased water temperatures are not advantageous for most organisms of the estuary. Some species can adapt, and some new species that are adapted to higher temperatures can enter the estuary. When temperatures fluctuate, however, with the nuclear reactor operating at different levels and at times shut down, both warm-adapted and cool-adapted species encounter temperatures unfavorable for them; and the over-all effect is detrimental to estuarine life.

There are, second, direct effects on marine populations in the sea itself. Ecologists know enough about the dynamics of fish populations to predict limits for sustained harvest. Knowledge is not always wisdom; and knowledge does not prevent the fishing fleets of competing companies and countries from increasing their catch until fish populations are forced into decline. Figure 7.9 represents the history of the Pacific sardine fishery, once the largest single fisheries catch of the North American continent. Overfishing has now affected many, or most, of the principal marine food species; and the indications of overfishing have spread geographically from the Baltic and North Atlantic near Europe to other areas, including the great fisheries offshore from eastern Canada and New England. At the same time some of the whales are being overharvested toward extinction; and for other marine populations—not only fish but lobster, shrimp, oysters and clams, and abalone—the catch is decreasing or must be taken from ever deeper water. Total fish landings from the seas increased steadily from the 1940's to 1968, and some increase in catch is still possible by still harder pressure upon fish populations. It is difficult to estimate the total sustainable catch that might be possible with good management of the marine fisheries. It is not difficult to recognize that without management they will be overfished, and to suspect that the maximum harvest sustainable for the long term may already have been reached or passed.

Other population effects are less direct, and we mention two cases.

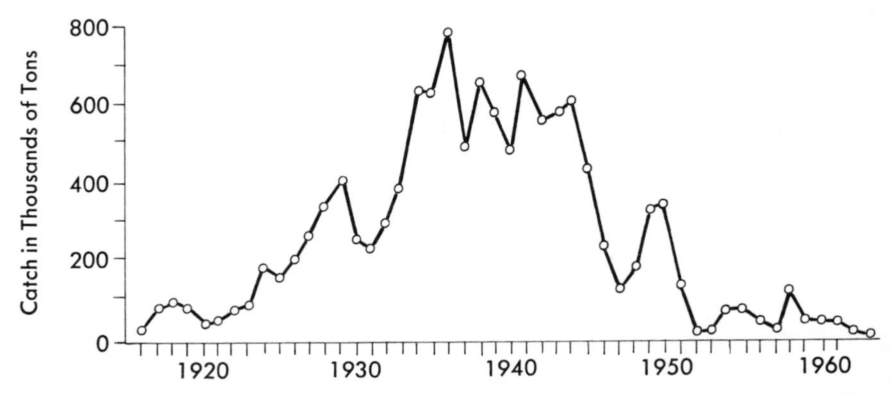

Figure 7.9. The rise and fall of a major fishery. Catches are shown for the sardine (*Sardinops caerulea*) along the Pacific Coast of North America as the catch rose to a peak and then declined, with increasing scarcity of the sardine, to collapse of the fishery. During the decline in the sardine population and the catch, overfishing reduced reproduction and left too few older individuals to carry the population through periods of reproductive failure. [Murphy, 1966.]

Along both the Atlantic and Pacific coasts of the United States "forests" of kelp, giant brown seaweeds, have been much reduced in area; off some of the southern California coast the kelp may have been reduced to one tenth of its former extent. The effect seems to be not from direct harvest of the kelp, on the California Coast, nor from direct effects of pollution on the kelp. Both areas are polluted, but within limits the kelps may be able to use the nutrients of pollution to increase their very high productivity. Consumption of the kelp, and particularly the bases of the plants, by erupting sea urchin populations appears to be responsible for reducing the kelp forests in both cases. The causes of increased numbers of sea urchins are uncertain. Direct effect of the pollution in increasing growth of sea urchins (in the absence of the sea otters that once controlled their populations) has been suggested for California; the effect of overfishing lobsters that are predators on the sea urchins has been suggested for the Atlantic Coast.

In the tropical Pacific Ocean coral reef communities include a giant starfish, the crown-of-thorns (*Acanthaster planis*). The starfish is a predator on coral, but its population is held in check by predation—by a giant snail on the adults and by corals on the young. Through extensive areas of the Great Barrier Reef of Australia, and many Pacific Islands, the crown-of-thorns population has erupted, denuding coral reefs and leaving them exposed to erosion and breakage by waves. Release from predation may have permitted the eruption. The giant snail population has been reduced in some areas; and disturbance by man has exposed extensive areas of coral where the crown-of-thorns can multiply without coral predation on its young, and from which the populations can emerge to denude and expose other coral areas. The coral reefs of southern Florida are reported to be dying and eroding in some areas. The cause, which in this case is not the crown-of-thorns, is unknown but likely to be some effect of pollution on the coral-forming animals.

Chemical influences are spreading in the oceans and may produce direct or indirect effects on ocean populations. Effects are most evident in semienclosed seas, where man's pollution is most concentrated. The Baltic Sea is bounded by several nations with their populations and industries concentrated on its shores. The Baltic is a great estuary, fed by many rivers, with waters ranging from nearly fresh to brackish. It is connected with the North Sea through narrow passages between Denmark and Sweden, but its waters are largely stagnant with a residence time of 22 years. As the rivers are polluted, so is the Baltic. The pollution input of phosphorus is estimated at 14,000 t/yr, versus natural inputs of about 3400 t/yr by rivers and 3000 t/yr from the atmosphere. Although it is about fifteen times the size of Lake Erie, the Baltic is subject to cultural eutrophication and to periods of extensive oxygen depletion and accumulation of hydrogen sulfide in the deeper waters. The bottom communities have been altered and the fisheries affected, and some of the beaches are noticeably polluted. Concentrations of DDT and polychlorinated biphenyls 8 to 10 times those in the Atlantic Ocean have been observed in Baltic animals.

The Mediterranean Sea is a water body also shared by many nations and their industry and sewage. The beauties of the Mediterranean coasts, that have been known to more civilizations than now survive in the world, are seriously affected. A survey of fishing and bathing areas of the Italian Coast indicated that 95 per cent had inadequate treatment of sewage and industrial wastes, 24 per cent were polluted by oil from ships, 35 per cent by other oil, 39 per cent by industry, and 68 per cent by sewage. It is appropriate for the multiple, convergent pressures of pollution on water bodies that these per cents total much more than 100. Some of the shores are doubly degraded by pollution and extensive commercial development. The whole water mass of the Mediterranean Sea is now affected by pollution, and declines in fish populations and shore life are observed.

The Mediterranean is more like a little ocean, with limited connection with the other oceans, than an estuary. Its condition is a forecast of effects on the larger oceans, given time and with continued industrial growth. The kinds of pollution effects that are evident in semienclosed bodies are beginning to occur in the larger areas of open coastal waters. The coastal fisheries, including fish species that do not spawn in estuaries, are thus subject to a pincers of overharvest and pollution. The open ocean, for all its immensity, is not invulnerable. We have mentioned that DDT is global, and that lead has been increased in the surface waters of the oceans by man. The oceans are also increasingly polluted by petroleum. Severe local effects on sea birds and shore life have been reported; but there is also a less intense and more general contamination of much of the ocean surface with oil. With time the petroleum on the surface is reduced in mass and converted into lumps of tar. Such tar balls are now a conspicuous, unwelcome catch in plankton nets in the Mediterranean and North Atlantic; they are widespread along with other wastes in the Sargasso Sea, and they occur in all the oceans.

The pollution history of the oceans thus suggests that of smaller water bodies, written more slowly. It is not evident that significant damage to the open oceans has resulted from their contamination. We mention, however, two further points on marine pollution. The first is its relentless quality. Because the oceans are the great sink for the ecosphere, a great share of man's wastes and poisons must find their way to them. Limitation of this input by mutual agreement among, and self-discipline and expense by, many nations is not easy. Characteristic lag effects apply—long after DDT use has stopped in a given country, some of that used is still being carried by air and water to the seas. The second is the importance of the seas for food. The fraction of man's food that comes from the seas is only 3 to 5 per cent, but the importance of the marine fisheries as a protein source is much more than this. Not only are marine fisheries affected by disturbance of estuaries, overharvest, and pollution, but there are also proposals for industrial use of the seas and mining of their waters and bottom deposits for minerals. The likely implication of continued industrial growth and invasion of the seas is accelerating accumulation of toxic materials in the seas. Not only may this toxication produce widespread effects on marine populations, it may make difficult the development of new marine food resources. There is thus conflict of interest between man's need for food from, and his industrial effects on, the seas.

Retrogression

If man pollutes ecosystems, how does the pollution change communities? We have observed that through evolutionary time communities may be enriched in species, and that this enrichment is made possible by evolution of niche difference and accommodation of species to one another. Severe environmental stress—of a kind that is difficult for organisms of a given group to adapt to—implies that few species of that group will survive and evolve accommodation to other species in that environment. We thus expect some stresses to be expressed in lower species diversities. In the course of succession the number of species in the community usually increases, even if the maximum number of species may not be in the climax. In the course of many successions productivity, biomass, and community height and structural complexity also increase, and the maxima for these may be in the climax. Many of the pollutants that concern us are toxic, and a toxin applied to an organism or a community is a stress. Our question about pollutant effects becomes: How are communities affected by new stresses, stresses that are exerted by man and that are of kinds or degrees that species and communities have not evolved adaptations to? Furthermore: How are community responses to stress related to succession?

Suggestions toward an answer can be found in the effects of overgrazing, a stress that antedates recent concern with pollution and radiation. A certain level of herbivore grazing is natural to a community, but overgrazing by man's livestock can be unnatural in the degree and kind of grazing pressure. Grazing animals that are foreign to the community are used to harvest a larger share of its primary productivity than under natural conditions. Some of the plant species native to a grassland are unable to survive this increased harvest. Effects of overgrazing appear first in the decline of these species (decreasers). Meanwhile other species (increasers), that are more tolerant of grazing and are now relieved of the competition of the decreasers, expand their coverage. At the same time the biomass, height, and total coverage of the grassland decrease. Continued overgrazing can overharvest and reduce the increaser species, while still other species (invaders) that are not part of the undisturbed community appear. These may then increase in coverage, and in time the grassland may become a weed field dominated by them. With still continued overgrazing and trampling the weeds, in turn, can be reduced in coverage and the soil further exposed to erosion. The final result of overgrazing to its limit may be a virtual noncommunity—a mudfield or eroded rocky slope, depending on location.

One notes reductions in community characteristics: biomass and productivity, coverage and structural complexity, species diversity, and environmental modification and nutrient control by the community. Grazing effects should not be generalized too freely; for in many areas the principal increasers or invaders are not weedy herbs but shrubs, and grazing converts a grassland into a shrub community. In the effects that we have described, however, we see reversal of the trends that characterize many successions. We can consequently speak of the effects as a kind of succession in reverse, a *retrogression*. The term does not imply a literal backwards retracing of the stages of a normal succession. It does imply the community's reduction—in some of those properties that tend to increase through successional time—in consequence of stress.

Lichens on the bark of trees form plant communities in miniature, and these are affected by air pollution. Lichens are especially vulnerable to air pollution because of their direct exposure to change in the composition of rainfall and the particles settling on them. Figure 7.10 shows the response to increasing pollution on the part of four species, three of them decreasers and one an increaser, toward the center of the city of Newcastle

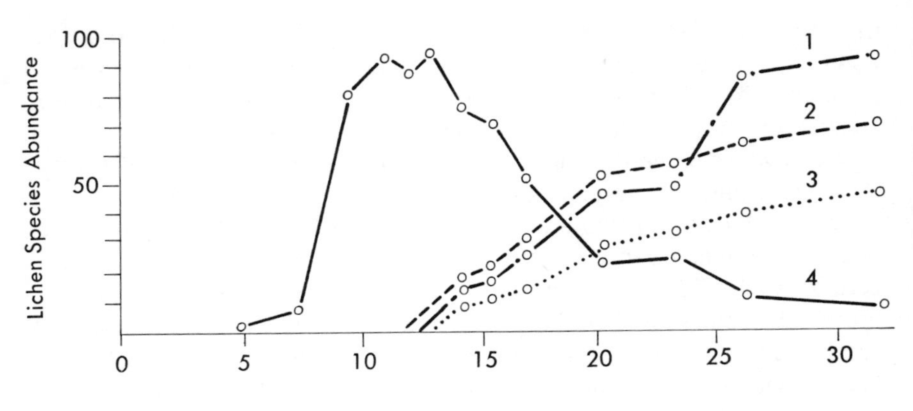

Figure 7.10. Response of lichen populations to air pollution. Most species populations decrease toward the center of the city, with increasing exposure to air pollution. One lichen species (number 4, *Lecanora conizaeoides*) acts as an increaser, becoming denser at intermediate pollution levels as other species are suppressed. The center of the industrial city (Newcastle upon Tyne, England) is a lichen and moss desert. [Gilbert, 1969.]

upon Tyne, England. There are no invader species in this case. Pollution effects are also correlated with the heights of lichen species. Those lichens that grow upward from the bark surface (fruticose and foliose species) are more sensitive to pollution, and the species most tolerant of pollution are some of those that grow as a crust on the bark surface. Species diversity, total coverage, and "community" height and structure thus decrease with increasing pollution. When the pollution sensitivities of lichen species and communities are known, lichen communities can be used to map intensities of air pollution (Figure 7.11).

Vascular plants are also affected by air pollution, though as a group they seem less vulnerable than lichens. Effects on vascular plant communities are most easily studied where a heavily polluting industry affects natural vegetation. An iron sintering plant at Sudbury, Ontario, releases sulfur dioxide in an area of spruce forest. About 25 km (16 miles) from the plant, the dominant spruce trees of the forests are killed by the SO_2. From 25 to 7 km the species diversities of vascular plants decrease progressively as the concentrations of SO_2 to which they are exposed increase. From 7 to 2 km only two plant species remain in most samples. The diversity gradient is illustrated in Figure 7.12; with increasing intensity of SO_2 exposure community height, biomass, production, and coverage decrease along with diversity.

The effects of pollution on aquatic communities are more varied because of the wide range of different kinds of pollution involved. Reductions in productivity, biomass, and diversity of communities are to be expected

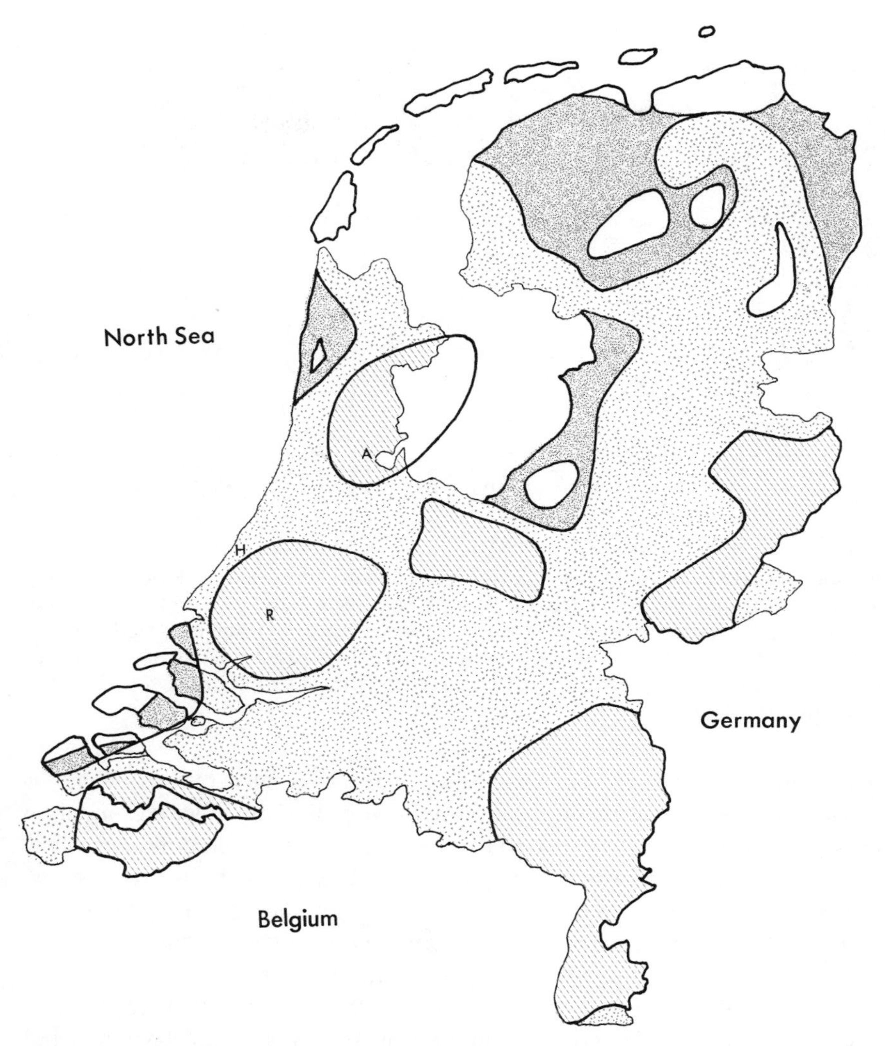

Figure 7.11. A moss and lichen map of air pollution intensity in the Netherlands. Epiphytic moss and lichen communities are well developed and rich in species on islands and along the coast of the Netherlands, where air reaching them from the west is least affected by air pollution. Within the rest of the country major urban areas produce a pattern of differing exposures to pollution that can be mapped on the basis of moss and lichen communities. Unstippled areas have rich epiphytic vegetation, fine stippled areas normal epiphytic vegetation for the Netherlands, and coarse stippled areas impoverished epiphyte vegetation, whereas the hatched areas are epiphyte deserts subject to heavy air pollution. Major cities: A Amsterdam, R Rotterdam, and H The Hague. [Barkman in ten Houten, 1969.]

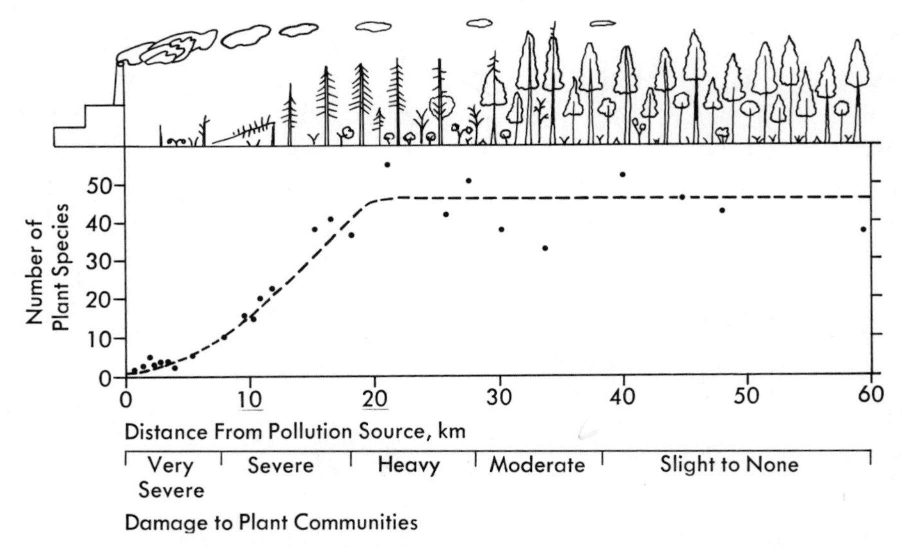

Figure 7.12. Reduction of plant species diversity by air pollution. With increasing exposure to sulfur dioxide, numbers of plant species are reduced from an average above 40 per sample to 0 to 2, close to an iron sintering plant in Ontario. Along the gradient the vegetation is reduced from spruce forest, to a mixture of shrubs and herbs with standing dead trees, to a meager cover of two herb species. [Gordon and Gorham, 1963.]

when large amounts of toxic materials are released into streams, lakes, or coastal waters of the oceans. Much aquatic pollution, however, involves sewage in which not toxic materials but organic wastes predominate. This wealth of organic material can increase secondary productivity, while altering profoundly the character of the aquatic community. In many temperate streams, for example, a natural community of bottom animals (midge larvae, mayfly and dragonfly nymphs, clams and snails, and so on) is replaced by a mat of worms (*Tubifex*) or rat-tailed maggots (*Syrphus*). These animals are adapted to feeding on organic wastes and bacteria, and they are tolerant of the low levels of oxygen in their environment produced by bacterial decomposition of the wastes. The blood of *Tubifex* contains a red hemoglobin pigment that aids it in obtaining oxygen from an environment low in oxygen, and the *Syrphus* larvae extend their tails to the water surface to obtain oxygen from the air. These species can sometimes form fur-like mats dense with thousands of individuals on the stream bottom. Organic pollution can support single-species animal communities that are highly productive, and highly unattractive. Most fish, especially the species desired as food by man, are among the more sensitive species that disappear with less intense pollution.

The reduction in species diversity can be shown also for diatoms in streams. To obtain controlled, comparable samples from different streams, glass microscope slides may be suspended in the water. After a period of two weeks or more, the films of diatoms and other organisms that develop on the glass surfaces can be studied as microcommunities and their species diversities measured. Figure 7.13 shows samples of such diatom communities from an unpolluted stream, above, and a polluted one, below. In each case some thousands of diatoms have been counted and recorded by species. The figure uses the lognormal plot of Preston (see Figure 3.13) to show the numbers of species of diatoms in octaves or doubling units for numbers of individuals in species. The curves are cut short on the left by the limits of the samples, but one can estimate the total numbers of diatom species by extending the curves to the left. The polluted stream has a lower number of species per octave, a wider dispersion of species through the octaves, stronger species dominance and lower equitability, and a much lower total number of species.

One other study of retrogression may be mentioned. The oak-pine forest at Brookhaven National Laboratory, New York, was exposed to gamma irradiation from a cesium-137 source that could be raised above the ground to irradiate the forest, and lowered into the ground to permit investigators to enter the area and study the effects. Circular zones of

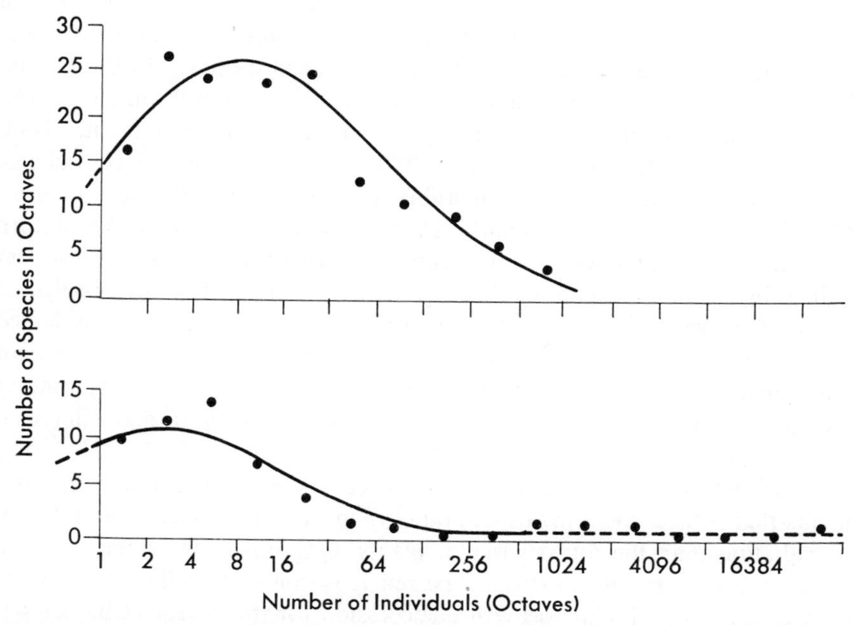

Figure 7.13. Effect of pollution on diatom species diversity in streams. Diatom species and individuals were counted on glass slides that had been suspended in streams and plotted as lognormal distributions for an unpolluted stream above (Ridley Creek, Pennsylvania) and a polluted stream below (Back River, Maryland). [Patrick, 1963.]

vegetation showing different intensities of radiation effects developed around the cesium source. Proceeding inward toward the source, the pines were killed first with least radiation, and then the oaks, leaving a middle zone of shrubs and the skeletons of trees. With more intense exposure the tall shrubs were killed, leaving lower shrubs (blueberries and huckleberries). With still more intense exposure these in turn were killed, leaving an inner zone in which vascular plants were reduced to one, a sedge. Still closer to the source the sedge was killed, and there remained only lichens on the bark of the dead trees. The radiation stress produced characteristic community reductions—of living biomass, productivity, height, structure, and diversity. The input of destructive energy, as radiation, was trivial compared with the energy of primary productivity that it disrupted.

The Brookhaven study permits comparison of radiation effects on three groups of plants—the forest vascular plants, forest lichens, and the herbs of an old-field succession that was separately studied. For study of stress effects we want a means of measuring community sensitivity, as distinguished from sensitivity of individual species. If reduction in species diversity is considered the most general response to stress, we can compare communities by the radiation intensity that reduces species diversity by 50 per cent from that of control, unirradiated samples. (We can also ask the stress intensity that produces a 50 per cent reduction in coefficient of community from that between replicate control samples. These two measures will be the same only if there are no invader species.) In the Brookhaven studies diversity was reduced by 50 per cent of control samples at 150 Roentgen units of irradiation per day for the forest vascular plants, 1000 R/day for the old-field herbs, and 2700 R/day for the forest lichens. Lichens as a group are more vulnerable to air pollution, but more tolerant of radiation, than vascular plants. In general the lower a plant's stature and the smaller the amount of its nonphotosynthetic support tissue, the less vulnerable it is to radiation. Among herbs the saturation- or *K*-selected forest herbs were more vulnerable than the weedy, exploitation- or *r*-selected old-field herbs. Underlying these sensitivity differences was a less evident plant characteristic—size of chromosomes. The larger the mean volume of the chromosomes, and consequently the larger the genetic targets offered to radiation, the more vulnerable the plant.

It is interesting that the weeds and lichens, adapted to extreme environments that subject them to stresses other than radiation, were more tolerant of radiation than the forest vascular plants. One notes also three parallels in community response to stress. The retrogression with radiation exposure at Brookhaven is similar to the fire succession for this forest (Chapter 4), in reverse. The physiognomic expression of increasing radiation—from forest to shrubland to sedge meadow—is suggestive of that along major climatic gradients, such as that from Brookhaven north to the arctic tundra. There are thus rough similarities in community relations to stress in the different circumstances of: climatic gradients toward increasing environmental severity, retrogressive gradients with increasing unnatural stress, and successional gradients (considered from the climax, with its moderated environment, back to the pioneers).

Our purpose has been to describe, not to project or evaluate, these effects on communities. However, it may be worthwhile to contemplate a projection. If industrial civilization continues its exponential growth and expansion around the world, then the natural communities of the world will be subjected to ever-increasing stresses by increasing quantities of increasing numbers of pollutant substances. Accelerating retrogressive effects on communities should be expected. Despite the variety of pollutants and particular effects, some general trends can be suggested. One of these clearly is reduction in species diversity.

Several interrelated effects on diversity may occur. Civilized man's occupation of a land tends to convert natural communities into islands, separated by agricultural and developed areas. Island species are vulnerable to extinction because they are small populations with limited habitats, and because survival by migration in the island, or emigration and re-immigration, may be impossible. If natural areas become separated islands in a world heavily occupied by industrial civilization, then many species will be both cornered in these "islands," and subject in them to increasing stress from widespread pollution. Some forms of pollution are subject to food-chain concentration and strike top carnivores with special intensity. Reduction in carnivore populations may imply reduction in the stability, and consequently the feasible niche differentiation, of the prey populations they feed on. It may also imply increasing tendency to eruption of herbivore populations, which then exert stress on plant populations and may press some of these toward extinction if they exist as small populations in isolated natural areas. The plants, themselves subject to widespread pollution, would be increasingly vulnerable to grazing stress and disease. Thus the unstabilizing effects of pollution can be transmitted downward through the community, from carnivores to herbivores and herbivores to plants. The trends toward destabilization of communities imply selective advantage for relatively unstable exploitation-selected and generalist species. They imply increased prevalence of species with the attributes of weeds and pests, and the evolution of new such species.

Apart from diversity effects, the cutting and clearing of forests imply a reduction of the biomass of land vegetation. The nutrient stocks held in terrestrial communities are thereby reduced; at the same time soil is exposed to erosion in some areas, and increasing amounts of fertilizer are applied to farm fields. A general consequence is the accelerated movement of nutrients from land into water bodies, which are thereby overloaded with these nutrients. A sufficient increase in general pollution on land would reduce the productivity of land communities, as the productivities of some farmlands subject to urban air pollution are even now reduced. A sufficient increase in the general pollution of water bodies might increase their productivity in some forms, while reducing the yield of food useful to man.

The general effect of industrial growth, sufficiently extrapolated beyond the present, would be a raveling of the biosphere's functions as regards both species in communities and the steady-state mechanisms of ecosystems and the ecosphere. The effects would not be welcome. This projection is not a prediction. It may, however, give basis for the suggestion that it will not profit our species to press the retrogression of the biosphere too far.

SUMMARY

Pollution effects of man are to be understood through ecosystem function and biogeochemical patterns. When pollutants are released into an ecosystem they may be concentrated many thousand times into organisms and along food chains, from low levels in environment to high and potentially toxic levels in organisms. At the same time pollutants are transferred between ecosystems in varied, and sometimes unforeseen, ways. Hazards from pollutants thus result from combinations of ecosystem function and biogeochemical transfer. Radioisotopes released by weapons tests and from nuclear reactors illustrate wide dispersal through environment and concentration into organisms. Parallel behavior appears in certain poisons, the persistent pesticides, that are used against agricultural pests and are now global contaminants. Some radioisotopes and pesticides are progressively concentrated up food chains, so that they affect top carnivores more strongly than other organisms. A number of carnivorous bird and fish populations have declined because of such concentration of persistent pesticides. Along with pesticides certain metals, notably lead and mercury, are now widely distributed through the environment and concentrated into organisms.

Radioisotopes, pesticides, and heavy metals are all transported about the world in the atmosphere as well as in moving water. The pollutants released by combustion—sulfur and nitrogen oxides, carbon monoxide, and others—also are transported in the atmosphere, but their ecological behavior is different. For these pollutants problems arise not so much by concentration from environment into organisms, as from accumulation to toxic levels in the environments of urban areas. Increases in sickness and death rates in human beings are resulting, and damage to natural communities can be recognized in some polluted areas. The most widespread problems are not acute pollution episodes but subtle effects on human health that are delayed in recognition, statistical in expression, and complexly correlated with different aspects of air pollution, exposure to metals, and other urban stresses. Despite the short residence times of pollutants in the atmosphere, air pollution is now in transition from a local problem to an often subcontinental condition and, increasingly, an ecosphere phenomenon. The increases in global carbon dioxide levels and in acidity of rainfall over wide areas illustrate the latter.

Water bodies are affected not only by toxic materials, but also by accelerating flows of nutrients into them. Lake productivity is governed in large part by nutrients; and many lakes are showing abnormal increases in productivity that distort community function and decrease their attractiveness and usefulness to man. This process of unnatural fertilization, or cultural eutrophication, is affecting also semienclosed and coastal waters of the oceans. Because the oceans are the world sink, into which most of the materials of the ecosphere may move in the end, the full variety of persistent pollutants from human industry may accumulate there. Despite the great volume of the oceans, some pollutant accumulation in the surface waters of the open ocean is occurring, and some marine species populations may be affected.

The effects of pollutants on different kinds of communities are varied, but some effects can be characterized as retrogression: reduction of communities in some of the characteristics that tend to increase through succession. These characteristics include species diversity, productivity, biomass, and height and structural complexity; but the most general effect of pollution seems to be reduction in species diversity. In some areas species are affected both by increasing pollution stress, and by increased "cornering" in limited preserves of natural communities surrounded by lands used and developed by man. Extinction of some of these species will result, and with this reduction in the stability of communities and increase in the prevalence of relatively unstable species such as weeds. Pollution may thus both alter the steady-state functions of the ecosphere and reverse the tendency of communities to evolve toward higher species diversity, refined niche differentiation, and relative stability of some of their populations.

References

General

BERRY, JAMES W., D. W. OSGOOD and P. A. ST. JOHN. 1974. *Chemical Villains: A Biology of Pollution.* St. Louis: Mosby. vii + 189 pp.

HODGES, LAURENT. 1973. *Environmental Pollution.* New York: Holt, Rinehart & Winston. xii + 370 pp.

JACKSON, WES, editor. 1971. *Man and the Environment*. Dubuque, Iowa: Brown. xxiii + 322 pp.

†*MURDOCH, WILLIAM W., editor. 1971. *Environment, Resources, Pollution and Society*. vii + 440 pp.

SCEP. 1970. *Man's Impact on the Global Environment:* Report of the Study of Critical Environmental Problems (SCEP). Cambridge: Mass. Inst. Technol. xxii + 319 pp.

SINGER, S. FRED., editor. 1970. *Global Effects of Environmental Pollution*. New York: Springer. xii + 218 pp.

TIE. 1972. *Man in the Living Environment:* Report of the Workshop on Global Ecological Problems. The Institute of Ecology. 267 pp.

WAGNER, RICHARD H. 1971. *Environment and Man*. New York: Norton. xiii + 491 pp.

Radioisotopes

ABERG, BERTIL and F. P. HUNGATE, editors. 1967. *Radioecological Concentration Processes:* Proceedings of the International Symposium, Stockholm, 1966. London: Pergamon. xiv + 1040 pp.

AUERBACH, S. I. 1965. Radionuclide cycling: current status and future needs. *Health Physics* **11**:1355–1361.

DAVIS, J. J. and R. F. FOSTER. 1958. Bioaccumulation of radioisotopes through aquatic food chains. *Ecology* **39**:530–535.

HANSON, W. C. 1967. Cesium-137 in Alaskan lichens, caribou, and eskimos. *Health Physics* **13**:383–389.

NELSON, DANIEL J. and F. C. EVANS, editors. 1969. *Symposium on Radioecology:* Proceedings of the Second National Symposium, Ann Arbor, 1967. Washington, U. S. Department of Commerce (CONF-670503). xii + 774 pp.

POLIKARPOV, G. G. 1966. *Radioecology of Aquatic Organisms*. New York: Reinhold. xxviii + 314 pp.

POLLARD, E. C. 1969. The biological action of ionizing radiation. *American Scientist* **57**:206–236.

REICHLE, D. E. 1967. Radioisotope turnover and energy flow in terrestrial isopod populations. *Ecology* **48**:351–366.

SCHULTZ, VINCENT and A. W. KLEMENT, JR., editors. 1963. *Radioecology:* Proceedings of the First National Symposium on Radioecology, Fort Collins, 1961. New York: Reinhold. xvii + 746 pp.

SCHULTZ, VINCENT and F. W. WHICKER, editors. 1972. *Ecological Aspects of the Nuclear Age: Selected Readings in Radiation Ecology*. Springfield, Va.: Natn. Techn. Informn. Serv. (TID-25978). 588 pp.

SEYMOUR, ALLYN H., chairman. 1971. *Radioactivity in the Marine Environment*. Washington: National Academy of Sciences. ix + 272 pp.

*WOODWELL, G. M. 1967. Toxic substances and ecological cycles. *Scientific American* **216**(3):24–31.

WOODWELL, G. M. 1969. Radioactivity and fallout: the model pollution. *Bio-Science* **19**:884–887.

Pesticides and Heavy Metals

*ACKEFORS, H., G. LÖFROTH, and C.-C. ROSÉN. 1970. A survey of the mercury pollution problem in Sweden with special reference to fish. *Oceanography and Marine Biology, an Annual Review* **8**:203–224.

ANDERSON, J. M. and M. R. PETERSON. 1969. DDT: sublethal effects on brook trout nervous system. *Science* **164**:440–441.

KLEIN, D. H. and E. D. GOLDBERG. 1970. Mercury in the marine environment. *Environmental Science and Technology* **4**:765–768.

LUNDIN, F. E., JR. et al. 1969. Mortality of uranium miners in relation to radiation exposure, hard-rock mining and cigarette smoking—1950 through September 1967. *Health Physics* **16**:571–578.

MATSUMURA, FUMIO, G. M. BOUSH, and T. MISATO, editors. 1972. *Environmental Toxicology of Pesticides:* Proceedings of a U.S.-Japanese seminar, 1971. New York: Academic. xiv + 637 pp.

MILLER, MORTON W. and G. C. BERG, editors. 1969. *Chemical Fallout: Current Research on Persistent Pesticides.* Springfield, Ill.: Thomas. xxii + 531 pp.

O'BRIEN, RICHARD D. 1967. *Insecticides: Action and Metabolism.* New York: Academic. xi + 332 pp.

*PATTERSON, C. C. 1965. Contaminated and natural lead environments of man. *Archives of Environmental Health* **11**:344–360.

*PEAKALL, D. B. 1970. Pesticides and the reproduction of birds. *Scientific American* **222**(4):72–78.

PIMENTEL, DAVID D. 1971. *Ecological Effects of Pesticides on Non-Target Species.* Washington: Office of Science and Technology. 220 pp.

RUDD, ROBERT L. 1964. *Pesticides and the Living Landscape.* Madison: Univ. Wisconsin. xiv + 320 pp.

WOODWELL, G. M., P. P. CRAIG, and H. A. JOHNSON. 1971. DDT in the biosphere: where does it go? *Science* **174**:1101–1107.

*WOODWELL, G. M., C. F. WURSTER, JR. and P. A. ISAACSON. 1967. DDT residues in an East Coast estuary: a case of biological concentration of a persistent pesticide. *Science* **156**:821–824.

WURSTER, C. F., JR. and D. B. WINGATE. 1968. DDT residues and declining reproduction in the Bermuda petrel. *Science* **159**:979–981.

Air Pollution

BRYSON, R. A. 1974. A perspective on climatic change. *Science* **184**:753–760.

*DIXON, J. P. and J. P. LODGE. 1965. Air conservation report reflects national concern. *Science* **148**:1060–1066.

HOUTEN, J. G. TEN, chairman. 1969. *Air Pollution:* Proceedings of the First

European Congress on the Influence of Air Pollution on Plants and Animals, Wageningen 1968. Wageningen: Centre for Agric. Publ. and Docmt. 415 pp.

JUNGE, CHRISTIAN E. 1963. *Air Chemistry and Radioactivity*. New York: Academic. xii + 382 pp.

KELLOGG, W. W., R. D. CADLE, E. R. ALLEN, A. L. LAZRUS and E. A. MARTELL. 1972. The sulfur cycle. *Science* **175**:587–596.

*LAVE, L. B. and E. P. SESKIN. 1970. Air pollution and human health. *Science* **169**:723–733.

LEWIS, HOWARD R. 1965. *With Every Breath You Take*. New York: Crown. xvii + 322 pp.

LIKENS, G. E., F. H. BORMANN, and N. M. JOHNSON. 1972. Acid rain. *Environment* **14**(2):33–40.

MILLER, P. R. 1969. Air pollution and the forests of California. *California Air Environment* **1**(4):1–3.

NEWELL, R. E. 1971. The global circulation of atmospheric pollutants. *Scientific American* **224**(1):32–42.

SKYE, E. 1968. Lichens and air pollution: a study of cryptogamic epiphytes and environment in the Stockholm region. *Acta Phytogeographica Suecica.* **52**:1–123.

STERN, ARTHUR C., editor. 1968. *Air Pollution: A Comprehensive Treatise*. 2nd ed., 3 vols. New York: Academic.

STERN, ARTHUR C. 1973. *Fundamentals of Air Pollution.* Academic, New York. xiv + 492 pp.

WOODWELL, GEORGE M. and ERENE V. PECAN, editors. 1973. Carbon and the Biosphere. *Brookhaven Symposia in Biology* 24, vii + 392 pp. Springfield, Va.: National Technical Information Service (CONF-720510).

Lake Production and Eutrophication

BEETON, A. M. 1965. Eutrophication of the St. Lawrence Great Lakes. *Limnology and Oceanography* **10**:240–254.

BRYLINSKY, M. and K. H. MANN. 1973. An analysis of factors governing productivity in lakes and reservoirs. *Limnology and Oceanography* **18**:1–14.

CLARK, J. R. 1969. Thermal pollution and aquatic life. *Scientific American* **220**(3):19–27.

*EDMONDSON, W. T. Fresh water pollution, pp. 213–229 in Murdoch 1971.

GOLDMAN, C. R. The role of micronutrients in limiting the productivity of aquatic systems, pp. 21–33 in Likens 1972.

GORHAM, E. 1961. Factors influencing supply of major ions to inland waters, with special reference to the atmosphere. *Bulletin of the Geological Society of America* **72**:795–840.

HASLER, A. D. Man-induced eutrophication of lakes, pp. 110–125 in Singer 1970.

HYNES, HUGH B. N. 1963. *The Biology of Polluted Waters*. Liverpool Univ. 202 pp.

IMBODEN, D. M. 1974. Phosphorus model of lake eutrophication. *Limnology and Oceanography* **19**:297–304.

LIKENS, GENE E., editor. 1972. *Nutrients and Eutrophication: The Limiting-Nutrient Controversy*. American Society of Limnology and Oceanography, Special Symposium **1**, x + 328 pp.

NATIONAL ACADEMY OF SCIENCES. 1969. *Eutrophication: Causes, Consequences, Correctives.* Washington: National Academy of Sciences–National Research Council. vii + 661 pp.

PATRICK, R., B. CRUM, and J. COLES. 1969. Temperature and manganese as determining factors in the presence of diatom or blue-green algal floras in streams. *Proceedings of the National Academy of Sciences, U. S. A.* **64**:472–478.

SCHINDLER, D. W. et al. 1972. Atmospheric carbon dioxide: its role in maintaining phytplankton standing crops. *Science* **177**:1192–1194.

*SCHINDLER, D. W. 1974. Eutrophication and recovery in experimental lakes: implications for lake management. *Science* **184**:897–899.

VOLLENWEIDER, R. A. 1969. Möglichkeiten und Grenzen elementaren Modelle der Stoffbilanz von Seen. *Archiv für Hydrobiologie* **66**:1–36.

VOLLENWEIDER, R. A. 1971. *Scientific Fundamentals of the Eutrophication of Lakes and Flowing Waters, with Particular Reference to Nitrogen and Phosphorus as Factors in Eutrophication.* Paris: Organization for Economic Cooperation and Development (DAS/CSI/78.27). 159 pp.

Man and the Oceans

*HOLT, S. J. 1969. The food resources of the ocean. *Scientific American* **221**(3):178–194.

†HOOD, DONALD W., editor. 1971. *Impingement of Man on the Oceans.* New York: Wiley-Interscience x + 738 pp.

*MACINTYRE, F. and R. W. HOLMES. Ocean pollution, pp. 230–253 in Murdoch 1971.

MURPHY, G. I. 1966. Population biology of the Pacific sardine (*Sardinops caerulea*). *Proceedings of the California Academy of Science* **34**:1–84.

†RUIVO, MARIO, editor. 1972. *Marine Pollution and Sea Life.* London: Fishing Books and FAO. xxiv + 624 pp.

RYTHER, J. H. and W. M. DUNSTAN. 1971. Nitrogen, phosphorus, and eutrophication in the coastal marine environment. *Science* **171**:1008–1013.

Retrogression

DYKSTERHUIS, E. J. 1949. Condition and management of range land based on quantitative ecology. *Journal of Range Management* **2**:104–115.

ELLISON, L. 1954. Subalpine vegetation of the Wasatch Plateau, Utah. *Ecological Monographs* **24**:89–184.

GILBERT, O. L. 1969. The effect of SO_2 on lichens and bryophytes around Newcastle upon Tyne, pp. 223–235 in ten Houk 1969.

GORDON, A. G. and E. GORHAM. 1963. Ecological aspects of air pollution from an iron-sintering plant at Wawa, Ontario. *Canadian Journal of Botany* **41**:1063–1078.

GORHAM, E. and A. G. GORDON. 1963. Some effects of smelter pollution upon aquatic vegetation near Sudbury, Ontario. *Canadian Journal of Botany* **41**:371–378.

PATRICK, R. 1963. The structure of diatom communities under varying ecological conditions. *Annals of the New York Academy of Sciences* **108**:359–365.

WHITTAKER, R. H. and G. M. WOODWELL. 1973. Retrogression and coenocline distance. *Handbook of Vegetation Science* **5**:53–73.

WOODWELL, G. M. 1962. Effects of ionizing radiation on terrestrial ecosystems. *Science* **138**:572–577.

WOODWELL, G. M. 1967. Radiation and the patterns of nature. *Science* **156**:461–470.

*WOODWELL, G. M. 1970. Effects of pollution on the structure and physiology of ecosystems. *Science* **168**:429–433.

WOODWELL, G. M. and R. H. WHITTAKER. 1968. Effects of chronic gamma irradiation on plant communities. *Quarterly Review of Biology* **43**:42–55.

8

Conclusion

Community Evolution

The preceding chapters have sought in various ways to give meaning to the concept of community introduced in Chapter 1. A major question remains: how communities evolve. This conclusion draws together from other chapters some implications for the way species relate to one another in community evolution, and the way characteristics of communities as wholes evolve.

One observation is that communities can accumulate increasing numbers of species through evolutionary time. Community species diversity tends to increase because on the whole, with certain steady-state exceptions, more species are added to communities than become extinct in them if there is no major change in environment. To survive, a species population entering a community from an adjacent community or a different area must be adapted in two ways; it must possess (1) some form of population buffering or regulation that protects it from fluctuating to extinction, and (2) some accommodation to interaction with other species in the community, especially difference from other species in niche (Chapters 2 and 3). These are not clearly separate adaptations. For a given species, interactions with other species may be part of the pressures toward population extinction, or part of the basis of population survival, or both. But the community evolves as an assemblage of interacting, buffered, niche-differentiated species populations to which other species can add themselves through evolutionary time in niches different from those of species already present.

Communities are the contexts in which species survive and evolve. The statement of context is incomplete, however, if we do not consider the full range of environments and communities in which a species occurs. The communities of a landscape form a pattern that corresponds to the landscape's pattern of environments (Chapter 4). The kind or kinds of environments in which a species occurs define its habitat. For a species new to the area to survive in the landscape without displacing another species, the new species must differ from other species in the landscape in either its niche, or its habitat, or both. It may survive in the same habitat and community as another species because it differs from that species in niche. It may survive with a niche much like that of another species—too close for both to persist in the same stable community—because it occurs in a different habitat from that species. In place of saying that the species differs from other species in either niche or habitat or both, we can say it differs in ecotope. By *ecotope* we mean the species' response to the full range of environmental factors affecting it, including both the intensive or intracommunity factors that define its niche, and the extensive or intercommunity factors that define its habitat. For clarity in discussing species relationships it is important to distinguish niche and habitat factors, but some of these factors are closely related to one another (as temperature change with elevation in mountains is a habitat factor, whereas temperature change with seasons in a community is a niche factor). Species populations evolve in adaptation to both kinds of factors at the same time, and many adaptations (for example an optimum temperature for reproduction) affect both niche and habitat.

Gradient analysis, which seeks to comprehend and represent landscape patterns, shows that species differ in habitats and positions in patterns and that undisturbed communities mostly intergrade continuously along environmental gradients (Chapter 4). The pattern of communities has been characterized as a complex population continuum. The pattern's character as such results from the manner of species evolution. Because species evolve toward difference in habitat, their population centers are scattered in the community pattern. Because they also evolve toward difference in niche, their populations can in most cases overlap broadly, forming the tapering distributions shown by gradient analysis. Because species evolve toward difference in both niche and habitat, they can form a community pattern in which species distributions are diverse and broadly overlapping, and the species form together a complex population continuum.

This manner of evolution suggests opposing tendencies. On the one hand, species evolve toward accommodation to or dependence on interactions with other species, and therefore toward membership in communities as systems of interacting species. On the other hand, they evolve toward difference in habitat from other species, as expressed in the principle of species individuality: Each species is distributed according to the way its own population responds to habitat factors and interactions with other species, and hence differently from any other species. In this we note a paradox of species evolution in communities. Species evolve both toward relatedness to other species (in the sense of interactions with them) and toward apparent unrelatedness (in the sense of distributional individuality). The paradox is resolved by considering the kinds of relationships among species that prevail in communities, and the distributional implications of these relationships. We allow, first, for the occurrence of some strict dependences of one species on another, because of which these species may have the same distribution. However, most species that depend on other species as predators or symbionts depend on several or many other species. Apart from strict dependences on one other species, predators and symbionts need not have the same distributions as any of the species on which they depend. Competition between species tends to produce difference in species' habitats and distributions. The direction of evolution is thus predominantly toward loose relationships among species that permit both terms of the paradox: species interaction in communities but species individuality in distribution.

Species can accumulate in the landscape's pattern of communities through evolutionary time. The addition of species differing in niche to communities implies increase in alpha diversity, whereas addition of species along habitat gradients implies increase in beta diversity. We should expect alpha and beta diversity to increase in parallel through evolutionary time. Yet there is reason to distinguish these two aspects of diversity as partly independent products of evolution. As noted in Chapter 4, tropical bird communities are only moderately higher in alpha diversity, but much higher in beta diversity, than temperate bird communities. Also, in some cases these two aspects of diversity have evolved somewhat differently in the Northern and Southern Hemispheres. Bird communities in Chile are of similar or somewhat higher alpha diversity than those in comparable climates of California. These similar diversities are based on generally similar niche relationships among the birds of comparable communities in the two areas, but this niche correspondence is not exact. In some cases there is a one-to-one correspondence in size, behavior, food, and other niche characteristics between species in Chile and California (or these and similar communities in South Africa). In most cases two or three birds that represent a guild in one area correspond to several birds of the comparable guild in the other area; but there is no close, species-to-species correspondence. In contrast to their similarity in alpha diversities, the Chilean communities are lower in beta diversity, but higher in geographic differentiation, than the California communities. The differences in beta and geographic diversities probably result from differences in the histories of the areas and differences in the relations of these to other areas of the continents from which bird species evolved into shrublands and related communities of Chile and California.

In general, however, through evolutionary time (without major climatic change) a landscape should be gradually enriched in species as more species are added to it than become extinct; and this enrichment should be expressed in both alpha and beta diversity. The consequence is increase in the landscape's total diversity. Striking contrasts in species diversities of plant communities of similar climates and physiognomies do occur, and probably result at least in part from differences in the times during which species have accumulated in landscapes and communities. The California chaparral (Plate 10) is relatively poor in species and recent (it has apparently evolved since Miocene time and been subject to extensive climatic change and geographic displacement). Comparable shrub communities in South Africa and southern Australia are far richer in species (Figure 8.1). The South African Cape scrub or fynbos (Plate 11) in particular is rich in species beyond Northern Hemisphere shrub communities, with both high alpha and beta diversities and a striking degree of floristic differentiation from one local area to another. The South African and southern Australian shrub communities are believed to be much older in the sense of continuous evolution without subjection to major change in climate.

Although species may be added to communities, they are not committed in their evolution to particular communities. Most species occur in a range of communities in which they interact with different combinations of other species. Many species include genetically different ecotypes, adapted to occurrence in different habitats and communities. Through evolutionary time ecotypes can be added to and lost from a species, and the range of habitats and communities in which the species occurs then changes. Species change their associations with other species in evolutionary time. A community observed in the present includes species of diverse histories in different other communities. The species of the California chaparral probably evolved into it from ancestral species that occurred in various

Figure 8.1. Vegetation convergence and nonconvergence between hemispheres. A temperate shrubland is shown on sandstone at 1550 m elevation on north-facing slopes (which in the Southern Hemisphere are dry) near Lady Grey in the Orange River Valley, South Africa. Community dominants include rosette shrubs (*Aloe ferox*) and other shrubs (*Rhus erosa, Rhus undulata* var. *burchellii, Olea africana, Diospyros lycioides* subsp. *lycioides,* etc.) in an open upper stratum 1–4 m tall above a lower shrub canopy about 0.75 m tall of other species (*Asparagus virgatus, Diospyros austro-africana,* etc.). Grasses (*Heteropogon contortus, Eragrostis curvula, Aristida diffusa* var. *burkei,* etc.) predominate in the herb stratum and contribute to a total vegetation cover of about 55 per cent. In broad physiognomic characterization—as open evergreen-sclerophyll shrubland—the community is convergent with chaparral on dry south-facing slopes in southern California and Arizona (Plate 10). In its more detailed physiognomic characteristics the community is unlike anything in these states or, probably, the Northern Hemisphere. Floristically the community shares a number of genera with California shrublands (part of the chaparral also includes two sclerophyll species of *Rhus,* sumac), but it is far richer in species. The mean number of vascular plant species in samples from this community type is 42; mean numbers in reasonably comparable American samples are 7 in chamise (*Adenostoma fasciculatum*) chaparral, 17 in mixed chaparral in the San Jacinto Mountains, California, and 24 in open sclerophyll scrub with pygmy conifers and rosette shrubs in Arizona. Other South African shrublands are even richer (see also Plate 11). [Photo courtesy of M. J. A. Werger; data on the Rhoo-Aloetum ferocis from Werger, *Phytosociology of the upper Orange River Valley, South Africa,* Pretoria (1973); California and Arizona data of R. H. Whittaker and W. A. Niering, unpublished, and *Ecology,* **46:**429 (1965).]

kinds of forest and semiarid vegetation. The evolution of communities is not branching or divaricate, in the sense that the whole assemblage of species in a community at one time divides into two different assemblages of species at a later time. Because species change their distributional relations to one another and communities, the evolution of communities is net-like or reticulate (Figure 8.2).

This reticulate evolution (together with the continuity of communities with one another in the present) has important implications: (1) Present and past communities are not arranged by ecologists in evolutionary trees,

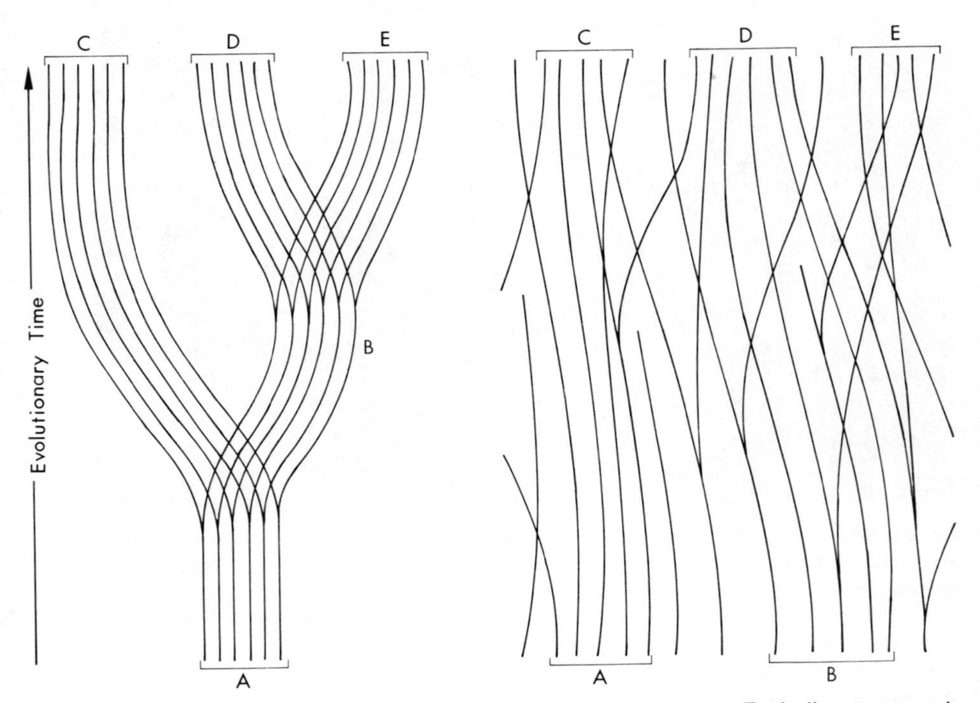

Figure 8.2. Two views of species association in evolution. Each line represents a species. Left: If community evolution were branching or divaricate, species would evolve largely in parallel. Three present communities (C, D, and E) are shown derived from a single ancestral community (A). Right: Actual community evolution is net-like or reticulate. Ancestral species are shown with their population centers scattered along an environmental gradient; among these species the groups labeled A and B are considered to characterize kinds of communities. Through evolutionary time species change their distributions somewhat independently. Some evolve in parallel; but some divide into two or more species, some become extinct, some evolve out of the range of communities represented, and some new species evolve into this range from other communities. After a period of evolution with some net increase in number of species, three groups characterizing kinds of communities (C, D, and E) are recognized along the gradient in the present. The species in each of these groups have diverse evolutionary histories of association with other species.

as present and extinct species can be arranged in some cases. (2) Community evolution cannot be made the underlying principle of community classification as evolution is, so far as possible, an underlying basis of the taxonomy of individual organisms. (3) Communities are not selected as wholes, with their species either all surviving or all going extinct together, as a basis of community evolution. (4) Care is needed in discussing evolution of characteristics of communities as a process separate from the evolution of their species.

Communities are, however, living systems and as such should have some characteristics in common with organisms. F. E. Clements and some other ecologists in the past regarded a community as a kind of superorganism and sought understanding based on the analogy between organisms and communities. Communities as living systems should possess emergent characteristics that apply to the whole and not only to its parts—characteristics including structural differentiation, growth and maturity, energy flow and material turnover, steady-state function and homeostasis, adaptive optimization, and organization. Some of these can be recognized in communities; the succession and climax of communities seem to resemble the growth and maturity of organisms. The resemblance is not close. In succession species replace one another in quite variable ways, much influenced by accidents of species dispersal, and not according to a genetic master plan like that governing an organism's development. Characteristics of the climax are not determined by an inherited design, but determined by characteristics of the environment and of the species that are able to establish themselves and maintain populations in the community.

It is possible to discuss adaptation at the level of the community—with due caution. We have used the plankton and the desert to illustrate community-wide adaptation (Chapter 4). Physiognomy of land communities is adapted to environment in ways that produce the convergences of biomes on different continents. This convergence expresses community adaptation; yet it is not, in a simple sense, an adaptive response of the community as a whole. Environment determines the potential productivity of the community and implies that certain growth-forms are most likely to become dominant. A climax community develops that is a mixture of species able to use the environmental resources and to survive together. But selection acts on the species, and community physiognomy is the expression of the kinds of species that make up the community and dominate its strata. Within broad limits physiognomy is predictable from climate. But the limits are broad; stable evergreen and deciduous forests can occur in the same humid continental climate, and grasslands, shrublands, and woodlands can occur in the same drier maritime climate. Some of the Southern Hemisphere, and especially Australian, communities are different in physiognomy from their closest equivalents in the Northern Hemisphere because of their separate evolution of different dominant species and growth-forms. Environment sets limits on the kinds of physiognomy that may develop in an area. Within these limits the development of community physiognomy is to some extent labile—different kinds of dominant species are equally able to use environmental resources and form stable communities of different physiognomies.

The labile quality of community structure bears upon the question of optimization. Some characteristics of the community should be fit to environment, in the sense of the best match of the community characteristics to those of environment, or the best compromise of different adaptive needs. This kind of optimal adaptation is often observed in the characteristics of individual species. Optimal qualities of communities might be sought, for example, in such characteristics as leaf surface area and chlorophyll content. Within broad limits these characteristics are consistent for communities of a given kind of physiognomy and level of productivity (Chapter 5). The range of leaf surface area of 4 to 6 m^2/m^2 in temperate deciduous forests presumably represents an optimum—leaf surface sufficient to use the light available, but not so great as to produce a loss in the shading of many leaves that would respire more than they produced. The range is not determined by community-wide selection. In some cases the community leaf area index is that of the dominant tree population itself, while the leaf areas of the remaining species are trivial. In other cases the community leaf area index is the sum of the leaf areas of many species from trees to herbs, some adapted to function at low light intensities. On the whole, forests of moist environments have higher undergrowth coverage than comparable forests with full canopy cover in drier environments. For a forest with a well-developed undergrowth the size of the leaf area index is determined in part by the extent in the community of species adaptated to low light intensities. Many community characteristics are consequences, or summations, of the adaptations of species. There may be respects in which community designs are optimized that have escaped notice. Because of the labile and species-dependent quality of community characteristics, however, the search for community-wide characteristics that are optimal and predictable has not been very successful.

Application of the concept of homeostasis to the community is affected by similar limitations. Homeostasis refers to the maintenance, by control mechanisms and feedback, of relative constancy of conditions within a system; for organisms it refers to the stabilization of an internal environment for the cells of the organism that is favorable for their function. A community modifies the environment of its organisms. Characteristics of a terrestrial community affect microclimate and soil properties, and function of an aquatic community determines the content of solutes and particulate organic matter in the water. In some cases—such as a forest's retention of its nutrients—community function works in a way that makes possible higher community productivity (Chapter 6). These effects are not really comparable to the homeostasis of an organism, for which specific regulatory mechanisms have evolved. The way the community affects its environment is largely an incidental consequence of other characteristics of the ecosystem, and not a homeostatic mechanism that has itself been selected for.

It is not clear that any characteristic of communities is selected for as such. Diversity increases by the addition of species to communities, not by selection of richer communities over poorer ones. The rates at which species are added are influenced by the kinds of environmental stresses affecting the species of different taxonomic groups. Subjection of communities to new stresses—such as pollution—usually reduces species diversity (Chapter 7). The reduction is not a community-wide adaptation to stress, but is a consequence of the elimination of the more vulnerable species from the community. Higher species diversity implies more diverse use of resources by species of the community. It is apparently not the case, however, that higher species diversity makes possible higher community productivity. It also seems not the case that communities are selected for their productivities.

It has been suggested that communities might be selected for stability, with more stable communities having selective advantage over, and tending to replace, less stable ones. But the species in a community differ in their responses to environmental fluctuation. A whole community of species will not fluctuate into extinction to be replaced by another community of species. Selective processes appear to lead toward different conditions of relative stability and instability, with different means of surviving adversity, in the different species of a given community. Communities differ widely in the relative stability of their dominant species—contrasting say, a redwood forest and a grassland, the marine benthos and the plankton. The different relative stabilities of communities (or their dominant species, at least) represent different adaptations to environment.

Some community characteristics are determined by environmental resources. Among the factors governing productivity, the flow of nutrients into and through the ecosystem may be most important for an aquatic community, and the flow of water from precipitation or other sources through the community by transpiration may be most important for a terrestrial community (Chapter 5). Because these resource flows are relatively stable from year to year (even if they are seasonably variable), the community's over-all function or productivity comes into balance with its resource flow. From some combination of productivity with rates of death and decomposition, in turn, results the climax community's steady-state biomass. The community's productivity and biomass are relatively stable in the two senses of constancy in the absence of disturbance, and return to the constant mean value after disturbance. The steady state of annual productivity can be achieved by either long-lived and relatively stable populations (the redwood forest) or short-lived and rapidly changing species populations (the plankton). The two aspects of relative stability—species populations and community productivity—are separate phenomena based on different kinds of regulatory mechanisms. (Resource flows contribute to the relative stability of some species populations, but various other mechanisms also do so.) Neither phenomenon results from selection of communities as wholes for their stability.

The evolution of communities seems to offer only a few general principles:

1. Diversity increases by addition of species differing from one another in niche and habitat.
2. Adaptation to environment is expressed in some characteristics of community structure and function. These adaptations seem to be loosely suggested by environment, and differently realized by particular combinations of species that evolve in different areas.
3. Communities evolve toward different degrees of stability in their structure and function, and species toward different kinds of population function and degrees of population stability in the same community. Evolution has produced some combinations of species (in climaxes) adapted to self-maintenance in a community of steady-state function.
4. Species associations with other species are predominantly loose and changeable, and community evolution is net-like in the sense that species are variously combined and recombined into communities in evolutionary time.
5. Since the community has no central control system, and no inherited genetic message for the community as a whole, community evolution is largely consequent on evolution of the species that make up communities.
6. Natural communities have evolved a kind of organization that is distinctive.

By organization we refer to the means by which the functional complexity of a system is maintained, or changes more or less harmoniously as part of a growth process. To the extent that the species of a community are organized, that organization depends on the interplay of various species differently affecting one another in competition for resources and in other interactions they have evolved in communities. This kind of organization resulting from the interplay of loosely related components, the species, with separate inheritances is unlike that of any other biological system. Analogies between organisms and communities have not been rewarding. Communities are living systems of very different sorts from organisms, and it is because they are so different that the kinds of evolutionary statements about communities are different, and more limited, than those about organisms.

Human Ecology

A theme can be traced from populations through communities and ecosystems to the biosphere: contrasting conditions of stability and instability. We note in general three modes of behavior, or time-relations, in living systems: (1) steady states, in which an approximate constancy of the system is maintained, superimposed on a flow of matter and energy through it; (2) limited growth, in which the system expands but is subject to negative feedbacks that slow growth as some ceiling or limit is approached; and (3) unregulated growth that, with continued exponential increase and positive feedback, expands past its resource limits to a reduction or crash of the system. Biological cases of these modes of behavior at the levels of organisms, populations, and communities include: (1) the steady states of mature organisms with determinate growth, stable populations with equal birth and death rates, and the climax of natural communities; (2) the regulated growth of individual organisms, either determinate, as in most higher animals, or indeterminate, as in most higher plants, the sigmoid stabilization of a population at its limit, and community succession to the climax; and (3) cancer in the organism, and eruption and collapse of some populations.

Communities, since their productivities are regulated by resource flows, do not widely overshoot their limits and crash; but many populations do. Our final question is the stability of one population and an interaction that is now altering the world—man's population and the ecosphere. The question is not only fundamental, but so formidable that we shall offer no conclusive answer.

Figure 8.3 plots the world population of man through three thousand years. The form of this curve is both typical of exponential growth, and dramatic as a case of such. It is this growth that casts across our time a curvilinear shadow of ascending problems in the relation of man to environment and of nations to one another. Although it is typical of exponential growth in appearance, the curve is distinctive in a way that Figure 8.3 does not show. The curve is not simply one of exponential growth at a constant rate; for the rate itself has been increasing. A constant rate of exponential growth, r, implies a constant doubling time for a population regardless of the population's size. For man's population the rate has been increasing, and the doubling time decreasing. Before the Christian era the doubling time for the world population of man was probably of the order of 2000 years; between 1650 and 1850 it was 200 years, by 1800 100 years, by 1910 50 years, and by 1950 40 years. The increased rate of growth is the result of positive feedbacks—the expansion of resources to support human populations, and the reduction of death rates by increased resources, improved living conditions, and medicine—consequent on human enterprise and intelligence. This is, however, unusual behavior for a population.

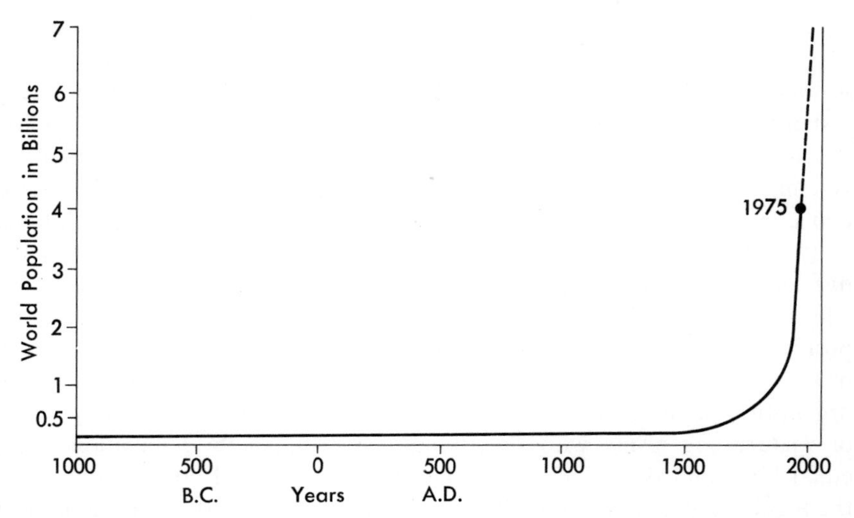

Figure 8.3. Growth of the human population through three millennia. The dramatic steepening of growth in the last three centuries coincides with the industrial and medical revolutions; reasonable projection implies a population around 7 billion by the year 2000. [Dorn, *Science,* **135:**283 (1962).]

We have mentioned effects of overgrowth or crowding that affect the populations of some other organisms. There are hazards for man in population overgrowth, among them:

1. Food shortage. Through much of human history the increase in food supply has barely kept up with population growth. Periods of plenty following the discovery of new food resources have alternated with periods of privation as populations grew until resources were more than fully used. The rapid population growth since World War II in particular has threatened to outgrow world food resources. Some additional time before it does so has been obtained by the "green revolution"—the export to some of the poor countries of advanced agricultural techniques using highly productive strains of cereal crops together with the techniques of irrigation, fertilization, and pest control needed to realize their high productivities. This time has been bought at the potential cost of: (a) increasing growth of city populations in the poor countries, growth that may exceed means of employment and that will increase dependence on intensive agriculture and on transport and distribution of food, (b) adverse environmental effects from intensive fertilizer and pesticide use, and (c) the possible instability of increased food production that is dependent on specific crop strains vulnerable to disease and on technological inputs that are becoming increasingly expensive (fertilizer, pesticides, irrigation water, and fuel to produce and power agricultural machinery).

2. Environmental degradation. If the per capita expenditure of industrial energy remained constant, the growth of population in Figure 8.3 would imply accelerating exploitation of environment and release of toxic materials into environment. In this respect Figure 8.3 is, although dramatic, itself an understatement. Per capital energy use, the mean release of industrial energy per individual, has been increasing and continued increase in per capita energy use is the policy of most nations. The present doubling time for world population is about 35 years; but energy use has increased more rapidly than this, and the doubling time for world petroleum use has been only 10 years. As long as growth in energy use continues, accelerating release of by-products of industry to environment should be expected, efforts to control some of these by-products notwithstanding. Furthermore, the acceleration in release of materials to environment has included increase in the number of kinds of potential pollutants, as well as in the amounts of these.
3. Resource shortage. The exponential growth of industry (including technological agriculture) has led to the current, gluttonous demands on such nonrenewable resources as fossil fuels and metals. It is difficult to forecast the time and effects of resource shortage because of uncertainties about amounts of resources and possible substitutions and extractions from very low-grade sources. Exponential growth may imply not only accelerating use of a given resource toward relative exhaustion in a time that, in an historic perspective, is now close. It may also imply accelerating costs that will come into play before exhaustion. These rising costs may result from competition for limited resources or from increasing energy expenditure for extraction from increasingly low-grade sources. The competition and costs could increase international tensions and economic stresses on both rich nations and poor.
4. Psychological effects. A human society is a living system made to work by a complex and partly invisible organization, one relating economic and political practices to individual psychology by way of culture and the acceptance of community purposes and individual restraints. The effects of population and industrial overgrowth bear upon individual psychology and thereby on the organization of the society and its ability to deal with its problems. For poor countries the psychological effects may include the prospect or the reality of famine, increasing concentration in cities with increasing exposure to poverty and unemployment, awareness of deepening relative national poverty, and discouragement before the population problem itself. These effects occur at the same time as a decline in the support offered to human life by traditional culture and value systems. For the rich countries effects of continued growth may include submergence of the individual in very large urban areas and social systems uninfluenced by him, concentration in cities of increasing poverty and pollution, and awareness of the declining quality

of environment and individual life. At the same time the effects of wealth, comfort, and commercialism may weaken culture and the sense of individual participation in and obligation to the society. For both rich and poor countries the energy of hopefulness and the will to contribute to building the society and tending the institutions by which it works could be eroded by a sense of entrapment in population problems, food shortage or environmental degradation, economic decline, cultural decline, or some combination of these.

The central problem is that of overgrowth—growth beyond the means of long-term support (including means of dispersing wastes or by-products). The potential for overgrowth is common to man's population and industry, natural and pest populations, and some other open systems. The hazard of overgrowth for man's population specifically is often referred to as the Malthusian problem, because of its recognition by Thomas Malthus in 1798. The rise of industry and the use of industrial power to increase and transport food have permitted human populations to grow far beyond the limits assumed by Malthus. Natural populations can be regulated by dependence on a stable resource flow, density-dependent limitation by negative feedback, and self-limitation; but industrial man has for a time been "free" from such limits. The Malthusian problem has not thereby been escaped, but delayed, changed in implication, and probably intensified. The signs of the present suggest that effects of the overgrowth are now being felt by world society, and that limits on population and industrial growth will come into play. The time, and manner, and effect of different potential limits on growth seem not predictable.

The effects may not involve collapse but prospects for the humane objectives of societies. These are objectives that use, support, and defend human values such as truth, freedom, equity, constructiveness, responsibility, and service, and the arts. One of the general effects of overgrowth may be the overloading of a society with problems, until these can no longer be faced and acted on, and their pressures threaten the humane traditions of the society. The question raised by Malthus, and by those who sought to alert the world to population problems from 1948 into the early 1950's, is not the survival but the condition of human life. The power of science and technology once offered man a vision: a world with a limited and stable population that had escaped real poverty to live in a peaceful and rational sharing of the world's resources. Overgrowth, beyond a point that cannot be stated, implies the impossibility of escape from history's following of periods of relative well-being by times of turmoil, travail, and tragedy. The consequence of failing to act in time on the Malthusian problem could be not merely food shortage but renewed subjection to history's grinding of human hopes.

The reason for overgrowth does not lie in any particular religion or political belief, but in man's identity as a gifted organism: an organism with the capacity for over-reproduction on the one hand, while gifted enough both to create civilizations and to grasp for ever-increasing wealth on the other. There are cases of self-stabilization of human populations, but control is difficult and achieved only by unwelcome measures. Normally, given their capacity for growth and the desires of parents, human populations grow. The need for limits on industrial growth was only recently recognized, is very difficult to achieve, and is in conflict with intense human desires for increasing prosperity and the human benefits that this can bring. Separate nations thus accept (if they do not welcome) population growth and seek maximal industrial growth. The effect, in a limited world, is that characterized by Garrett Hardin as "the tragedy of the commons."

In a village a shared grazing meadow or commons may serve all while the number of cattle owners and cattle using it is small. When the number of owners and cattle is larger, the commons is overgrazed and degraded; yet it seems still to the advantage of each cattle owner to increase the number of his cattle if he can. Similarly, when population and industrial wealth are small in a world of abundant resources, the commons—in this case the resources and the pollution-dispersing means of the ecosphere—provides for all. Growth seems then not only acceptable, but desirable. When population and industry are large in a world of limited resources, growth comes to imply the exhaustion of resources, the overloading of the ecosphere, and intensifying competition among nations—hence detriment to the commons and all who use it. Even when detriments to the commons are occurring, it seems still to the advantage of individual nations to intensify their efforts. The effort toward material progress and prosperity thus tends to press on toward the degradation of the commons on which man's well-being is dependent. The problem was differently stated by William James: the trouble with man is that he cannot have enough without having too much.

It has been suggested that disturbance of environmental resources—by overcropping or overgrazing especially—may have contributed to the decline of some human societies. Given the limits of our knowledge and the complexities of historic processes, we cannot be too sure what brought the decline of the Mayan and other past societies. It is possible, however, to observe patterns of population change in particular societies, as distinguished from the world pattern of Figure 8.3. There are only a few human societies for which data are adequate to give a realistic picture of population change through some centuries of history. Figure 8.4 shows three of these for which the records are most effective.

1. Egypt has had the longest record of continuous occupation of an area by a major, civilized society of any land. This long occupancy has been supported by a great river that annually flooded its valley, adding water and nutrients, as well as silt, to the valley soils. A further, fortunate

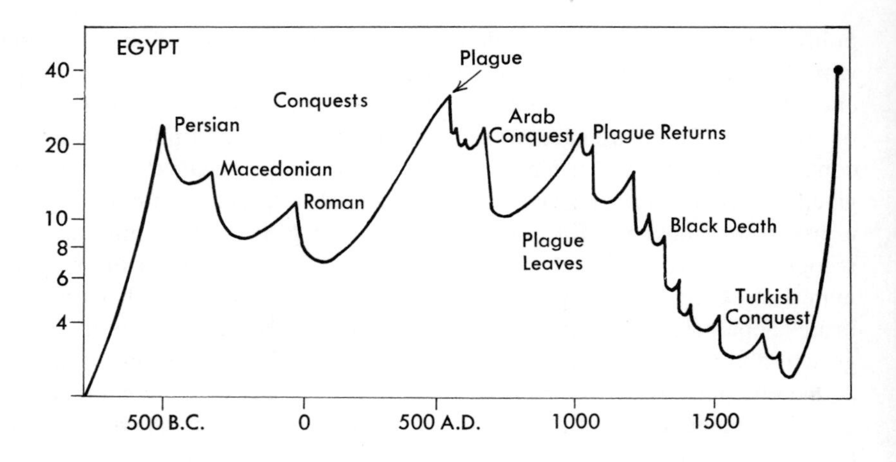
EGYPT
40
20
10
8
6
4
Plague
Conquests
Persian
Macedonian
Roman
Arab
Conquest
Plague Returns
Plague
Leaves
Black Death
Turkish
Conquest
500 B.C.
0
500 A.D.
1000
1500

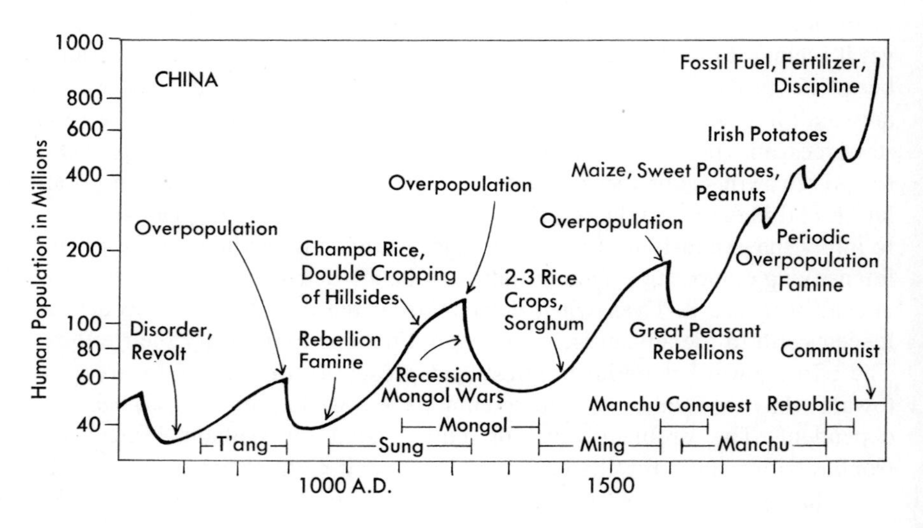
Human Population in Millions
1000
800
600
400
200
100
80
60
40
CHINA
Fossil Fuel, Fertilizer,
Discipline
Irish Potatoes
Maize, Sweet Potatoes,
Peanuts
Overpopulation
Overpopulation
Champa Rice,
Double Cropping
of Hillsides
Overpopulation
Periodic
Overpopulation
Famine
2-3 Rice
Crops,
Sorghum
Disorder,
Revolt
Rebellion
Famine
Recession
Mongol Wars
Great Peasant
Rebellions
Communist
Manchu Conquest
Republic
T'ang
Sung
Mongol
Ming
Manchu
1000 A.D.
1500

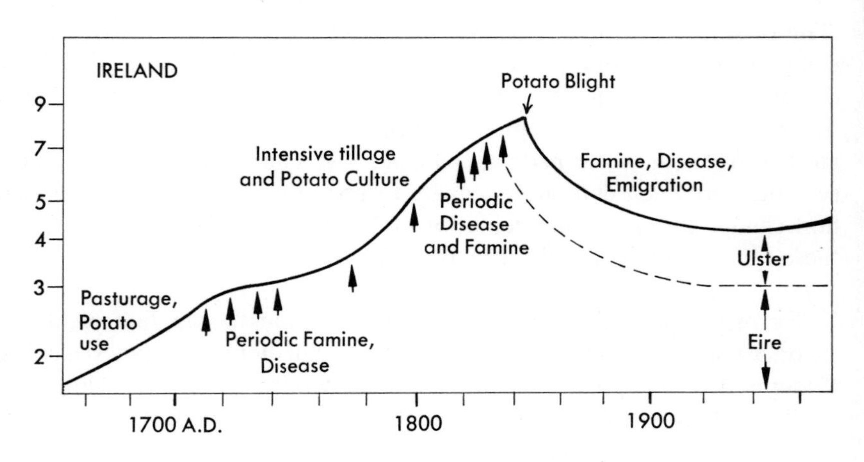
IRELAND
9
7
5
4
3
2
Potato Blight
Intensive tillage
and Potato Culture
Famine, Disease,
Emigration
Periodic
Disease
and Famine
Ulster
Pasturage,
Potato
use
Periodic Famine,
Disease
Eire
1700 A.D.
1800
1900

circumstance favored the Egyptians. Unlike other societies, they were not able to reach and exploit the lands from which the river supporting them flowed, and they were consequently unable to disrupt or destroy the basis of their livelihood. Their history records a human population supported by a stable resource and a slowly or scarcely changing technology. The population has fluctuated, and it suffered a tenfold decline from about 30 million in 540 A.D. to fewer than 3 million in the Nineteenth Century. Repeated conquests have produced population decreases lasting a century or more. (The population of central Mexico is believed to have collapsed from 25 million, supported by the Aztec civilization, in 1519 before Spanish conquest to 1 million in 1605.) Such effects of some, though not all, conquests may result from cultural and psychological disruption of a society, as well as from violence and the introduction of disease. Plague, including that of the period of the Black Death in Europe, affected Egypt even more severely than Europe. The recent rise in population with plague largely controlled is probably the most rapid in Egyptian history.

2. The record of China differs in its sawtooth ascent and more clearly Malthusian implications. Periods of well-being, population expansion, and dynastic continuity have each been made possible by new agricultural technology or expansion into new cultivated areas. Each has led to overpopulation for the resource base available at that time, and resulting famine, disorder, and dynastic change. Comparable fluctuations, including population decreases from about 16 million in 500 B.C. to 8 million in 200 B.C., and about 70 million to 40 million during the First Century A.D., can be inferred from Chinese history before the period illustrated in Figure 8.4.
3. The population of Ireland grew through the Seventeenth Century to around 3 million in 1720. It was probably then saturated for an agriculture based primarily on grazing and limited reliance on the "Irish" potato (introduced from the New World, perhaps in 1588). Periodic famine and disease (primarily smallpox) occurred in 1709 and thereafter. A change in agriculture to intensive cultivation of both cereals and potatoes, expansion of farming into less favorable lands, and heavy

Figure 8.4. Human population history in civilized societies. Estimates of population numbers, together with some events affecting them, are plotted for three nations; the population numbers are on logarithmic scales. Top: Egypt, 800 B. C. to the present, as interpreted from various historic records and estimates. [Hollingsworth, 1969.] Middle: China, A. D. 600 to the present, based on censuses in some periods and estimates from other historic records. [Cook, 1972; see also Durand, 1960; Ping-ti Ho, *Studies on the Population of China 1368–1953* (1959); and Colin Clark, *Population Growth and Land Use* (1967).] Bottom: Ireland, 1650 to the present, based on censuses from 1821 to the present and interpretation of less reliable household counts before 1800. [K. H. Connell, *The Population of Ireland 1750–1845* (1941); and Reinhard et al., 1968.]

> reliance on the potato as the principal food made possible more rapid population increase from about 1780 on. Population saturation for this resource base was expressed in periodic famine and disease from 1817 to 1841. A disease, the potato blight, caused failure of this crop beginning in 1845; and a major famine accompanied, as is often the case, by epidemic disease resulted. The population declined from about 8,175,000 in 1841 to 4,705,000 in 1890. The population of what is now Eire, the Irish Republic, stabilized at about 3 million from 1900 to the present (an additional 1.6 million live in Ulster, or Northern Ireland). The stabilization was apparently based on an unstated agreement: that the Irish would not again overpopulate their island, but would live within the support offered by a mixed agriculture with population limited by late marriage, non-marriage, small families, and emigration.

The implication of these population records is dual: human populations—especially civilized populations—are relatively unstable, but their self-stabilization is possible.

The relative instability is a consequence of growth processes without adequate feedback control. For a poor country, increasing entrapment in population problems is a gradual process. There is no particular time at which the effects are so clear, and the implications so inescapable, that the policy of population stabilization can establish itself over the great difficulties in its way and the great temptation to let the matter ride in hope and illusion. For the industrial countries pollution problems have appeared more rapidly; yet these are also progressive developments that specify no particular time at which growth must cease. A pollution process does not enter into public concern until it is already advanced and based on an extensive and apparently necessary industry. Control is delayed by a lag in the public perception of the problem, and by difficulty in obtaining political action after a problem is recognized. Control is made difficult also by the complexities in which population and environmental problems are embedded. Given complex and partly conflicting evidences, it is easy for men and governments to choose to believe the more favorable, if less valid, fraction of the evidence that asks of them less concern and self-discipline. The potential for tragedy lies not only in the Malthusian problem, but in the facility with which men use short-term evidences to deceive themselves about long-term implications. Control is difficult, finally, because of the conflict of short-term individual (or corporate or national) desires with long-term, delayed, indirect, social or global consequences.

Nevertheless, a reasonable future for man depends on the limitation of growth. Technology has provided great benefits, including humane contributions to health and culture as well as wealth. Yet it is futile to depend on technology alone for solution of the problems of the future. At the present level of their power and diversity, technology and industry produce or intensify problems more rapidly than solutions. It is also misleading to trust in great new sources of energy—primarily the possibility of nuclear fusion—for the solution of human problems. In theory, sufficient energy might accomplish almost any purpose, including control of environmental degradation. In actuality, the enormous industry supporting, and supported by, exponentially increasing use of energy from nuclear fusion would accomplish the degradation of the biosphere whether or not that industry itself released extensive pollutants. It is likely that abundant energy would be for the industrial nations, like the green revolution for the poor countries, a brief reprieve and false remedy—buying time for continued growth while the consequent problems expand to a new level of unmanageableness. In both cases problems are met again later, when continued growth and increasing dependence on technology have increased the society's potential instability.

Continued growth can prevent solving the problems of man's overgrowth and potential overpowering of the ecosphere. Piecemeal solutions of individual environmental problems are much to be desired, yet these as tactical measures cannot solve the long-term strategic problem. The only long-term solution is a strategy simple to state and very difficult to achieve: a steady-state function of human societies. A rational strategy for a long-term human future should include: (1) population stabilization, (2) steady-state economic function, with efforts toward improved equity of individual incomes, (3) tight circulation and conserving use of non-renewable resources, (4) industrial and individual restraints minimizing pollution and environmental alteration, (5) minimal disturbance and maximal conservation of the biosphere, combined with skillful management of part of the biosphere for human needs, (6) due concern for the deeper psychological needs of human life and the profound readjustments of individual psychology implied by a stable society. These may amount to history's greatest challenge to man, and success in meeting this challenge is not assured.

Point (5) is one of "conservation of nature" for reasons based on man's own best interests. Natural communities provide the research subjects through which evolution and the function of the living world may be understood, and a standard by which the behavior of ecosystems altered by man can be interpreted. Pollution processes are indicated by effects on some of the many natural species around us. These species, being often more sensitive than man, are monitors for man that warn of increasing toxication of man's own environment. The psychological contribution of natural environments to the well-being of societies may be greater than we know how to measure. The psychological value of natural and harmoniously cultivated, as opposed to urban and exploited, landscapes may increase in a time of increasing urban concentrations and social constraints. It is not possible now to foresee clearly when and how increasing disturbance of the biosphere might be seriously detrimental to man. Apart from management for food or other needs, minimal disturbance of the biosphere and of individual natural environments is the policy of wisdom. In the management of ecosystems, skill in management breaks down when change is too rapid, or when the ecosystem is disturbed or polluted to a point beyond which desirable species cannot be maintained, or management effects cannot be predicted. The reasons for conservation are humane. By protecting natural areas from destruction, man not only preserves resources of great, if not simply economic value, but affirms the possibility of a reasonable future. If man can restrain himself from degrading the biosphere, he may thereby restrain himself from degrading his own future.

The problems of the human future range far beyond ecology, yet ecology is an essential part of them. The relations of human populations to environment, their effects in changing environments, and the implications of these changes for human beings, are aspects of human ecology. The study of communities and ecosystems can serve human practical concerns in important ways. Ecological research can make possible knowledgeable and nondestructive harvest of ecosystems, contribute to planning that balances best use and minimal disturbance of environment, and clarify the meaning and potential dangers of pollution and retrogression. These are among the objectives of ecology. Those objectives include knowledge of adaptation and population process, appreciation and understanding of natural communities and the evolution of organisms in them, comprehension of the function of ecosystems and the ecosphere, and contribution to a wiser and more understanding long-term management of natural communities and environment in relation to human needs.

References

Community Evolution

ALLEE, W. C., O. PARK, A. E. EMERSON, T. PARK, and K. P. SCHMIDT. 1949. *Principles of Animal Ecology*. Philadelphia: Saunders. xii + 837 pp. (pp. 440, 695–729).

CLEMENTS, F. E. 1936. Nature and structure of the climax. *Journal of Ecology* **24**:252–284.

CODY, M. L. 1970. Chilean bird distribution. *Ecology* **51**:455–464.

CODY, M. L. 1974. Towards a theory of continental species diversities: Bird distributions over mediterranean habitat gradients. In *The Ecology and Evolution of Communities,* ed. J. M. Diamond and M. L. Cody. Cambridge: Harvard Univ. (in press).

DUNBAR, M. J. 1960. The evolution of stability in marine environments; natural selection at the level of the ecosystem. *American Naturalist* **94**:129–136.

MASON, H. L. 1947. Evolution of certain floristic associations in western North America. *Ecological Monographs* **17**:201–210.

MAY, ROBERT M. 1973. *Stability and Complexity in Model Ecosystems.* Princeton Univ. ix + 235 pp.

MAYNARD SMITH, J. 1974. *Models in Ecology.* Cambridge Univ. xii + 146 pp.

MCCREE, K. J. and J. H. TROUGHTON. 1966. Non-existence of an optimum leaf area index for the production rate of white clover under constant conditions. *Plant Physiology* **41**:1615–1622.

MCMILLAN, C. 1960. Ecotypes and community function. *American Naturalist* **94**:245–255.

MOONEY, H. A. and E. L. DUNN. 1970. Convergent evolution of mediterranean-climate evergreen sclerophyll shrubs. *Evolution* **24**:292–303.

WHITTAKER, R. H. 1957. Recent evolution of ecological concepts in relation to the eastern forests of North America. *American Journal of Botany* **44**:197–206.

WHITTAKER, R. H. 1969. Evolution of diversity in plant communities. *Brookhaven Symposia in Biology* **22**:178–196.

*WHITTAKER, R. H. and G. M. WOODWELL. 1972. Evolution of natural communities. In *Ecosystem Structure and Function,* ed. John A. Wiens. Oregon State University Annual Biology Colloquia **31**:137–159.

Human Ecology

*COOK, E. 1972. Energy for millenium three. *Technology Review* **75**(2): 16–23.

CROWE, B. L. 1969. The tragedy of the commons revisited. *Science* **166**:1103–1107.

DALY, HERMAN E. 1973. *Toward a Steady-State Economy.* San Francisco: Freeman. x + 332 pp.

DURAND, J. D. 1960. The population statistics of China, A.D. 2–1953. *Population Studies* **13**:209–256.

EHRLICH, PAUL R. and A. H. EHRLICH. 1970. *Population, Resources, Environment: Issues in Human Ecology.* San Francisco: Freeman. 383 pp.

EISELEY, LOREN C. 1971. *The Night Country.* New York: Scribner. xi + 240 pp.

FOERSTER, H. VON, P. M. MORA, and L. W. AMIOT. 1960. Doomsday: Friday 13 November, A. D. 2026. *Science* **132**:1291–1295.

*HARDIN, G. 1968. The tragedy of the commons. *Science* **162**:1243–1248.

HOLLINGSWORTH, T. H. 1969. *Historical Demography.* Ithaca: Cornell Univ. 448 pp.

ISTOCK, C. 1969. A corollary to the dismal theorem. *BioScience* **19**:1079–1081.

MALTHUS, THOMAS R. 1798. *First Essay on Population.* Reprint with notes by J. Bonar, 1926. London: Royal Economic Society and Macmillan. 396 pp.

MEADOWS, DONELLA H., DENNIS L. MEADOWS, JØRGEN RANDERS, and WILLIAM W. BEHRENS III. 1972. *The Limits to Growth.* New York: Universe. 205 pp.

*MILES, RUFUS E. 1971. Man's population predicament. *Population Bulletin* **27**(2):1–39.

MULLER, HERBERT J. 1952. *The Use of the Past: Profiles of Former Societies.* Reprint 1957. New York: Oxford Univ. xi + 394 pp.

ORTEGA Y GASSET, JOSE. 1932. *The Revolt of the Masses.* Reprint 1957. New York: Norton. 190 pp.

PADDOCK, WILLIAM AND PAUL PADDOCK. 1967. *Famine—1975! America's Decision: Who Will Survive?* Boston: Little, Brown. x + 276 pp.

PLATT, J. 1969. What we must do. *Science* **166**:1115–1121.

REINHARD, MARCEL R., ANDRÉ ARMENGAUD, et JACQUES DUPAQUIER. 1968. *Histoire général de la population mondiale.* 3rd ed. Paris: Montchrestien. ix + 708 pp.

THOMAS, WILLIAM L., editor. 1956. *Man's Role in Changing the Face of the Earth.* Univ. Chicago. xxviii + 1193 pp.

Index

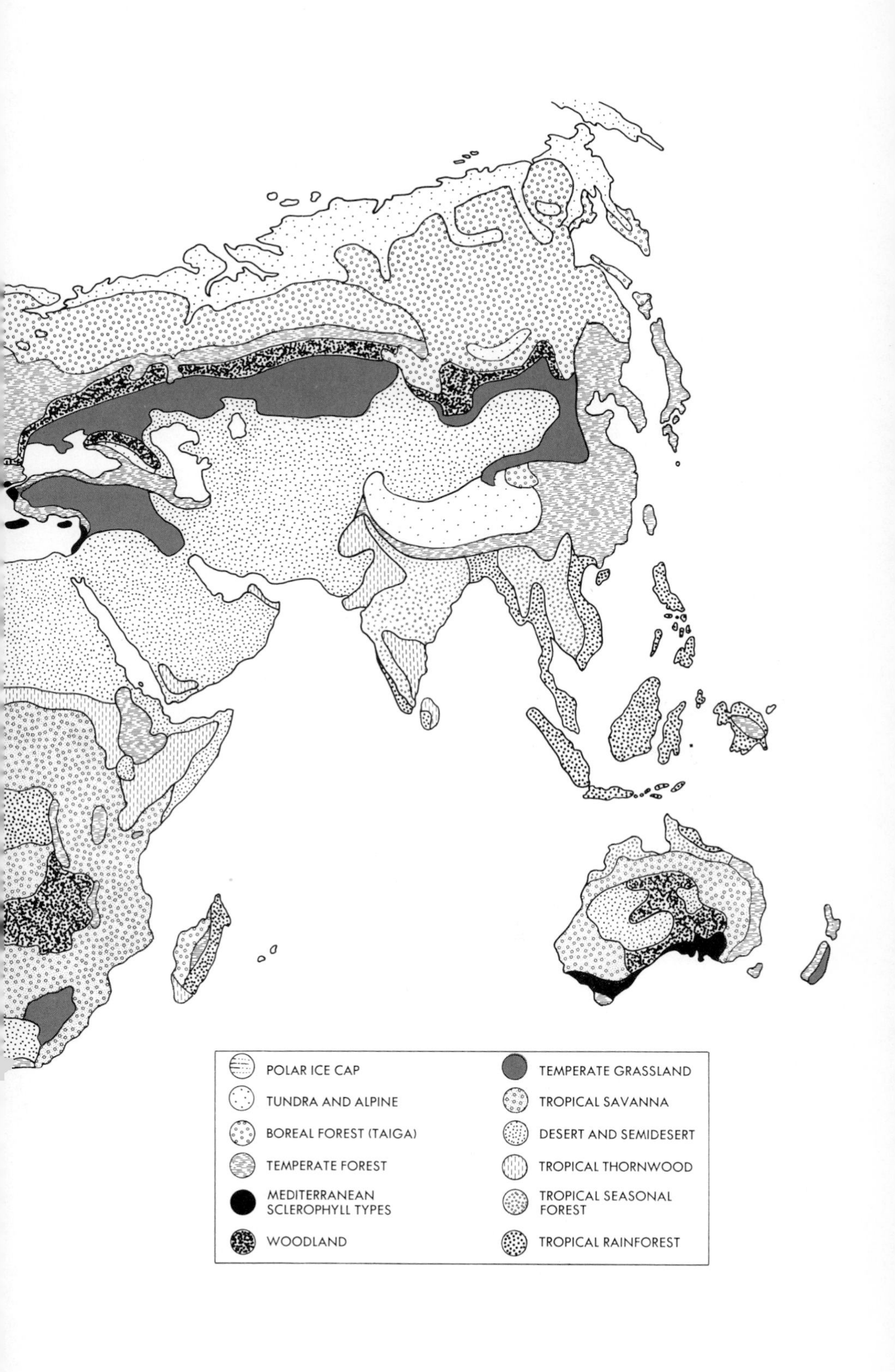

POLAR ICE CAP
TUNDRA AND ALPINE
BOREAL FOREST (TAIGA)
TEMPERATE FOREST
MEDITERRANEAN SCLEROPHYLL TYPES
WOODLAND
TEMPERATE GRASSLAND
TROPICAL SAVANNA
DESERT AND SEMIDESERT
TROPICAL THORNWOOD
TROPICAL SEASONAL FOREST
TROPICAL RAINFOREST

574.524 W617c 1975

Whittaker, Robert Harding,
1920-

Communities and ecosystems

574.524 W617c 1975

Whittaker, Robert Harding,
1920-

Communities and ecosystems

DISCARDED